Soil–Water–Root Processes:
Advances in Tomography and Imaging

Soil–Water–Root Processes:
Advances in Tomography and Imaging

Stephen H. Anderson and Jan W. Hopmans, Editors

SSSA Special Publication 61

Soil Science Society of America, Inc.
5585 Guilford Road, Madison, WI 53711-5801 USA
soils.org

dl.sciencesocieties.org
societystore.org

ISBN: 978-0-89118-958-9 (print)
ISBN: 978-0-89118-959-6 (electronic)
doi:10.2136/sssaspecpub61
Library of Congress Control Number: 2013939207

SSSA Special Publications
ISSN: 2165-9826 (online)
ISSN: 0081-1904 (print)

Cover design: Patricia Scullion
Cover photo: (Also Fig. 2–5) A partially transparent view of a pore, showing the complex nature of the pore-grain interfaces.

Printed in the United States of America.

Contents

Foreword

We should all enjoy seeing plants grow, flower, and fruit, but what we see aboveground is only part of the true beauty of plants. The belowground portion, or rhizosphere, is of equal importance and should be marveled at just as much as the aboveground portion. However, studying this realm requires different and novel methods. Over the past 20 years advanced techniques using sophisticated computer assisted imaging and other techniques have made great strides forward that now allow us to evaluate the root–soil interactions at multiple scales. This book updates the 1994 SSSA Special Publication 36, *Tomography of Soil–Water–Root Processes*, and contains details on new techniques to investigate pore and root architecture, small-scale hydraulic properties, soil–water–root transformations and translocations, and field-scale root interactions using geophysical methodology.

Consisting of 12 chapters written by world renowned researchers who are working at the cutting-edge of this field, this book will serve as a resource for years to come. It also makes suggestions on where future research should be directed. Its utility extends to all who not only work directly with these techniques but also to those who seek further knowledge of soil–water–root processes and interactions.

David Lindbo, 2013 SSSA President

PREFACE

During the past 20 years, significant advances have occurred in the use and application of X-ray computed tomography and other imaging methods to evaluate soil and root systems. *Soil–Water–Root Processes: Advances in Tomography and Imaging,* SSSA Special Publication 61, is a unique assemblage of contributions exploring applications of imaging and tomography systems in soil science; it provides an updated collection of X-ray computed tomography, synchrotron microtomography, neutron imaging, magnetic resonance imaging, geophysical imaging tools, and other tomography techniques for evaluating soils and roots. Exciting new procedures and applications have been developed over the past two decades which are discussed in this publication.

As a result of the rapid advances in tomography and imaging techniques and their successful application in soil and plant science, we proposed in 2010 to develop a new edition of the 1994 *Tomography of Soil–Water–Root Processes,* SSSA Special Publication 36. This new publication presents exciting studies using synchrotron microtomography to analyze rhizosphere fluid transport, changes of root-induced rhizosphere physical properties, and characterization of biofilm architecture in porous media. Synchrotron sites include the Advanced Photon Source at the Argonne National Laboratory near Chicago, IL and the National Synchrotron Light Source at Brookhaven, NY. In addition, X-ray microtomography was applied to study biopore architecture across spatial scales, and tomography and microtomography with varying spatial resolution provided enhanced understanding of the effects of management techniques on pore geometry. The revised edition includes a chapter highlighting the versatility of nuclear magnetic resonance imaging in analyzing root–soil interactions and soil hydraulic properties.

Neutron imaging is an even more recent method that is presented for studying root architecture and water distribution in the root–soil system, and shows much promise for complementary experiments as new neutron beams become accessible. We include a chapter on electrical resistivity tomography for mapping of plant root zones as an example to present new advances in geophysics and its capabilities at finer spatial resolution. The last chapter reviews additional geophysical methods such as spectral-induced polarization, electrical impedance tomography, electromagnetic induction, and ground penetrating radar. Various chapters focus on data analysis challenges of tomography and imaging methods in general. Specific chapters report on the need for image segmentation to allow quantitative characterization of pore features, and the modeling of both static and dynamic liquid behavior, and the use of semivariograms to characterize porous media structural properties. Undoubtedly, future studies of tomography and imaging will further explore and develop image segmentation techniques.

In addition to the publication, the Soil Physics Division of the Soil Science Society of America sponsored a special symposium entitled "Tomography and Imaging for Soil–Water–Root Processes" during the 2012 annual meetings in Cincinnati, OH. The objective of the symposium was to provide a forum to highlight research using tomography and imaging technologies to quantify soil and plant properties and processes. We thank all participants and authors of papers for their efforts in making the symposium and this publication a success.

Stephen H. Anderson and Jan W. Hopmans, Editors

CONTRIBUTORS

Anderson, S.H.	Dep. of Soil, Environmental and Atmospheric. Sciences, Univ. of Missouri, 302 ABNR Bldg., Columbia, MO 65211 (AndersonS@missouri.edu)
Aravena, J.E.	Dep. of Civil and Environmental Engineering, Univ. of Nevada, Reno, NV 89557 (jaravena@unr.edu)
Arnon-Zur, A.	Geological Survey of Israel, Jerusalem 95501, Israel
Assouline, S.	Dep. of Environmental Physics and Irrigation, Agricultural Research Organization, Volcani Center, Bet Dagan 50250, Israel (vwshmuel@volcani.agri.gov.il)
Armstrong, R.T.	School of Chemical, Biological and Environmental Engineering, Oregon State Univ., 103 Gleeson Hall, Corvallis, OR 97331 (armstror@onid.orst.edu)
Beckingham, L.	Princeton Univ., Princeton, NJ 08544
Berli, M.	Division of Hydrologic Sciences, Desert Research Institute, Division of Hydrologic Sciences, 755 E. Flamingo Rd., Las Vegas, NV 89052 (markus.berli@dri.edu)
Brown, G.O	Biosystems and Agricultural Engineering, Oklahoma State Univ., 118 Ag Hall, Stillwater, OK (gbrown@okstate.edu)
Carminati, A.	DNPW—Soil Hydrology, George-August-Universitat Gottingen, Busgenweg 2, Gottingen, 37077, Germany (acarmin@gwdg.de)
Chen, Y.-C.	Brookhaven National Lab., 75 Brookhaven Avenue, Bldg. 725B, Upton, NY 11973-5000
Davit, Y.	Institut de Mecanique des Fluides de Toulouse, Univ. of Toulouse, France; now at Mathematical Institute, Univ. of Oxford, 24-29 St. Giles, Oxford OX1 3LB, UK
Fink, W.	Dep. of Electrical and Computer Engineering, Dep. of Biomedical Engineering, Visual and Autonomous Exploration Systems Research Laboratory, Univ. of Arizona, 1230 E. Speedway Blvd., Room 521, Bldg. 104, P.O. Box 210104, Tucson, AZ 85721 (wfink@ece.arizona.edu)
Furman, A.	Technion Institute of Technology, Civil & Environmental Engineering, Technion IIT, Haifa 32000, Israel (afurman@tx.technion.ac.il)
Gantzer, C.J.	Dep. of Soil, Environmental, and Atmospheric Sciences, 330 AB Natural Resources Building, Univ. of Missouri, Columbia, MO 65211 (gantzerc@missouri.edu)
Ghezzehei, T.A.	Univ. of California, Merced, 5200 North Lake Road, School of Natural Sciences, Merced, CA 95344 (taghezzehei@ucmerced.edu)
Haber-Pohlmeier, S.	ITMC, RWTH Aachen Univ., Worringer Weg 1, 52074 Aachen, Germany (haber-pohlmeier@itmc.rwth-aachen.de)
Hopmans, J.W.	Land Air Water Resources Dep., Univ. of California-Davis, 123 Veihmeyer Hall, 1 Shields Ave., Davis, CA 95616 (jwhopmans@ucdavis.edu)
Horn, R.	Univ. Kiel, Faculty of Agricultural and Nutritional Sciences, Institute of Plant Nutrition and Soil Science, Hermann-Rodewald-Str. 2, 24118 Kiel, Germany (rhorn@soils.uni-kiel.de).
Huisman, J.A.	Agrosphere Institute, IBG-3, Forschungszentrum Jülich GmbH, D-52425 Jülich, Germany (s.huisman@fz-juelich.de)
Iltis, G.C.	School of Chemical, Biological and Environmental Engineering, Oregon State Univ., 103 Gleeson Hall, Corvallis, OR 97331 (iltis@engr.orst.edu)
Javaux, M.	Earth and Life Institute, Université Catholique de Louvain-la-Neuve, Croix du Sud 2, L7.05.02, Louvain-la-Neuve, B-1348, Belgium (mathieu.javaux@uclouvain.be)
Jones, K.W.	Brookhaven National Lab., 75 Brookhaven Ave., Bldg. 725B, Upton, NY 11973-5000 (jones@bnl.gov)
Lehmann, E.	Paul Scherrer Institut, 5232 Villigen PSI, Switzerland
Lindquist, W.B.	Dep. of Applied Mathematics and Statistics, Stony Brook, NY 11794-1401 (lindquis@ams.sunysb.edu)
Mandava, A.K.	Dep. of Electrical and Computer Engineering, Univ. of Nevada, 4505 S. Maryland Parkway, Las Vegas, Nevada 89154-4026 (mandavaa@unlv.nevada.edu)

Menon, M. Kroto Research Institute, Dep. of Civil and Structural Engineering, Univ. of Sheffield, Broad Lane, Sheffield S3 7HQ, UK (m.menon@sheffield.ac.uk)

Moradi, A.B. Dep. of Land, Air and Water Resources, Univ. of California, Davis, CA 95616 (amoradi@ucdavis.edu)

Newman, L. SUNY-ESF, 1 Forestry Drive, Syracuse, NY 13210 (lanewman@esf.edu)

Nico, P.S. Lawrence Berkeley National Laboratory, Earth Sciences Division, 90R1116, One Cyclotron Road, Berkeley, CA 94720 (psnico@lbl.gov)

Oswald, S.E. Univ. of Potsdam, Institute of Earth and Environment, Geoecology, Karl-iebknecht-Straße 24-25, Potsdam, 14476 Germany (sascha.oswald@uni-potsdam.de)

Pagenkemper, S.K. Univ. Kiel, Faculty of Agricultural and Nutritional Sciences, Institute of Plant Nutrition and Soil Science, Hermann-Rodewald-Str. 2, 24118 Kiel, Germany (s.pagenkemper@soils.uni-kiel.de)

Peters, C.A. Dep. of Civil and Environmental Engineering, Princeton Univ., Princeton, NJ 08544 (cap@princeton.edu)

Peth, S. Univ. of Kassel, Faculty of Ecological Agriculture, Department of Soil Science, Nordbahnhofstr. 1a, 37213 Witzenhausen, Germany (peth@uni-kassel.de)

Pillai, N.S. Dep. of Electrical and Computer Engineering, Univ. of Nevada, Las Vegas, NV 89154

Pohlmeier, A. IBG-3, Agrosphere Institute, Research Center Jülich, 52425 Jülich, Germany (a.pohlmeier@fz-juelich.de)

Porter, M.L. Earth and Environmental Sciences Division, Los Alamos National Laboratory, Los Alamos, NM 87544

Regentova, E.E. Dep. of Electrical and Computer Engineering, Univ. of Nevada, Las Vegas, NV 89154

Rivers, M.L. Dep. of Geophysical Sciences and Center for Advanced Radiation Sources, Univ. of Chicago, Chicago, IL 60637

Sabo-Attwood, T. Univ. of Florida, Gainesville, FL 32610

Steude, J. Dep. of Geological Sciences and Engineering, Univ. of Nevada, Reno, NV 89557

Tappero, R. Brookhaven National Lab., 75 Brookhaven Avenue, Bldg. 725B, Upton, NY 11973-5000

Tuller, M. Univ. of Arizona, Dep. of Soil, Water and Environmental Science, Tucson, AZ 85721 (mtuller@cals.arizona.edu)

Tyler, S.W. Dep. of Geological Sciences and Engineering, Univ. of Nevada, MS 0175, Reno, NV 89557

Udawatta, R.P. Center for Agroforestry, 203 ABNR Bldg., Univ. of Missouri, Columbia, MO 65211 (UdawattaR@missouri.edu)

Um, W. Pacific Northwest National Lab., MS P7-22Richland, WA 99352 (wooyong.um@pnl.gov)

Uteau Puschmann, D. Univ. Kiel, Faculty of Agricultural and Nutritional Sciences, Institute of Plant Nutrition and Soil Science, Hermann-Rodewald-Str. 2, 24118 Kiel, Germany (d.uteau@soils.uni-kiel.de)

van der Kruk, J. Agrosphere Institute, IBG-3, Forschungszentrum Jülich GmbH, D-52425 Jülich, Germany

Vanderborght, J. Agrosphere Institute, IBG-3, Forschungszentrum Jülich GmbH, D-52425 Jülich, Germany (j.vanderborght@fz-juelich.de)

Vereecken, H. Agrosphere Institute, IBG-3, Forschungszentrum Jülich GmbH, D-52425 Jülich, Germany (h.vereecken@fz-juelich.de)

Vogel, J.R. Biosystems and Agricultural Engineering, Oklahoma State Univ., Stillwater, OK

Wang, J. Brookhaven National Lab., 75 Brookhaven Ave., Bldg. 725B, Upton, NY 11973-5000

Wildenschild, D. School of Chemical, Biological and Environmental Engineering, Oregon State Univ., 103 Gleeson Hall, Corvallis, OR 97331 (dorthe@engr.orst.edu)

Young, M.H. Bureau of Economic Geology, Jackson School of Geosciences, Univ. of Texas, University Station, Box X, Austin, TX 78712-8924 (michael.young@beg.utexas.edu)

Yuan, Q. Brookhaven National Lab., 75 Brookhaven Ave., Bldg. 725B, Upton, NY 11973 5000

1

Using Synchrotron-Based X-Ray Microtomography and Functional Contrast Agents in Environmental Applications

Dorthe Wildenschild,* Mark L. Rivers, Mark L. Porter, Gabriel C. Iltis, Ryan T. Armstrong, and Yohan Davit

Abstract

Despite very rapid development in commercial X-ray tomography technology, synchrotron-based tomography facilities still have a number of advantages over conventional systems. The high photon flux inherent of synchrotron radiation sources allows for (i) high resolution to micro- or nanometer scales depending on the individual beam-line, (ii) rapid acquisition times that allow for collection of sufficient data for statistically significant results in a short amount of time as well as prevention of temporal changes that would take place during longer scan times, and (iii) optimal implementation of contrast agents that allow us to resolve features that would not be decipherable in scans obtained with a polychromatic radiation source. This chapter highlights recent advances in capabilities at synchrotron sources, as well as implementation of synchrotron-based computed microtomography (CMT) to two topics of interest to researchers in the soil science, hydrology, and environmental engineering fields, namely multiphase flow in porous media and characterization of biofilm architecture in porous media. In both examples, we make use of contrast agents and photoelectric edge-specific scanning (single- or dual-energy type), in combination with advanced image processing techniques.

Abbreviations: ALS, Advanced Light Source; APS, Advanced Photon Source; CAMD, Center for Advanced Microstructure and Devices; CCD, charge-coupled device; CLSM, confocal laser scanning microscopy; CMT, computed microtomography; CT, computed tomography; ESEM, environmental scanning electron microscopy; ESRF, European Synchrotron Radiation Facility; GSECARS, GeoSoilEnviroCARS; HASYLAB, Hamburger Synchrotronstrahlungslabor; ISA, ASTRID, Institute for Storage Ring Facilities; LuAG, lutetium–aluminum–garnet; MRM, magnetic resonance microscopy; NAPL, nonaqueous phase liquid; NSLS, National Synchrotron Light Source; REV, representative elementary volume; SLS, Swiss Light Source; S/N, signal-to-noise ratio.

D. Wildenschild, G.C. Iltis, and R.T. Armstrong, School of Chemical, Biological and Environmental Engineering, Oregon State Univ., Corvallis, OR 97331. M.L. Rivers, Dep. of Geophysical Sciences and Center for Advanced Radiation Sources, Univ. of Chicago, Chicago, IL 60637. M.L. Porter, Earth and Environmental Sciences Division, Los Alamos National Laboratory, Los Alamos, NM 87544. Y. Davit, Institut de Mecanique des Fluides de Toulouse, Univ. of Toulouse, France (now at Mathematical Institute, Univ. of Oxford, 24-29 St. Giles', Oxford OX1 3LB, UK). *Corresponding author (dorthe@engr.orst.edu).

doi:10.2136/sssaspecpub61.c1

Soil–Water–Root Processes: Advances in Tomography and Imaging. SSSA Special Publication 61. S.H. Anderson and J.W. Hopmans, editors. © 2013. SSSA, 5585 Guilford Rd., Madison, WI 53711, USA.

The focus of this chapter is on X-ray *absorption* tomography (we do not cover X-ray fluorescence tomography or X-ray phase-contrast tomography). We describe various elements of the imaging process and present examples of how synchrotron radiation can be used with great advantage for characterizing many different variables of interest to the soil science, hydrologic, and environmental engineering communities. The synchrotron-based X-ray tomography section provides a brief introduction to synchrotrons and how X-rays interact with matter and also contains a discussion of how to optimize image quality via a discussion of resolution, contrast sensitivity, and artifacts. In the contrast agents section, we discuss contrast agents and the use of monochromators, while the image processing and analysis section provides a brief overview of some of the steps involved in image processing. Finally, the environmental applications section contains example applications where the use of synchrotron radiation and contrast agents has been crucial to success. We wish to emphasize that this entire chapter is meant more as a how-to guide than an extensive review of all the remarkable work that has been accomplished over the years with tomography—synchrotron-based or not. The two applications we focus on are based in work performed by this group of people and focuses on applications where we have found the use of monochromatic radiation to be essential. There are many other areas of research where great strides are being made using synchrotron-based microtomography, but a full review of these applications is beyond the scope of this text.

Synchrotron-Based X-Ray Tomography

A synchrotron light source is typically an electron storage ring that uses radio-frequency cavities to accelerate electrons to near the speed of light and magnetic fields to steer and focus the beam. When high-speed electrons are accelerated by the magnetic fields of a bending magnet, wiggler, or undulator, they emit electromagnetic radiation. The radiation spans an enormous range in wavelength and includes radio waves, visible light, and X-rays. The resulting synchrotron radiation is millions of times brighter than sunlight or laboratory X-ray tubes (see Fig. 1 in Kinney and Nichols, 1992).

Synchrotron sources can be divided into insertion devices (undulators and wigglers) and bending magnets, the main differences being in the spatial distribution and spectrum of light that is emitted. Undulators emit light that is highly collimated in both the vertical and horizontal directions (typically less than 0.05 mrad) and is in a narrow wavelength bandwidth. Bending magnets (and wigglers) emit a wide horizontal fan (several mrad), with a narrower vertical opening angle (0.1–0.3 mrad) and a continuous broad band of wavelengths. What all these sources have in common is very high emitted flux, and this is what renders synchrotron tomography such a powerful imaging tool. The wide bending magnet X-ray spectrum is most often monochromated into a narrow energy bandwidth, and this is a crucial feature, since it allows for energy specific imaging. Specifically, this permits the separation of features in the images using dual-energy (above and below the photoelectric edge) imaging (see the photoelectric edge enhancement and monochromators section for more on monochromators and edge enhancement). Tomography beam-lines based on bending magnets are

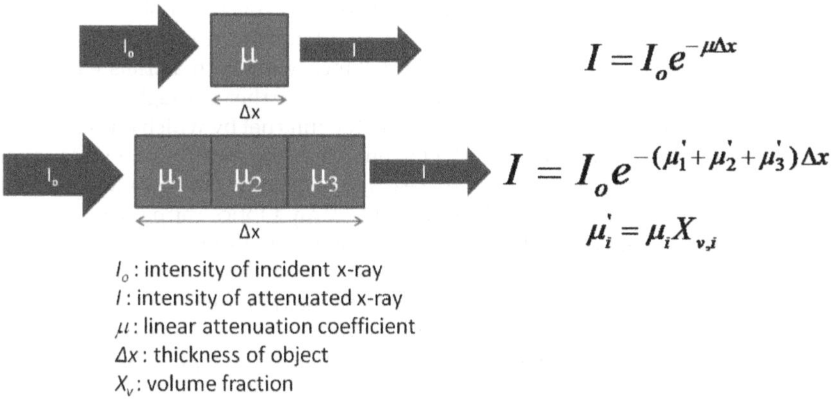

$$I = I_o e^{-\mu \Delta x}$$

$$I = I_o e^{-(\mu_1' + \mu_2' + \mu_3')\Delta x}$$

$$\mu_i' = \mu_i X_{v,i}$$

I_o : intensity of incident x-ray
I : intensity of attenuated x-ray
μ : linear attenuation coefficient
Δx : thickness of object
X_v : volume fraction

Fig. 1–1. X-ray attenuation of single and multi-component materials.

typically suitable for specimens in the 1- to 50-mm size range, while undulators are ideal for high-resolution images of samples under 2 mm in size.

X-Rays and Their Interaction with Matter

A synchrotron produces photons that can be used for a variety of analytical techniques and thus a very wide range of applications. Photons interacting with matter can be absorbed, scattered (elastic or inelastic), diffracted, or transmitted through the material. The absorption of photons can also stimulate emission of electrons, visible light, and X-rays. To better understand how to optimize tomographic imaging at synchrotrons, it is necessary to understand the basics of X-ray absorption (see Fig. 1–1).

Attenuation of X-rays as they pass through a solid object follows Lambert–Beer's law:

$$I = I_0 \exp(-\mu x) \tag{1}$$

where I_0 is the incident monochromatic radiation intensity, I is the attenuated intensity after the X-rays have passed through an object of thickness x, and μ is the linear attenuation coefficient, which depends on the bulk, and thus electron density of the material, and the energy of the radiation. For low X-ray energies in the typical synchrotron beam-line range (~5–50 keV), X-rays interact with matter predominantly by photoelectric absorption, which is strongly dependent on atomic number, and this dependence is what allows us to use contrast agents to enhance contrast in many environmental applications as described in the contrast agents section. A more detailed description of X-ray attenuation and its relative dependence on these various components as a function of X-ray energy can be found in, for example, the papers by McCullough (1975) and Wildenschild et al. (2002). It can be seen from Eq. [1] that I is a decreasing function of distance, x, since the exponential argument ($\mu \Delta x$) is negative. This reflects the fact that the incident X-rays, I_0, are attenuated as they pass through an object, and this decrease has a characteristic length of $1/\mu$, called the attenuation length. This is the distance traveled by the X-rays before they are absorbed by 1/e, or about 63% (Margaritondo, 2002). Thus, materials with a high attenuation coefficient will allow X-rays to penetrate only a relatively short distance, whereas materials with a low attenuation coefficient will allow X-rays to travel farther through the material. There are a number of resources available on the internet by which one can evaluate the absorption properties of both elements and compounds for a wide range of energies: NIST XCOM Photon Cross-sections Database (http://www.nist.gov/pml/data/xcom/index.cfm) and the Center for X-ray Optics (http://www.cxro.lbl.gov/). The X-ray Data Booklet available from http://xdb.lbl.gov/ is also an invaluable resource.

From Eq. [1], we see that by measuring the incident intensity as well as $I(x)$, we can calculate the average linear attenuation coefficient of the (composite) material that the X-rays have passed through. Because attenuation coefficients for composite materials add linearly (see Fig. 1–1), we can also compute linear attenuation coefficients for the composite parts (μ') if we know the volume fractions, X_v. What is required for favorable imaging is sufficient variation in attenuation coefficients to accurately identify or classify objects or materials of different composition and density.

To perform tomography, it is necessary to collect projection (radiographic) images at a large number of angles, and these are then "reconstructed" into an image that reveals the details of the internal structure of the object. This reconstruction process is generally performed by a mathematical back-projection algorithm, a technique originally developed in 1917 by Johann Radon (later translated in Radon, 1986), who derived a method based on calculus by which one can unfold projection images from an object and from these projections recover the object itself (filtered back-projection).

Figure 1–2 illustrates how this would work for a pencil beam, and for the case of only four projections (which would generate a single two-dimensional slice). In three dimensions (and using a parallel beam), the object is similarly rotated and two-dimensional radiographic projections are collected at a large number of angles such that the full three-dimensional distribution of attenuation coefficients can be mathematically back-calculated, that is, reconstructed. If a sufficiently large number of projections (500–1000) are collected at different angles, typically over at least 180°, the optimization problem becomes sufficiently well-

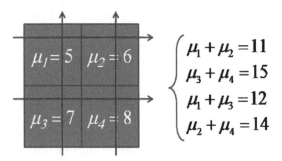

Fig. 1–2. Set of equations that needs to be solved for four projections, in this simplified case for four single ray paths, resulting in a two-dimensional slice of the object.

posed that it can be solved uniquely and with high accuracy, that is, producing a highly resolved three-dimensional distribution (image) of X-ray attenuation values. The rate of raw data generation from a synchrotron beam-line can exceed 10 Mb s^{-1}, and these data must be fairly rapidly normalized (to adjust for variation in white and dark currents), ordered into sinograms (all the projections put together to a volume makes up the sinogram), and then converted into image data during reconstruction. The typical back-projection reconstruction algorithm is an $O(N^3)$ problem, and with projections approaching 2000 by 2000 pixels, large computational resources are needed. Because it is possible to collect data 24 h a day, and at a very rapid rate, the amount of data generated is significant. A single reconstructed volume can vary in size from ~0.5 to 16 Gb depending on resolution, and during a typical 2- to 3-d run at the Advanced Photon Source (APS), we collect on the order of a quarter to half a terabyte of data. Processing, storing, and backing up these data locally require financial resources and incur network storage and administrative costs that are worth considering before starting a tomography-based project.

Synchrotron Facilities and Access

There are an increasing number of synchrotron radiation user facilities around the world; however, not all of these have beam-lines dedicated to X-ray tomography. Among the ones currently developed for tomography are the beam-lines at the

- Advanced Light Source (ALS) at Lawrence Berkeley National Laboratory, Berkeley, CA, USA
- Advanced Photon Source (APS) at Argonne National Laboratory, Argonne, IL, USA
- National Synchrotron Light Source (NSLS) at Brookhaven National Laboratory, Brookhaven, NY, USA
- Center for Advanced Microstructure and Devices (CAMD) at Louisiana State University, Baton Rouge, LA, USA
- European Synchrotron Radiation Facility (ESRF), Grenoble, France
- Swiss Light Source (SLS), Viligen, Switzerland
- Hamburger Synchrotronstrahlungslabor (HASYLAB) at DESY, Hamburg, Germany
- Diamond Light Source, Oxfordshire, United Kingdom
- Institute for Storage Ring Facilities (ISA, ASTRID), Aarhus, Denmark (energy range limited to image biological material only)
- MAX-LAB at Lund University, Lund, Sweden (proposed)
- Australian Synchrotron, Melbourne, Australia (commissioning)
- SPring-8 at Japan Synchrotron Radiation Research Institute, Hyogo, Japan
- Photon Factory, National Laboratory for High Energy Physics, Tsukuba, Japan
- Shanghai Synchrotron Radiation Facility, Shanghai, China

Most of these are run as user-facilities where individual users can apply for beam-time and, if successful, get access to the analytical equipment. An extensive review of the inner workings of an electron storage ring, the various types of synchrotrons (first to fourth generation), etc., is beyond the scope of this chapter and can be found in the works by Sham and Rivers (2002) and Margaritondo (2002), for example.

Experimental Constraints

A typical tomography beam-line setup is shown in Fig. 1–3. It generally involves an automated, micrometer-precision stage where the specimen is mounted and that can rotate through 360° as well as translate in three directions. After the X-rays have passed through the specimen, they strike a scintillator that converts the X-rays to visible light. At the GeoSoilEnviroCARS (GSECARS) bending magnet beam-line at the Advanced Photon Source (APS) at Argonne National Laboratory, the scintillator now consists of a single crystal lutetium–aluminum–garnet (LuAG) scintillation crystal. Various microscope objectives (or zoom and/or macro lenses) are then used to project the visible light onto a charge-coupled device (CCD) camera for image capture.

Depending on the brightness of the source, beam-line optics, and data collection parameters, acquisition times can be under 2 min per scan, but at some facilities, can also take hours, which bring the latter on par with modern polychromatic systems in terms of acquisition time. However, they still have the bright synchrotron light in common, which can be monochromated, and therefore used very favorably with contrast agents (see the contrast agents section), an approach that is not as optimally implemented when using a polychromatic radiation source. When choosing to perform research at a synchrotron facility, a potential user will therefore want to consider whether acquisition time is a constraint and which environmental settings can be accommodated (sample dimensions, vertical, and horizontal field of view, pressure and temperature control, etc.). Is equilibration time important for your experiment? Is it sensitive to temperature variations in the hutch?

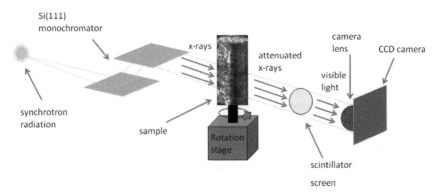

Fig. 1–3. Typical synchrotron tomography beam-line setup.

While the specific details of each microtomography beam-line may differ, all generally work roughly as described in the following which represents the setup at the GSECARS bending magnet beam-line (13 BMD).

The GSECARS bending magnet provides a fan beam of high-brilliance radiation, with an intrinsic vertical size of about 5 mm in the experimental station, about 55 m from the electron beam source point. When used with an Si [111] double-crystal monochromator, energies in the range from 7 to 65 keV can be obtained with a beam size up to 50-mm width and 5-mm height (Rivers et al., 1999). The stage setup facilitates automated translation and rotation of the object in the beam. Since the detector is two-dimensional, many slices (e.g., a complete three-dimensional data set) are obtained in a single 180° rotation (Fig. 1–3). After the X-rays are converted to visible light by the scintillator, they are imaged with a Nikon Macro lens, or 5×, 10×, or 20× Mitutoyo microscope objectives onto a high-speed 12-bit CCD camera (Photometrics CoolSnap HQ2), with 1392 by 1040 pixels, each 6.45 by 6.45 mm^2 in size. The raw data used for tomographic reconstruction are 12-bit images and a total of 720 to 1200 such images are typically collected as the sample is rotated from 0 to 180° in 0.25 to 0.15° steps. Reconstruction is accomplished either with filtered back-projection or the FFT-based Gridrec software, using the programming language IDL (Research Systems Inc.).

Resolution and Contrast

To obtain high-quality images, one must chose the X-ray energy such that there is 20 to 50% transmission through the more absorbing portions of the sample. This will result in good contrast and sufficient X-rays at the detector to obtain a good signal/noise (S/N) ratio. At the same time, the detector needs to have sufficient spatial resolution to discriminate between narrowly separated photon ray paths. High spatial resolution also depends on having a small source size, at a large distance, because source size contributes to image blur.

Compared to images obtained with many older conventional radiation sources, synchrotron-based tomography allows for superior resolution. However, polychromatic systems are developing rapidly, and it is now feasible to acquire data at similar resolutions with some of these newer systems, albeit using much longer acquisition times and generally with lower sensitivity to subtle attenuation differences. As a rule of thumb, the voxel resolution can be assumed to be approximately 1/1000 of the horizontal dimension of the specimen. This is mainly dictated by the field of view (sample size) divided by number of pixels of the detector (CCD camera), and the potential use of imaging optics. Resolution is also affected by the number of projections used, by S/N ratio (e.g., Stock, 1999), and by crystal resolution, scattering, and depth of field of the scintillator, the latter reducing it to 1 to 2 μm. Ultimately, resolution is constrained by the fact that the fundamental limit for a non-magnifying (parallel beam) technique is the diffraction limit for visible light, which is about 0.5 μm. For a non-parallel beam (i.e., cone or fan), the spot size of the source determines the physical limit of the resolution.

More recent developments using zone plates are capable of generating higher resolutions into the nanometer range, but for an increasingly small specimen size. It is worth noting that to obtain resolution on the order of nanometers, the specimen size needs to be so small that the chance of actually measuring REV (representative elementary volume)-appropriate flow and transport variables is diminishing.

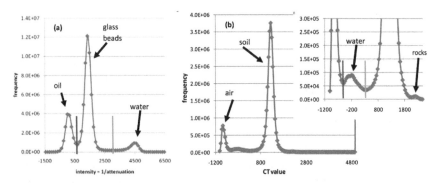

Fig. 1–4. (a) Synchrotron-based image of a partially saturated glass bead pack imaged at 10.8 μm. Using the new optics and lutetium–aluminum–garnet scintillator the different phases (oil, water, beads) are easily separated. (b) Typical medical computed-tomography (CT)-based histogram for a partially saturated clayey till soil column imaged at ~600 μm. Four different phases (air, water, soil, rocks) can be identified based on the histogram.

It should be noted here that for fine-grained soils, the resolution limit and resulting REV issue may become an issue for tomography at the micrometer resolution level as well because of the small scale heterogeneity and anisotropy of the pores.

Contrast is a measure of how well a feature can be distinguished from the surrounding background. It is often defined by the difference in attenuation between the feature and background, divided by the background attenuation. The ability to discriminate between two materials with closely similar linear attenuation values depends on the accuracy with which the values of μ, the linear attenuation coefficient, can be determined, and thus how well attenuation values for the phases of interest are separated (e.g., spatially in a histogram).

It should also be noted that there is a very large difference in what the human eye–brain cognitive unit can distinguish and what a computer can be programmed to automatically distinguish, especially for low-contrast objects.

Figure 1–4 shows example histograms for a partially saturated soil sample (20 cm tall by 20 cm in diameter), and a 7-mm diameter glass bead pack, partially saturated with water and oil. The samples were imaged with a medical CT (computed tomography) scanner at 600 μm resolution and with a synchrotron-based system at 10.8 μm resolution, respectively. It can be seen that the three phases of interest (solids, water, and air) can be distinguished in the histogram for the medical CT scan (Fig. 1–4b). Similar separation of features was also obtained using a tube X-ray microCT system (i.e., polychromatic radiation) by Tippköetter et al. (2009) so phase segmentation is straightforward and no contrast enhancement needed.

For data collected 10 yr ago with the synchrotron-based system, some of the different phases of interest would overlap, despite the use of a contrast agent; however, with recent upgrades of imaging optics and scintillator, the histograms are now similarly favorable (Fig. 1–4a) in terms of segmentation as those obtained for larger specimen size with medical CT or using a microCT system (see the experimental constraints and segmentation sections for further discussion of Fig. 1–4).

To optimize both spatial resolution and contrast sensitivity, it is necessary to select an X-ray energy that is appropriate for the material in question. The X-rays need to be sufficiently energetic to penetrate the sample, such that adequate counting statistics (S/N ratio) can be obtained. On the other hand, if the energy of the incoming radiation is too powerful, the relative attenuation will be low and the object becomes virtually transparent, with little or no contrast between the various phases.

Other Factors Affecting Image Quality

One of the advantages of (monochromatic) synchrotron radiation over polychromatic sources is that the images are not affected by beam-hardening artifacts. However, synchrotron sources are not free of ring artifacts, yet they can be minimized with careful correction algorithms in the data processing step. Ring artifacts are caused by local defects (drift and nonlinearities) in the scintillator or detection device, resulting in faulty low or high beam intensities, which then appear as rings in the reconstructed image. Furthermore, cosmic or scattered X-rays hitting the detector chip directly can cause anomalously bright pixels (zingers) and result in streak artifacts.

Other artifacts arise from movement in the object being imaged. If for instance, fluid interfaces in a porous medium are moving (equilibrating and redistributing) during the scan, a certain degree of blurring or streaking will result in the scanned image. The added "benefit" of this is that one can rather effectively ascertain whether quasi-equilibrium conditions have been achieved for the fluids involved. The term quasi-equilibrium is used here because minor adjustments in interface curvature may take longer and may not as readily show up as a motion artifact.

Partial-volume effects arise from the fact that a scanned object is often composed of a number of different substances, and the resulting discretized representation is therefore often an array of averaged values that cross interface boundaries. In other words, a scan of an object consisting of two different materials will likely produce a tomographic array of attenuation values that has a large fraction of voxels that can easily be classified as either material, but it will also have a number of voxels that have attenuation values that are fractional averages of the two materials. In addition, because of the inherent resolution limitations (of most imaging modalities), material boundaries are blurred to some extent, and the materials in a neighboring voxel may affect attenuation values in surrounding voxels.

All X-ray-based imagery is associated with some level of noise. In a perfect world, all voxel values for a uniform object should be identical; however, in reality, the voxel values are generally spread around a mean value. The magnitude of this variation is called image noise and arises because X-ray interaction and detection is a statistical process. Some noise can be successfully dealt with via image processing, but it is desirable to reduce the noise during image capture.

The time required to image a volume element or voxel with a certain statistical confidence increases drastically as the size of the voxel decreases. An object of smaller cross-section will absorb fewer photons and therefore requires longer exposure time to assure acceptable counting statistics. For instance, reducing the voxel size for a cube that is 100 μm on a side to 10 μm on a side will increase the exposure time by a factor of 10^4 (see Table 1 in Davis, 1999). Consequently, increasing spatial resolution requires larger incident photon intensity or longer integration times. Synchrotron-based radiation is well suited for high resolution imaging because of the extremely high photon flux available. However, because it is difficult to produce energies above approximately 50 keV with synchrotron radiation sources, maximum sample size is generally limited to a few centimeters for specimens with composition and densities in the typical range of soil and rock samples so that the beam can penetrate the sample, whereas larger samples can be examined in conventional systems that generally use higher energies (e.g., Wildenschild et al., 2002).

Advantages and Limitations Relative to Polychromatic Radiation Sources

Synchrotron radiation has several advantages over traditional X-ray sources. These include the high intensity (number of photons per second) and thus rapid scan times, parallel beam geometry (which leads to more accurate reconstructions compared to cone-beam), no beam-hardening artifacts, and the ability to tune the photon energy over a wide range using a monochromator for obtaining optimal image contrast and element-specific measurements (Kinney and Nichols, 1992). The trade-off is in specimen size that can be imaged, and that the data is relatively noisy for small objects (low S/N ratio for micrometer-sized objects, unless longer exposure times are used): as discussed in the resolution and contrast section, high resolution does not dictate high contrast and easy segmentation. The histograms in Fig. 1–4 illustrate how it is possible to distinguish different phases in the medical CT data; however, only features that can be detected at 600 μm are captured. In the past, much smaller glass bead samples that were scanned with synchrotron radiation were significantly more noisy (e.g., Culligan et al., 2004) and thus the phases were not as easily distinguished. As mentioned earlier, recent advances in instrumentation has alleviated that problem, and regardless, the ability to optimally use contrast enhancement and dual-energy scanning which allow for generation of truly high-resolution images, where micrometer-scale features can be measured, far outweighs noise-related segmentation issues.

Contrast Agents
Photoelectric Edge Enhancement and Monochromators

In the range of energies that most synchrotrons operate at, X-rays interact with matter predominantly by photoelectric absorption, which strongly increases with atomic number (e.g., McCullough, 1975; Wildenschild et al., 2002). Absorption of X-rays occurs when an incoming X-ray photon is absorbed, resulting in the ejection of electrons from the inner shell of the atom, and the subsequent ionization of the atom. The ionized atom consequently returns to the neutral state (filling the vacated spot in the inner shell) often with the emission of an X-ray characteristic of the atom. The photoelectric effect can be used to great advantage to enhance the contrast of a phase (e.g., a fluid phase) that would otherwise have a very low X-ray cross-section (low absorption) and therefore not be easily distinguished in an X-ray tomographic image. By adding a contrast agent (dissolvable salt or suspension of a high atomic number element) to such a fluid phase (or otherwise adding it to the specimen of interest; see the environmental applications

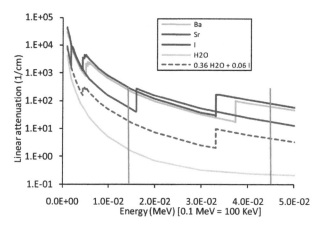

Fig. 1–5. Linear attenuation as a function of energy for select elements of interest in tomographic imaging. The curve for water and a 1:6 mixture of water and KI is also shown. The vertical orange bars depict the (15–45 keV) range of energies that it is feasible to use for soil and rock samples of ~5-mm diameter.

section), it is possible to enhance the contrast of this fluid phase, if the energy of the incident X-rays is tuned to enhance the absorption of the contrast agent.

As mentioned previously, bending magnet beam-lines are often used with a tunable monochromator that can be used to select a narrow energy band from the white synchrotron light. Following Bragg's law, each constituent wavelength that is directed at a single crystal of known orientation and d-spacing (also known as the interatomic spacing) will be diffracted at a discrete angle.

Monochromators make use of the Bragg relationship to selectively pass only the radiation of interest, that is, in a narrow tunable band of interest. The radiation outside of this energy range is thus removed. Because of the high photon flux of synchrotrons, the flux after this monochromatization is still sufficient for fast and high-resolution imaging, whereas a similar procedure would render a conventional radiation source rather depleted and result in very long scan times and very noisy images. Figure 1–5 shows the attenuation as a function of energy for several commonly used contrast agents (Ba, Sr, and I), as well as the attenuation for water, and for a 1:6 mass ratio of water and potassium iodide.

The abrupt increases in attenuation represent the point where the incoming X-rays exceeded the photoelectric (K-shell edge) energy for the element or compound of interest. From this figure, it is evident that two images collected immediately above and below the edge will result in very different absorption, and the two images can be subtracted to clearly bring out the phase that contains the element (or compound) of interest. The range of energies that can realistically be achieved at a synchrotron, and that will allow for penetration (and adequate counts) of a approximately 0.5-cm diameter object with the density of a soil or rock sample (~15–45 keV) is indicated by orange bars on the figure. As the sample thickness increases, it is necessary to use higher photon energies, and thus contrast agents (elements) in the higher energy range.

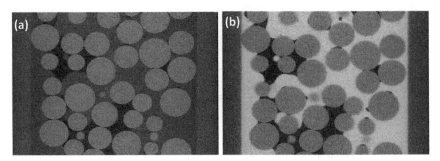

Fig. 1–6. (a) Below the Cs edge image; (b) The same image above the Cs edge.

Traditional Dissolved and Functionalized Contrast Agents

The photoelectric effect has been used to great advantage in X-ray tomography studies for decades; however, it can be optimally exploited using synchrotron-based systems. Among the contrast agents most commonly used are dissolved solutions of compounds such as KI, NaI, CsCl, RuCl, and iodized organics such as iodoheptane, iodobenzene, and iodononane. These compounds are, for instance, added to the fluid of interest (e.g., water or oil) in percentages necessary to produce good contrast (generally in the few-percent range). It is then fairly straightforward to tune the energy of the incident radiation to the relevant edge and produce images with high absorption in the doped fluid phase. The potential impact of the dopant on fluid properties and behavior can be a concern. Initial examination of this issue indicated no measurable impact of the dopant on measured properties (saturation, blob volumes, and surface areas) for the systems investigated by Schnaar and Brusseau (2005). This issue should be examined for each application of interest. Figure 1–6 shows images obtained with energies below and above the Cs-edge, respectively. When comparing the two images, the photoelectric effect is obvious. If two fluids need to be separated in the resulting images, two different contrast agents can be used, and scans can be performed above each element-specific edge to allow for subsequent subtraction, which then brings out the separate phases. This is useful when studying three-phase-fluid systems where an organic, a water, and a gas phase need separation and quantification (Brown et al., 2011).

In addition to the more traditionally used dissolved contrast agents, other means of achieving contrast can also be implemented as illustrated by the example provided below on imaging biofilms in porous media. Key to the process is to choose contrast agents that speciate, dissolve, attach, adsorb, or are excluded from the phase of interest.

Image Processing and Analysis

Ideally, tomographic images would be easy to segment into the desired phases of interest, but generally a number of processing steps of varying difficulty need to be accomplished to arrive at that point. As a general rule, it is much easier to obtain good image quality by optimizing the process at the point of capture, so adjusting the settings (energy level, exposure time, number of projections, etc.)

before starting a scan is well worth one's time. As described above, much can also be done at the point of image capture with respect to contrast enhancement using energy selective dopants. This drastically simplifies the amount of processing needed, and is often necessary to generate quantitative results that are accurate enough to draw technically reliable conclusions. A few of the most common aspects of image processing are described in the following section; however, several of these steps can be fairly complex and deserve attention beyond the scope of this chapter.

Pixelation and Discretization Effects

All images consist of pixilated versions of reality and no matter how good a resolution one can achieve, that is ultimately the case. That said, higher resolution will obviously produce more reliable quantification of small features, but also at the cost of the size of the imaged region, and therefore REV. Use of numerical models similarly force us to discretize the world we wish to investigate, and one can argue quite strongly that a discretized version of pore-scale phenomena is better than no characterization.

Registration

In many quantitative applications, it is desirable to be able to track a specimen as it goes through some sort of alteration, such as changing fluid saturations, precipitates forming or dissolving, or colloids attaching to an interface. For this purpose, it is necessary to register all the images to the same coordinate system. This can be accomplished quite readily because most beam-lines are equipped with very accurate rotation and translation stages that can be adjusted to below micrometer resolution. If for some reason the object has moved, there are now many types of commercial (and some free) software available that provide algorithms for auto-registration.

Segmentation

Following a close second to optimal image capture, segmentation (also often referred to as thresholding or binarization) is the most important step in producing high-quality results from a scan. If sufficient effort has been made to generate data with good contrast and high resolution, segmentation can be achieved using simple histogram thresholding. Two examples were shown in Fig. 1–4 that illustrate the relative ease with which thresholding can be performed when the different phases of interest have well-separated attenuation values; the two segmentation thresholds are indicated by the red and green vertical lines. However, even if a histogram looks less favorable than the ones in Fig. 1–4, it is often possible, using a number of segmentation steps, to generate good quantitative results. In a situation where the solid and water phases overlap, a simple approach is to scan the dry porous medium and then, relying on perfect registration, subtract the "dry" image from the partially saturated image (e.g., Culligan et al., 2004). A number of approaches also exist for handling more complex segmentation problems, such as the watershed segmentation approach (e.g., Sheppard et al., 2004), K-means cluster analysis (e.g., Porter and Wildenschild, 2009), and indicator kriging (e.g., Oh and Lindquist, 1999), the latter incorporating spatial information through the two-point covariance of the image. A full description of the many different types of segmentation algorithms is beyond the scope of this chapter; we refer to Iassonov et al. (2009) for a detailed overview of segmentation techniques. In summary, much more attention has been paid to the extraction of pore network, porosity, mineral phases, etc. than to extracting information about fluid phases, interfacial characteristics, and their evolution. A number of segmentation algorithms and processing steps leading to estimates of interfacial area from tomographic images were tested by Porter and Wildenschild (2009) to assess their accuracy against measured values. The particular geometries were fluid–fluid interfaces (menisci) in capillary tubes of varying sizes that were imaged with microtomography. Most of the algorithms, except for two-point correlation functions and voxel counting approaches, produced estimates that were between 2 and 15% of the actual values. In general, one can assume that the more complex the scheme, the more computationally intensive it is.

Phase Quantification and Distribution

Once the image is segmented, the data analysis can commence. Some of the simpler measurements that can be made from tomographic images are sample porosity, fluid saturations and their spatial distributions, classification of pores into matrix and macropores, etc. Porosity and saturation can be estimated by simply counting the number of voxels assigned to each phase during segmentation. In many cases, medial-axis based network generating algorithms such as 3DMA Rock (Lindquist, 1999; Lindquist and Venkatarangan, 1999; Prodanović et al., 2006) are used to produce statistical representations of the imaged pore space for use in numerical network models. More recently (e.g., Lehmann et al., 2006; Vogel et al., 2010), mathematical morphology and Minkowski functionals have also been employed to quantify various properties of the porous medium, such as pore volume (porosity), surface area, curvature, and the Euler characteristic, which quantifies the connectivity of a porous medium.

Surface Generation and Curvature Estimation

For more complex variables such as solid surface area, fluid–fluid interfacial area, and curvature, it is highly recommended to use more sophisticated surface generating techniques such as the commonly implemented marching cubes algorithm (Lorensen and Cline, 1987), as opposed to using voxel-counting techniques. The marching cubes technique is a high-resolution three-dimensional surface construction algorithm that produces a triangle mesh by computing isosurfaces from discrete data. By connecting the patches from all cubes on the isosurface boundary, a surface representation is produced. From this surface, it is possible to measure variables such as surface area and fluid–fluid interfacial area (e.g., Porter and Wildenschild, 2009), and by approximating the surface locally by a quadratic polynomial (and using the principal curvatures at the point on the graph of such a quadratic polynomial as the approximation of the principal curvatures at the original surface point), interfacial curvature can be calculated (Armstrong et al., 2012; Armstrong and Wildenschild, 2012).

Environmental Applications
Multiphase Flow

In the following section, the term multiphase shall refer to both air-water and oil-water fluid systems in porous media. Computed microtomographic (CMT) imaging has been widely used in the fields of soil science, hydrology, petroleum engineering, and environmental engineering. In petroleum engineering, the focus has often been on extraction of porosity, pore morphology, network information, and relative permeability estimates for use in pore network simulators (e.g., Coles et al., 1998; Lindquist and Venkatarangan, 1999; Turner et al., 2004; Prodanović et al., 2007), whereas in soils and hydrology research, more work has focused on multiphase variables and on estimating properties such as fluid saturation and distribution (e.g., Clausnitzer and Hopmans, 1999, 2000; Perret et al., 2000); on describing soil structural features such as macropores (e.g., Anderson et al., 1990; Peth et al., 2008; Luo et al., 2010), root structure (e.g., Kaestner et al., 2006; Tracy et al., 2010), and plant uptake mechanisms (e.g., Scheckel et al., 2007). We refer to Taina et al. (2008) for an extensive review of tomography applications in soil science, and to Werth et al. (2010) for contaminant hydrology-type applications. In environmental engineering, *synchrotron*-based microtomography has been widely used to describe multiphase variables such as nonaqueous phase liquid (NAPL) characteristics (blob morphology, e.g., Al-Raoush and Willson, 2005a, 2005b; Schnaar and Brusseau, 2005, 2006) and on measuring multiphase variables such as fluid saturations and distribution (e.g., Wildenschild et al., 2005), fluid–fluid specific interfacial area (e.g., Culligan et al., 2004, 2006; Brusseau et al., 2006, 2007; Costanza-Robinson et al., 2008; Porter et al., 2009, 2010) and curvatures (Armstrong et al., 2012).

While porosities and fluid saturations are relatively easy to obtain with conventional sources, the more recent measurements of variables such as fluid–fluid interfacial area and curvature could not have been accomplished without the use of synchrotron beam-lines, either because of insufficient resolution and/or contrast, but also because of the rapid acquisition times that allow for collection of sufficient data for statistically significant results in a short amount of time. The rapid scan times also prevent temporal changes that would take place during scans with longer acquisition time. These capabilities have, for instance, allowed for concurrent measurement of capillary pressure (P_c), saturation (S), and fluid–fluid interfacial area (a_{nw}), a relationship that has received much attention in recent years as researchers are trying to generate data in support of new thermodynamics-based theories of multiphase flow.

Hassanizadeh and Gray (1993) expanded the traditional functional dependence of the P_c–S relationship to include a_{nw}, which explicitly accounts for the numerous fluid–fluid interfacial configurations that may exist for any given saturation value. In addition, Hassanizadeh and Gray (1993) hypothesized that the inclusion of a_{nw} in the macroscale formulation of P_c would account for hysteresis observed in the traditional P_c–S relationship. In recent work using synchrotron-based measurements of P_c–S–a_{nw} and lattice-Boltzmann simulations, Porter et al. (2009) were able to show that this appears to be the case because they found that hysteresis was virtually nonexistent in the P_c–S–a_{nw} relationship as opposed to the hysteretic P_c–S plane.

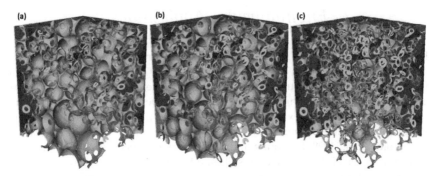

Fig. 1–7. Computed microtomography image of glass bead pack with continuous wetting phase (blue) and isolated pendular rings (green). Solid and gas phases have been removed. Image resolution is 13.0 μm and the imaged region is 5.5 mm tall. (a) drainage S_w = 0.38 (b) imbibition S_w = 0.37 (c) imbibition S_w = 0.09.

Similarly, high-resolution imaging has also provided tools that allow us to better understand how fluid phases and interfaces evolve as saturation and capillary pressures change—for instance as illustrated in Fig. 1–7. In this particular sequence of images, we are able to observe how the wetting phase (blue) is disconnected and left behind as pendular rings (green) as the sample changes saturation, and those disconnected pendular rings are then subsequently absorbed into the bulk wetting fluid on imbibition. The images also illustrate the significant difference in fluid distribution during drainage (Fig. 1–7a) and imbibition (Fig. 1–7b) at similar saturations (connected + disconnected). Additionally, the images illustrate that even at the low wetting fluid saturation of 9% there is still connected fluid from top to bottom of the imaged region.

The availability of multiphase data sets with this type of pore-scale detail has also allowed for detailed comparison and evaluation of various types of numerical models (pore network models, lattice-Boltzmann, pore morphology-based), for instance by Vogel et al. (2005), Schaap et al. (2007), Joekar-Niasar et al. (2007, 2010), Sukop et al. (2008), and Porter et al. (2009).

Biofilm Architecture

Imaging biofilms in porous media without disturbing the natural spatial arrangement of the porous medium and associated biofilm is challenging, primarily because porous media generally precludes conventional imaging. Conventional techniques for imaging biofilm include light microscopy (e.g., Yang et al., 2000; Sharp et al., 2005), environmental scanning electron microscopy (ESEM) (e.g., Davis et al., 2009), and confocal laser scanning microscopy (CLSM) (e.g., Leis et al., 2005; Rodriguez and Bishop, 2007), all of which are useful for examining biofilm on surfaces or in two-dimensional or quasi-two-dimensional porous systems. Imaging porous-media-associated biofilm using these techniques requires that model porous media systems either be constrained to a few particle diameters, that the porous medium and fluid be index-matched, or that samples be extracted and prepared, thereby disrupting the pore scale structure (Iltis et al., 2011). Thus, new techniques that allow for direct visualization of biofilm in situ are required to characterize biofilm surface architecture, and spatial distribu-

Fig. 1–8. (a) Glass bead pack (yellow) with biofilm (purple) delineated using the strained Ag particle approach (*Deinococcus radiodurans* imaged at 9.8 μm); (b) Glass bead pack (yellow) with biofilm (purple) delineated using the BaSO$_4$ suspension approach (*Escherichia coli* imaged at 11.3 μm). Flow is upward in both experiments.

tion within porous media. One such three-dimensional technique is magnetic resonance microscopy (MRM) (e.g., Seymour et al., 2004, 2007). Yet, thus far, the method has been limited in resolution (resolving features on the order of 50–100 μm), and acquisition time is significant.

Synchrotron-based X-ray tomography offers a potential alternative because it renders the solid phase transparent. The main obstacle to using X-rays for biofilm visualization and characterization is the fact that biofilms and their aqueous environment have very similar X-ray absorption capacities and are therefore difficult to separate in a reliable and quantitative manner. To overcome this problem, two different approaches have been developed based on novel use of X-ray contrast agents. At this point, we have tested two different techniques:

(i) Physical straining or adhesion of an X-ray contrast agent on the outer surface of the biofilm, and

(ii) Physical separation of biofilm and aqueous solution based on size-exclusion of a suspension of an X-ray contrast agent.

Examples of biofilm geometries imaged with each of these two approaches are shown in Fig. 1–8. For the first approach, silver-coated hollow glass microspheres were added to the fluid phase to function as an X-ray contrast that does not diffuse into the biofilm mass, but attaches to the outer surface of the biofilm. Using this approach, biofilm imaging in porous media was accomplished with sufficient contrast to differentiate between the biomass- and fluid–filled pore spaces (see Fig. 1–8a). The method was validated by using both light microscopy and CMT imaging to image biofilm in a two-dimensional micromodel flow cell by Iltis et al. (2011). Additional work is required to optimize this imaging approach; specifically, we find that the quality of the images are highly dependent on the coverage of the biofilm by the dopant microspheres, which could be problematic for dead-end pore space and for very low density biofilms. This technique is, however, particularly well-suited for outlining the biofilm surface that is well-connected to flow paths and thus well-supplied with nutrients and active in transformation. The second approach is based on use of a $BaSO_4$ suspension, which functions as a very good X-ray contrast agent and is size-excluded from entering the biofilm (see Davit et al., 2011). Using a commercially available, polychromatic tomography system and packs of polystyrene beads as their porous medium, Davit et al. (2011) used a second aqueous phase dopant (KI) which also diffused into the biofilm to help separate the low-contrast (polystyrene) beads from the other phases. Before imaging, the aqueous solution was replaced with the $BaSO_4$ suspension to effectively separate the two phases of interest (aqueous phase via Ba contrast and biofilm phase via I contrast) from the polystyrene beads. This approach obviously lends itself very nicely to implementation using a synchrotron-based system because of the edge-specific imaging that the monochromatic light allows for. In addition, we have found that when using glass beads as the solid phase, it is not necessary to use the second dopant since the solid phase (glass beads) and proxy aqueous phase (the $BaSO_4$ suspension) uniquely defines the biofilm as the remaining phase once these other two are delineated (see Fig. 1–8b).

Conclusions

X-ray tomography has been used for many decades now to generate three-dimensional information about geological, biological, and manufactured objects of interest. A large number of commercial systems have been developed over the years, and along with those many very capable software programs (both commercial and freeware) that can perform almost any type of analysis, regardless of complexity. The days of having to write processing algorithms from scratch when analyzing the accompanying large (and ever-growing) volumes of data are fortunately over, as long as one can afford the not insignificant cost involved in purchasing such software or if one has the resources to learn how to use some of the excellent freeware programs.

The availability of synchrotron-based tomography to a general population of users has brought about tremendous gains in knowledge in a vast number of fields, not the least in the petroleum engineering, hydrologic, soil science, and environmental engineering areas of research. Data obtained with synchrotron-based microtomography has helped advance our understanding of how pore-scale mechanisms and interactions take place, and has established a platform for evaluation of how pore-scale processes affect continuum-scale flow, transport, and transformation. The availability of detailed pore-scale information has helped, and will continue to help, advance pore-scale modeling efforts by providing realistic input information against which numerical models can be tested.

New developments in instrumentation and data acquisition speeds, such as new CMOS cameras which can be run at very high speeds, up to 1200 frames per s for 2048 by 2048 cameras, allow an entire three-dimensional dataset to be collected in under 1 s. When used with filtered white beam from a bending magnet, it will allow for dynamic imaging. Similarly, the rapidly evolving field of phase contrast tomography will likely also open up new avenues of exploration—in particular in areas where materials with low X-ray attenuation are of interest, but where differences in refractive index can be taken advantage of, say for imaging biological materials in a harder surrounding matrix (e.g., biofilms in porous media).

Acknowledgments

This work was supported by NSF-EAR-06101108 and NSF-EAR-0337711. A portion of this work was conducted at GeoSoilEnviroCARS (Sector 13), Advanced Photon Source (APS), Argonne National Laboratory. GeoSoilEnviroCARS is supported by the National Science Foundation-Earth Sciences (EAR-0622171) and Department of Energy-Geosciences (DE-FG02-94ER14466). Use of the Advanced Photon Source was supported by the U.S. Department of Energy, Office of Science, Office of Basic Energy Sciences, under contract DE-AC02-06CH11357. We thank the entire staff at GSECARS for experimental support. This work was also partially supported by the Environmental Remediation Science Program (DE-FG02-09ER64734) under the Department of Energy, Office of Biological and Environmental Research (BER), grant ER64734-1032845-0014978. Some of the work was also performed at Beam-line 8.3.2 at the Advanced Light Source (ALS), LBNL.

References

Al-Raoush, R.I., and C.S. Wilson. 2005a. Extraction of physically realistic pore network properties from three-dimensional synchrotron x-ray microtomography images of unconsolidated porous media systems. J. Hydrol. 300:44–64. doi:10.1016/j.jhydrol.2004.05.005

Al-Raoush, R.I., and C.S. Wilson. 2005b. A pore-scale investigation of a multiphase porous media system. J. Contam. Hydrol. 77:67–89. doi:10.1016/j.jconhyd.2004.12.001

Anderson, S.H., R.L. Peyton, and C.J. Gantzer. 1990. Evaluation of constructed and natural soil macropores using X-ray computed tomography. Geoderma 46:13–29. doi:10.1016/0016-7061(90)90004-S

Armstrong, R.T., M.L. Porter, and D. Wildenschild. 2012. Linking pore-scale interfacial curvature to column-scale capillary pressure. Adv. Water Resour.46:55–62.

Armstrong, R.T., and D. Wildenschild. 2012. Microbial enhanced oil recovery in fractional-wet systems: A pore-scale investigation. Transp. Porous Media 92:819–835.doi: 10.1007/s11242-011-9934-3

Brown, K.I., D. Wildenschild, W.G. Gray, and C.T. Miller. 2011. Interfacial area measurements for robust models of multiphase flow in porous media. Proceedings of the 2011 Goldschmidt Conference, August 14–19, 2011, Prague, Czech Republic, abstract.

Brusseau, M.L., S. Peng, G. Schnaar, and M.S. Costanza-Robinson. 2006. Relationship among air–water interfacial area, capillary pressure and water saturation for a sandy porous medium. Water Resour. Res. 42:W03501. doi:10.1029/2005WR004058

Brusseau, M.L., S. Peng, G. Schnaar, and A. Murao. 2007. Measuring air–water interfacial areas with X-ray microtomography and interfacial partitioning tracer tests. Environ. Sci. Technol. 41:1956–1961. doi:10.1021/es061474m

Clausnitzer, V., and J.W. Hopmans. 1999. Determination of phase volume fractions from tomographic measurements in two-phase systems. Adv. Water Resour. 22:577–584. doi:10.1016/S0309-1708(98)00040-2

Clausnitzer, V., and J.W. Hopmans. 2000. Pore-scale measurements of solute breakthrough using microfocus X-ray tomography. Water Resour. Res. 36:2067–2079. doi:10.1029/2000WR900076

Coles, M.E., R.D. Hazlett, P. Spanne, W.E. Soll, E.L. Muegge, and K.W. Jones. 1998. Pore level imaging of fluid transport using synchrotron X-ray microtomography. J. Petrol. Sci. Eng. 19:55–63. doi:10.1016/S0920-4105(97)00035-1

Costanza-Robinson, M.S., K.H. Harrold, and R.M. Lieb-Lappen. 2008. X-ray microtomography determination of air–water interfacial area-water saturation relationships in sandy porous media. Environ. Sci. Technol. 42:2949–2956. doi:10.1021/es072080d

Culligan, K.A., D. Wildenschild, B.S. Christensen, W.G. Gray, and M.L. Rivers. 2006. Pore-scale characteristics of multiphase flow in porous media: A comparison of air–water and oil–water experiments. Adv. Water Resour. 29:227–238. doi:10.1016/j.advwatres.2005.03.021

Culligan, K.A., D. Wildenschild, B.S. Christensen, W.G. Gray, M.L. Rivers, and A.B. Tompson. 2004. Interfacial area measurements for unsaturated flow through porous media. Water Resour. Res. 40:W12413. doi:10.1029/2004WR003278

Davis, D., L. Pyrak-Nolte, E. Atekwana, and D. Werkema. 2009. Microbial- induced heterogeneity in the relationship among acoustic properties of porous media. Geophys. Res. Lett. 36:L21405. doi:10.1029/2009GL039569

Davis, G.R. 1999. Image quality and accuracy in X-ray microtomography. Proc. SPIE Dev. X-ray Tomogr. 3772:147.doi:10.1117/12.363716

Davit, Y., G. Iltis, G. Debenesta, S. Veran-Tissoiresa, D. Wildenschild, M. Gerinoc, and M. Quintard. 2011. Imaging biofilms in porous media using X-ray computed microtomography. J. Microsc. 242:15–25.doi: 10.1111/j.1365-2818.2010.03432.x

Hassanizadeh, S.M., and W.G. Gray. 1993. Thermodynamic basis of capillary pressure, saturation, interfacial area and relative permeability using in porous media. Water Resour. Res. 29:3389–3405.

Iassonov, P., T. Gebrenegus, and M. Tuller. 2009. Segmentation of X-ray computed tomography images of porous materials: A crucial step for characterization and quantitative analysis of pore-network modeling. Transp. structures. Water Resour. Res. 45:W09415. doi:10.1029/2009WR008087

Iltis, G., R.T. Armstrong, D.P. Jansik, B.D. Wood, and D. Wildenschild. 2011. Imaging biofilm architecture within porous media using synchrotron-based X-ray computed microtomography. Water Resour. Res. 47:W02601.doi: 10.1029/2010WR009410

Joekar-Niasar, V., S.M. Hassanizadeh, and A. Leijnse. 2007. Insights into the relationships among capillary pressure, saturation, interfacial area and relative permeability using pore-network modeling. Porous Med. 74(2): 201–19. doi:10.1007/s11242-007-9191-7

Joekar-Niasar, V., M. Prodanovic, D. Wildenschild, and S.M. Hassanizadeh. 2010. Network model investigation of interfacial area, capillary pressure and saturation relationships in granular porous media. Water Resour. Res. 46:W06526. doi:10.1029/2009WR008585

Kaestner, A., M. Schneebeli, and F. Graf. 2006. Visualizing three-dimensional root networks using computed tomography. Geoderma 136:459–469. doi:10.1016/j.geoderma.2006.04.009

Kinney, J.H., and M.C. Nichols. 1992. X-ray tomographic microscopy (XTM) using synchrotron radiation. Annu. Rev. Mater. Sci. 22:121–152. doi:10.1146/annurev.ms.22.080192.001005

Lehmann, P., P. Wyss, A. Flisch, E. Lehmann, P. Vontobel, M. Krafczyk, A. Kaestner, F. Beckmann, A. Gygi, and H. Flühler. 2006. Tomographical imaging and mathematical description of porous media used for the prediction of fluid distribution. Vadose Zone J. 5:80–97. doi:10.2136/vzj2004.0177

Leis, A.P., S. Schlicher, H. Franke, and M. Strathmann. 2005. Optically transparent porous medium for nondestructive studies of microbial biofilm architecture and transport dynamics. Appl. Environ. Microbiol. 71(8):4801–4808. doi:10.1128/AEM.71.8.4801-4808.2005

Lindquist, W., and A. Venkatarangan. 1999. Investigating 3d geometry of porous media from high resolution images. Phys. Chem. Earth 25(7):593–599.

Lindquist, W.B. 1999. 3DMA-Rock, A Software Package for Automated Analysis of Pore Rock Structure in 3D Computed Microtomography Images; http://www.ams.sunysb.edu/~lindquis/3dma/3dma_rock/3dma_rock.html.

Lorensen, W.E., and H.E. Cline. 1987. Marching cubes: A high resolution 3D surface construction algorithm. Comput. Graph. 21:163–169. doi:10.1145/41997.41999

Luo, L., H. Lin, and S. Li. 2010. Quantification of 3-D soil macropore networks in different soil types and land uses using computed microtomography. J. Hydrol. 393:53–64. doi:10.1016/j.jhydrol.2010.03.031

Margaritondo, G. 2002. Elements of synchrotron light for biology, chemistry, and medical research. Oxford Univ. Press, Oxford, UK.

McCullough, E.C. 1975. Photon attenuation in computed tomography. Med. Phys. 2:307–320. doi:10.1118/1.594199

Oh, W., and B. Lindquist. 1999. Image thresholding by indicator kriging. IEEE Trans. Pattern Anal. Mach. Intell. 21(7):590–602. doi:10.1109/34.777370

Perret, J., S.O. Prasher, A. Kantzas, and C. Langford. 2000. A two-domain approach using CAT scanning to model solute transport in soil. J. Environ. Qual. 29:995–1010. doi:10.2134/jeq2000.00472425002900030039x

Peth, S., R. Horn, F. Beckmann, T. Donath, J. Fischer, and A.J.M. Smucker. 2008. Three-dimensional quantification of intra-aggregate pore-space features using synchrotron-radiation-based microtomography. Soil Sci. Soc. Am. J. 72:897–907. doi:10.2136/sssaj2007.0130

Porter, M.L., M.G. Schaap, and D. Wildenschild. 2009. Comparison of interfacial area estimates for multiphase flow through porous media using computed microtomography and lattice-boltzmann simulations. Adv. Water Resour. 32:1632–1640.doi:10.1016/j.advwatres.2009.08.009, Advances in Water Resources.

Porter, M.L., and D. Wildenschild. 2009. Validation of an image analysis method for computed microtomography image data of multiphase flow in porous systems. Comput. Geosci. 14:15–30.doi:10.1007/s10596-009-9130-5

Porter, M.L., D. Wildenschild, G. Grant, and J.I. Gerhard. 2010. Measurement and prediction of the relationship between capillary pressure, saturation, and interfacial area in a NAPL-water-glass bead system. Water Resour. Res. 46:W08512. doi:10.1029/2009WR007786

Prodanović, M., W.B. Lindquist, and R.S. Seright. 2006. Porous structure and fluid partitioning in polyethylene cores from 3D X-ray microtomographic imaging. J. Colloid Interface Sci., 298, 282–297.

Prodanović, M., W.B. Lindquist, and R. Seright. 2007. 3d image-based characterization of fluid displacement in Berea core. Adv. Water Resour. 30:214–226. doi:10.1016/j.advwatres.2005.05.015

Radon, J. 1986. On the determination of functions from their integral values along certain manifolds. IEEE Trans. Med. Imaging 5(4):170–176. doi:10.1109/TMI.1986.4307775

Rivers, M.L., S.R. Sutton, and P. Eng. 1999. Geoscience applications of X-ray computed microtomography. Proceedings of the SPIE Conference on Developments in X-ray Tomography II, July 1999, Denver, CO. p. 78–86.

Rodriguez, S.J., and P.L. Bishop. 2007. Three-dimensional quantification of soil biofilms using image analysis. Environ. Eng. Sci. 24(1):96–103. doi:10.1089/ees.2007.24.96

Schaap, M.G., M.L. Porter, B.S.B. Christensen, and D. Wildenschild. 2007. Comparison of pressure-saturation characteristics derived from computed tomography and Lattice Boltzmann simulations. Water Resour. Res. 43:W12S06. doi:10.1029/2006WR005730

Scheckel, K.G., R. Hamon, L. Jassogne, M. Rivers, and E. Lombi. 2007. Synchrotron X-ray absorption-edge computed microtomography imaging of thallium compartmentalization. Plant Soil 290:51–60.doi:10.1007/s11104-006-9102-7

Schnaar, G., and M.L. Brusseau. 2005. Pore-scale characterization of organic immiscible-liquid morphology in natural porous media using synchrotron X-ray microtomography. Environ. Sci. Technol. 39:8403–8410. doi:10.1021/es0508370

Schnaar, G., and M.L. Brusseau. 2006. Characterizing pore-scale configuration of organic immiscible liquid in multi-phase systems using synchrotron X-ray microtomography. Vadose Zone J. 5:641–648. doi:10.2136/vzj2005.0063

Seymour, J.D., J.P. Gage, S.L. Codd, and R. Gerlach. 2004. Anomalous fluid transport in porous media induced by biofilm growth. Phys. Rev. Lett. 93(19):198103. doi:10.1103/PhysRevLett.93.198103

Seymour, J.D., J.P. Gage, S.L. Codd, and R. Gerlach. 2007. Magnetic resonance microscopy of biofouling induced scale dependent transport in porous media. Adv. Water Resour. 30(6–7):1408–1420. doi:10.1016/j.advwatres.2006.05.029

Sham, T.K., and M.L. Rivers. 2002. A brief overview of synchrotron radiation. In: P.A. Fenter et al., editors, Reviews in mineralogy and geochemistry, Vol. 49: Applications of synchrotron radiation in low-temperature geochemistry and environmental science. Mineralogical Society of America, Chantilly, VA. p. 117–148.

Sharp, R.R., P. Stoodley, M. Adgie, R. Gerlach, and A. Cunningham. 2005. Visualization and characterization of dynamic patterns of flow, growth and activity of biofilms growing in porous media. Water Sci. Technol. 52(7):85–90

Sheppard, A.P., R.M. Sok, and H. Averdunk. 2004. Techniques for image enhancement and segmentation of tomographic images of porous materials. Phys. A 339:145–151. doi: 0.1016/j.physa.2004.03.057

Stock, S.R. 1999. X-ray microtomography of materials. Int. Mater. Rev. 44:141–164. doi:10.1179/095066099101528261

Sukop, M.C., H. Huang, C.L. Lin, M.D. Deo, K. Oh, and J.D. Miller. 2008. Distribution of multiphase fluids in porous media: Comparison between lattice Boltzmann modeling and micro-x-ray tomography. Phys. Rev. E Stat. Nonlin. Soft Matter Phys. 77:026710. doi:10.1103/PhysRevE.77.026710

Taina, I.A., R.J. Heck, and T.R. Elliot. 2008. Application of X-ray computed tomography to soil science: A literature review. Can. J. Soil Sci. 88:1–20. doi:10.4141/CJSS06027

Tippkötter, R., T. Eickhorst, H. Taubner, B. Gredner, and G. Rademaker. 2009. Detection of soil water in macropores of undisturbed soil using microfocus X-ray tube computerized tomography (mCT). Soil Tillage Res. 105:12–20. doi:10.1016/j.still.2009.05.001

Tracy, S.R., J.A. Roberts, C.R. Black, R. Colin, A. McNeill, R. Davidson, and S.J. Mooney. 2010. The X-factor: Visualizing undisturbed root architecture in soils using X-ray computed tomography. J. Exp. Bot. 61:311–313. doi:10.1093/jxb/erp386

Turner, M., L. Knüfing, C. Arns, A. Sakellariou, T. Senden, A. Sheppard, R. Sok, A. Limaye, W. Pinczewski, and M. Knackstedt. 2004. Three-dimensional imaging of multiphase flow in porous media. Physica A 339:166–172. doi:10.1016/j.physa.2004.03.059

Vogel, H.-J., J. Tölke, V. Schulz, M. Krafczyk, and K. Roth. 2005. Comparison of a lattice-Boltzmann model, a full-morphology model, and a pore network model for determining capillary pressure–saturation relationships. Vadose Zone J. 4:380–388. doi:10.2136/vzj2004.0114

Vogel, H.-J., U. Weller, and S. Schlüter. 2010. Quantification of soil structure based on Minkowski functions. Comput. Geosci. 36:1236–1245. doi:10.1016/j.cageo.2010.03.007

Werth, C.J., C. Zhang, M.L. Brusseau, M. Oostrom, and T. Baumann. 2010. A review of non-invasive imaging methods and applications in contaminant hydrogeology research. J. Contam. Hydrol. 113(1–4):1–24. doi:10.1016/j.jconhyd.2010.01.001

Wildenschild, D., J.W. Hopmans, A.J.R. Kent, and M.L. Rivers. 2005. A quantitative study of flow-rate dependent processes using X-ray microtomography. Vadose Zone J. 4:112–126. doi:10.2113/4.1.112

Wildenschild, D., J.W. Hopmans, C.M.P. Vaz, M.L. Rivers, and D. Rikard. 2002. Using X-ray computed tomography in hydrology: Systems, resolutions, and limitations. J. Hydrol. 267(3–4):285–297. doi:10.1016/S0022-1694(02)00157-9

Yang, X.M., H. Beyenal, G. Harkin, and Z. Lewandowski. 2000. Quantifying biofilm structure using image analysis. J. Microbiol. Methods 39(2):109–119. doi:10.1016/S0167-7012(99)00097-4

2

Tomographic Investigations Relevant to the Rhizosphere

Keith W. Jones,* Jun Wang, Yu-chen Chen,
Qingxi Yuan, W. Brent Lindquist, Lauren Beckingham,
Catherine A. Peters, Wooyong Um, Lee Newman,
Tara Sabo-Attwood, and Ryan Tappero

Abstract

The rhizosphere is a complex system that requires knowledge of fluid transport of nutrients and contaminants in the subsurface region, interaction of water, soils, microbes, and plant tissues at the soil–water–root interface, and transport of metals and organic compounds through the plant tissue. Tomography is a powerful analytical method that can be applied to all components of the rhizosphere and across the range of size scales of interest. Here, we describe relevant synchrotron computed microtomography experiments on soil and root structures that were performed at the Brookhaven National Synchrotron Light Source.

The rhizosphere consists of soil, root, water, and microbial components that exist together and that are related by a complex set of interactions. Understanding these interactions is difficult, requiring many types of experiments and the employment of sophisticated analytical tools for measurements from the molecular to the macro scale.

In particular, X-ray computed microtomography (CMT) has come into increasing use over the past 20 yr for experiments on the root–soil–water interface [see Taina et al. (2008) for a literature review]. Investigations have used both laboratory type scanners based on conventional X-ray tube sources and scanners

---

Abbreviations: BNL, Brookhaven National Laboratory; BSE, back-scattered-electron; CCD, charge-coupled device; CMT, computed microtomography; NSLS, National Synchrotron Light Source; SEM, scanning electron microscopy; TEM, transmission electron microscopy; TXM, transmission X-ray microscope.

K.W. Jones, R. Tappero, J. Wang, Y.-C. Chen, and Q. Yuan, Brookhaven National Lab., Upton, NY 11973-5000. W.B. Lindquist, Stony Brook Univ., Stony Brook, NY 11794. L. Beckingham and C.A. Peters, Princeton Univ., Princeton, NJ 08544. W. Um, Pacific Northwest National Lab., Richland, WA 99352. L. Newman, SUNY-ESF, Syracuse, NY 13210. T. Sabo-Attwood, Univ. of Florida, Gainesville, FL 32610. *Corresponding author (jones@bnl.gov).

doi:10.2136/sssaspecpub61.c2

Soil–Water–Root Processes: Advances in Tomography and Imaging. SSSA Special Publication 61. S.H. Anderson and J.W. Hopmans, editors. © 2013. SSSA, 5585 Guilford Rd., Madison, WI 53711, USA.

based on synchrotron X-ray sources. Experiments have been performed on both dry and wet samples and to determine soil characteristics, water uptake in roots and other parameters. Early synchrotron work at Brookhaven National Laboratory (BNL) was described by Spanne et al. (1994). Other interesting experiments selected from an extensive literature include the following: Hainsworth and Aylmore (1983), Gregory and Hinsinger (1999), Hamza et al. (2001), Gregory et al. (2003), Schmidt et al. (2010), Tracy et al. (2010), and Vaz et al. (2011).

Our objective is to illustrate different ways that we are looking at features of the soil–water–root rhizosphere topic using the commonality of the CMT technique as a unifying feature of the discussion. The experiments are our specific examples of the analytical power that CMT brings to the field. Other ongoing work at other laboratories is also evidence for the power of the CMT approach.

Specifically, we describe applications of synchrotron-computed microtomography to precipitation–dissolution reactions in soil samples, transport of Au nanoparticles in root samples, and uptake of Fe and As in water into root structures. Supplementary measurements with scanning electron microscopy and X-ray fluorescence maps are discussed. Finally, improvements to the analytical techniques based on upgrades to synchrotron and computational hardware capabilities now in progress will be mentioned.

Materials and Methods

The CMT technique is a common factor in the work on the several different types of materials investigated. Therefore, we first describe the CMT equipment and supplementary scanning electron microscopy (SEM) apparatus and then follow with a presentation of the sample details. The SEM was used to provide data that supplemented the CMT data by providing data with more than order-of-magnitude better spatial resolution. The applicability of SEM is limited since it is a destructive analytical method and does not give volume information.

CMT experiments were performed at beam lines X2B and X8C at the National Synchrotron Light Source (NSLS) at BNL. Tomography at both beam lines is based on measurement of the sample attenuation coefficient with an area X-ray detector as a function of rotation angle.

The X2B equipment employs a silicon monochromator to select X-ray energies from 10 to 35 keV. Slits are then used to define the beam size to roughly 5 by 5 mm. Samples were mounted on a rotator that could be adjusted in all three dimensions for selection of the region of the sample to be investigated. The X-rays passing through the sample were detected with a CsI (Tl) scintillator viewed by a magnifying lens used to focus the emitted light onto a 16-bit charge-coupled device (CCD) camera producing image sizes of 1340 by 1300 pixels. We used pixel sizes of 2 and 4 μm during the present experiments, adjusting the field-of-view to define the portion of the sample to be investigated. Acquisition times were generally between 2 and 3 h. Tomograms were based on 1200 views of the sample collected over a rotation range of 180°. The volume images were produced using several proprietary and nonproprietary software programs. The ExxonMobil Corporation provided the proprietary reconstruction software specific to the beam line. Nonproprietary software included tomo_display by Rivers (2010) ImageJ (National Institutes of Health, 2011) and Drishti (Limaye, 2012).

The nano-scale tomography was performed at X8C, a new dedicated imaging beamline at NSLS equipped with a transmission X-ray microscope (TXM) manufactured by Xradia, Inc. The data were collected with an X-ray beam energy of 8.4 keV with a 20 by 20 µm field of view. The size of the CCD camera image was 2048 by 2048 pixels. A spatial resolution of 30 nm was achieved for each projection. The tomogram was obtained based on 361 views of the sample collected over a rotation range of 180°. The reconstructions and visualization of the experimental data was done with proprietary software from Xradia, Inc.

Back-scattered-electron (BSE) images were taken using the FEI Quanta 200 FEG Environmental-SEM at the Princeton Institute for Science and Technology of Materials (PRISM) Imaging and Analysis Center. Additional images were taken in low vacuum, or environmental SEM, mode with electron energy of 15 keV and working distance of 10 mm. Energy dispersive X-ray spectroscopy was performed to determine mineral identification.

We investigated four samples in the present work:

1. Soil samples were packed in 10-cm-long polyetheretherketone (PEEK) tubes with an inside diameter of 3.1 mm. Sequential measurements were made of the soil structure following exposure to high pH fluids over a period of several months between measurements. The soils were sieved samples from a site at Pacific Northwest National Laboratory. To prepare for SEM imaging, the sample column was flushed with ethanol and dried with desiccated air before flowing a low-viscosity epoxy through the column. Sections were cut from the column for SEM analysis.

2. An ordinary laboratory wooden stirring rod (Solon Mfg. Co, Solon, ME. EV-Wood 7340–753–5565) was used as a convenient test sample plant fibrous material.

3. Tomato plants (*Solanum lycopersicum* L., 'Brandywine') were grown in a soil-less media, and then transferred to 10-mL tubes containing half-strength Hoagland's solution after the emergence of the first true leaves. After 2 wk of growth, plants were transferred to tap water for 2 d, and then the ta water was replaced with a solution containing Au nanoparticles. Plants remained in the nanoparticle solution for 3 d, with transpired solution being replaced with tap water. Control plants were transferred and held in tap water. Established and published protocols were used for the synthesis and characterization of the 3.5-nm gold particles (Sabo-Attwood et al., 2011). Characterization was done using UV-Vis, transmission electron microscopy (TEM), and electrophoretic mobility in water. Prior studies with various capping agents showed that the citrate agent had the lowest level of toxicity to the plants, so this was used for all plant-based studies (Sabo-Attwood et al., 2011).

4. Rice plants (*Oryza sativa* L. ssp. *Japonica*) were grown hydroponically for 30 d in magenta boxes aseptically in a biology lab under growth lights. The nutrient solution was prepared using a Hoagland's nutrient packet altered by adding Fe at 11 ppm and As at 400 ppb.

Results and Discussion
Soils

Tomographic measurements were made on the samples at varying time intervals during the treatment process with caustic waste solutions, as described in the previous section. In general, one measurement was made where the caustic liquid entered the column; two measurements were made at points about 3 mm apart, and one where the fluid flowed from the column. The results were analyzed using the Stony Brook 3DMA-Rock program (http://www.ams.sunysb.edu/~lindquis/3dma/3dma_rock/3dma_rock.html). It was used to segment each three-dimensional image in the time sequence into pore and grain phases and to derive porosity and distributions for pore size, throat size, and tortuosity.

The tomography work was complemented by BSE measurements on sections taken from the column after the CMT work was completed. The spatial resolution and elemental detection sensitivity were much better than those delivered by the CMT method. However, to make the BSE measurements, it was necessary to section the sample and was, therefore, not suitable for making sequential measurements. In contrast, CMT is not only nondestructive, but also delivers measurements over an extended volume giving a better assessment of the sample variability. Such volume data also provides a better basis for a detailed validation of the results of numerical computations of reactive flow.

Results from the SEM–BSE scans are shown in Fig. 2–1 and 2–2. Figure 2–1 is a scan over an entire cross-sectional area of the soil column. Figure 2–2 shows grain details at higher magnifications. Different minerals are present, as are a variety of very small particles. Both images show the rich variety in the structures of the mineral grains and also the existence of micro cracks in the grains. SEM–BSE analyses of the section data showed regions of precipitation, and gave a two-dimensional estimate distribution of pore and throat sizes. The fine structure observed on the grain surfaces shown in Fig. 2–2 and its effects on flow, if any, should be further investigated.

A typical section through the sample when imaged using CMT is shown in Fig. 2–3. Here the pixel size is 4 μm, much larger than that employed for the SEM–BSE work. However, the general features are the same. A representation of the volume image is shown in Fig. 2–4. Analysis of the data was performed using the 3DMA-Rock software to give time dependent values for the pore volume, throat area, effective pore radius and tortuosity (Cai et al., 2009; Crandell et al., 2012). Good agreement was obtained between the SEM–BSE results when compared to the results from the final image in the CMT time sequence of the column. For example, total porosity changes were found to be 0.04 while values for the SEM–BSE work ranged from 0.068 to 0.11. The values are derived from experiments with different spatial resolutions and with different types of data analysis. This leads to differences in the results found through evaluation of pore and throat sizes and other parameters. The experiment also showed that it is possible to preserve sample integrity despite many shipments between the PNNL treatment location and the BNL imaging location.

Further visualization work was performed using the three-dimensional immersive cube facility at the Center for Excellence in Wireless and Information Technology at Stony Brook University. The software enables a complete three-dimensional visualization of the CMT volume; the immersive cube makes possi-

Fig. 2–1. Scanning electron microscopy–back-scattered-electron (SEM–BSE) image of a section of the entire sample of Hanford sand used in the sequential tomography imaging experiments. Digital processing was used to remove secondary mineral precipitation.

Fig. 2–2. Left panel. Scanning electron microscopy–back-scattered-electron (SEM–BSE) image of Hanford sand reacted with simulated tank waste showing cancrinite (C) precipitation on grain surface. Identified mineralogy includes plagioclase feldspar, P, amphibole, A, and an iron titanium compound, T. The image resolution is 0.1 μm. Right Panel. Reacted Hanford sand taken using SEM–BSE imaging depicting sub-grain scale mineralogy and micro-fractures at a resolution of 0.24 μm. Mineralogy consists of cancrinite precipitation on grain surfaces, C, plagioclase feldspar, P, iron titanium compound, T, quartz, Q, potassium feldspar, K, and amphibole, A.

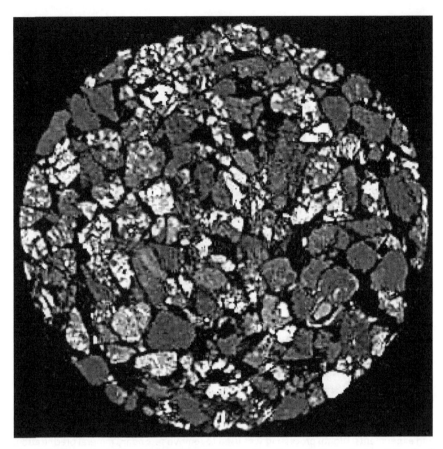

Fig. 2–3. Computed microtomography (CMT) section through the column of Hanford soil. It is not taken at the same location as the scanning electron microscopy–back-scattered-electron (SEM–BSE) image shown in Fig. 2–1. The pixel size is 4 µm. The differences in the different spatial resolutions are clear.

ble a fly through of the data so that the complexity of the pore structures become apparent and can be compared with the computer models. Our goal is to combine a flow model that incorporates dissolution and/or precipitation processes and can show points where they are important. It will also be possible to deliver experimental and theoretical visualization results to a computer at remote locations such as offices or beam lines. In particular, this will be much needed at new synchrotron facilities such as NSLS II where data acquisition will be substantially higher than at present. As an example, a partially transparent view of a pore is shown in Fig. 2–5. It can be seen that the structures are very complex and likely could demand more refined consideration of the definitions of pore and throat sizes.

Wood Test Sample

We imaged an ordinary laboratory stirring stick to determine the feasibility of measuring fibrous materials with the X-ray beams provided at beam line X2B. In

Fig. 2–4. A three-dimensional view of a typical computed microtomography (CMT) tomographic volume is displayed. The voxel size is 4 μm.

this case, we used a pixel size of 4 μm and energy of 12.0 keV. A section through the rod is shown in Fig. 2–6. A partial view of the volume is shown in Fig. 2–7. The results show that it is possible obtain useful images at this X-ray energy and to see changes (small) in the wood structures as shown by variation in attenuation coefficients. This could be a useful analytical method for characterizing different types of woods in a search for improved biomass as an energy source.

Root Structures and Composition

We investigated samples of roots from tomato and rice seedlings. The aim was to image the morphology of the root and stem structures and to investigate the possible uptake from the hydroponic growing medium of gold nanoparticles into the tomato plant and of Fe and As into the rice plant.

Uptake of Gold Nanoparticles in Tomato Roots

Gold nanoparticles have been proposed for use in biomedical applications for targeted drug delivery and to increase the efficiency of treatment of cancer tumors. It has been accepted thought that the gold nanoparticles are inert, and exposure to them does not pose a health risk from exposure. If the gold particles are to be used for this and other applications to health and in beauty products (e.g., skin creams), the gold particles will move into our waste water stream, and from there into the soils and sediments. Thus, plants will be exposed to the particles, and it is important to know if they are, in fact, as inert as claimed. Prior work done by

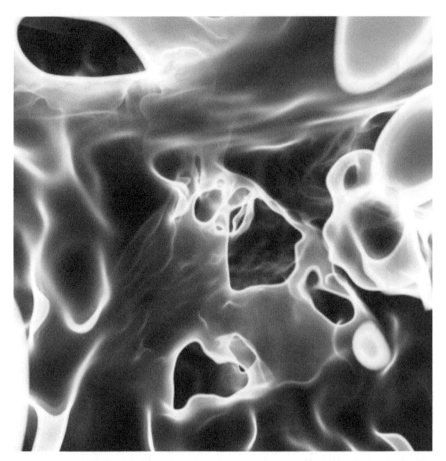

Fig. 2–5. Expanded view of the data shown in Fig. 2–4 showing the complex nature of the pore-grain interfaces. The interior can be explored in real time using the immersive three-dimensional cube at Stony Brook University.

some of our group has shown that the particles are toxic to the plants, and ongoing studies are elucidating the mechanism of toxicity.

Prior studies performed with tobacco (*Nicotiana tobacum* L. 'Xanthi') plants have shown that within 3 to 5 d, gold nanoparticles move into the plants, enter the transpiration stream, and move to the aerial portions of the plants (Sabo-Attwood et al., 2011). Additionally, nanoparticles within the plant root cells will occasionally form aggregates that are visible using brightfield microscopy (100×). Analysis of the aggregates by TEM with elemental analysis confirmed that the aggregates were of gold nanoparticles. For these and other nanoparticle studies, it is important to know where in the plant they are localized to focus the toxicological studies. Thus, working with methodology that can both do localization and speciation becomes important.

Transverse and longitudinal cross-sections of the tomato root sample are shown in Fig. 2–8. Two images are shown for each section. The image on the left shows all attenuation values and the other image on the right shows only values

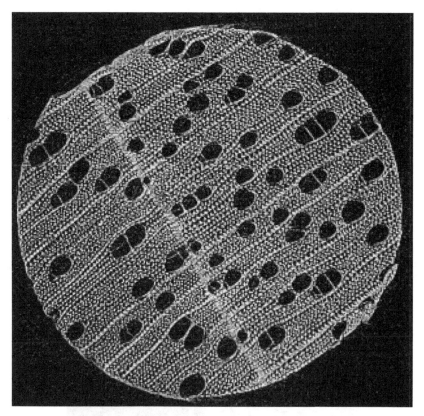

Fig. 2–6. Section through a computed microtomography (CMT) volume of a wooden stirring rod showing the morphology of the wood at the micrometer scale. The pixel size is 4 μm.

above the limit for organic material. The latter values are attributed to the presence of Au nanoparticles. A rough estimate of the mass of the gold can be made by assuming that the attenuation coefficient should be twice that of the organic material. Inspection of the attenuation values for Au at the L-edge gives values of 0.002 and 0.001 g cm^{-3} for below and above the edge as a rough estimate for the detection limits for Au in a single voxel. The bulk concentration for this slice, as estimated from a histogram showing the fraction of pixels with Au, is in the ppm range. Gold can also be detected by subtracting a tomogram taken below the Au L-edge at 11.92 keV from a tomogram taken above the edge. The difference section through the stem displayed in Fig. 2–9 is consistent with the interpretation of the attenuation coefficients since it shows high-value pixels, especially around the periphery.

One tomogram on a root tip was collected with the TXM instrument working with a 30-nm pixel size. One view of the tomogram is given in Fig. 2–10; the high value voxels at the left are on the surface of the root. Other regions to the right are in the interior of the root and show uptake in discrete regions. This first measure-

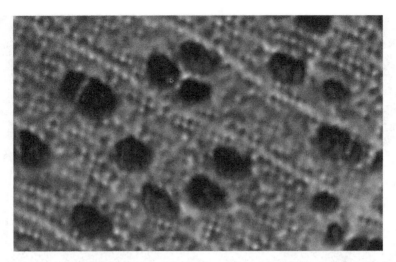

Fig. 2–7. Expanded view of the computed microtomography (CMT) stirring rod volume showing the fine structure of the material.

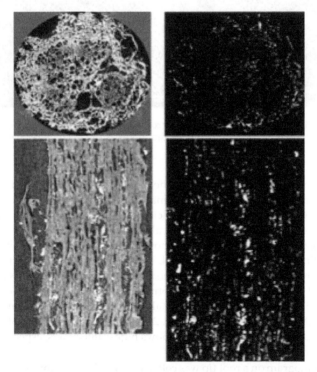

Fig. 2–8. Transverse (top row) and longitudinal (bottom row) sections through a tomato root stem. All attenuation values are shown in the left column and attenuation values chosen to exclude organic compound s are shown in the right column. This is evidence for the detection of Au nanoparticles. The pixel size is 2 μm.

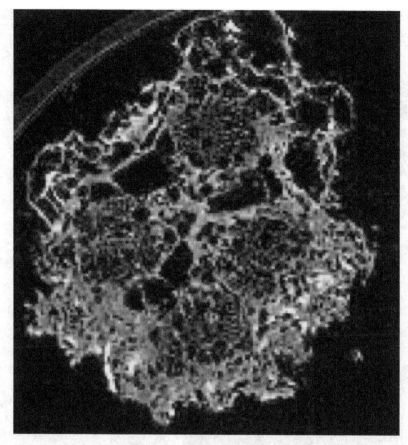

Fig. 2–9. The tomato stem section shown is the difference between data taken above and below the Au L-edge at 11.92 keV and also demonstrates that there are Au nanoparticles widely distributed through the sample.

ment indicates that tomography can be used to detect Au nanoparticles in particular cells and help delineate the pathways of the particles into the root structures.

Uptake of Iron and Arsenic in Rice Roots

Arsenic is a common contaminant in soils and water in many regions around the world and represents a substantial threat to public health. In particular, food crops such as rice grown in areas with high concentrations of As in soil or irrigation water could be a pathway resulting in human exposure. Investigation of the ways that As is incorporated into rice roots will yield information that is needed to define the magnitude of the problem and guide work designed to mitigate the uptake.

Experimentally, characterization of the As content in plant roots can be done across size scales by use of X-ray microfluorescence experiments, and a spatial association for Fe (plaque) and As on rice roots was observed initially using this technique (Moyer et al., 2010). CMT based on X-ray attenuation can also be useful tool. Specifically, the association of As with Fe in the rice roots is interesting since

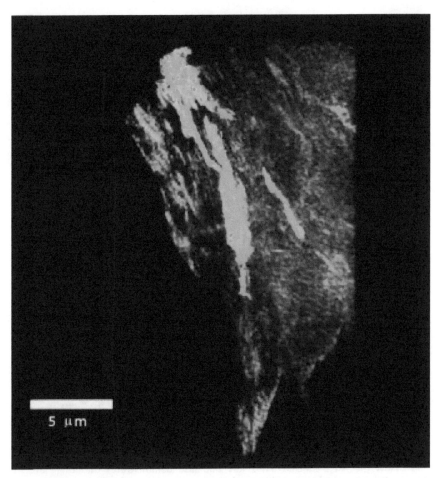

Fig. 2–10. A tomographic volume of a tomato root tip taken with 30-nm resolution with the National Synchrotron Light Source (NSLS) X8C TXM instrument. The image shows Au nanoparticles accumulated on the root surface and also in the interior of root. This is in agreement with the data shown in Fig. 2–8 and 2–9.

the Fe could provide a way to bind As and thereby reduces the transport of As to the rice grains. The present experiment explores detection of Fe by a CMT measurement below the absorption edge of the As K X-ray. A search for the presence of As was made by subtracting tomograms made below the As X-ray absorption K-edge from a tomogram made above the edge. We assume that the native metal concentrations in the root are small compared to uptake of Fe from the growing medium.

A CMT image of the rice roots acquired at an energy of 12.0 keV, above both the As and Fe K X-ray absorption edges, is shown in Fig. 2–11. It includes the tube used to hold the roots during the measurement and all values of the attenuation coefficients. If only values of the attenuation coefficients that exclude organic material are presented, then only voxels containing metals (e.g., Fe and As) will be visible. A visualization of these voxels is given in Fig. 2–12. It can be seen

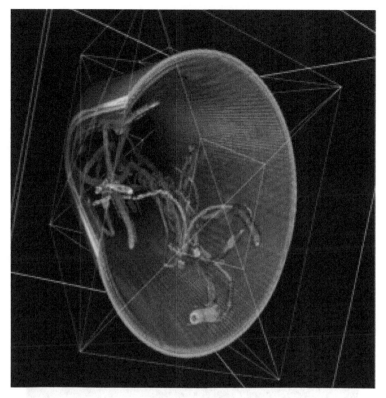

Fig. 2–11. Computed microtomography (CMT) volume taken at 12.0 keV, above the As K X-ray absorption edge, showing multiple rice roots contained in a polyimide tube. The pixel size is 2 μm.

that metals are readily taken up by the plant. Figure 2–13 shows cross-sections through the roots obtained at an X-ray energy of 11.5 keV, below the As K X-ray absorption edge. The image on the left in Fig. 2–13 shows the results for all values of the attenuation coefficients. The image on the right excludes the values for the organic materials and shows that Fe and other metals are present in the roots. Using CMT, no evidence was found for the presence of As in roots because the detection limits for the technique are too high. In general, studies of metal uptake in plants will be enhanced by combining X-ray microprobe measurements on single sections with the CMT technique to trace the transport through a specimen volume.

Fluorescence CMT

The fluorescence CMT technique has been used for many years (Vincze et al., 2002; Hansel et al., 2001; De Nolf and Janssens, 2010; Lombi et al., 2011) and is being vigorously applied at present. It gives multi-element detection combined with good spatial resolution. However, the attenuation length of the characteristic X-rays in the sample material limits its use. Nevertheless, it is a powerful technique that will be widely used at the new synchrotron facilities coming into operation.

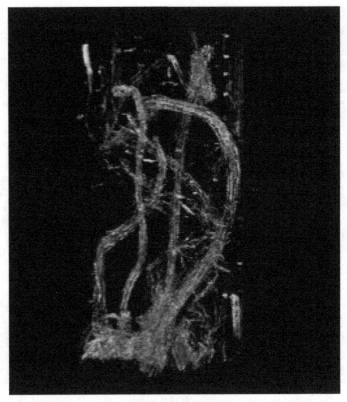

Fig. 2–12. Data shown in Fig. 2–11 showing only voxels with high attenuation values corresponding to metals (e.g., Fe/As). The pixel size is 2 μm.

Conclusions and Future Directions

We described different experiments that applied the technique of attenuation-coefficient based synchrotron CMT to rhizosphere related topics. First, we described an experiment that looked at changes to soil properties caused by transport of a caustic waste solution as a function of treatment time. Values of porosity and other parameters were obtained for the same regions of the sample since CMT is a nondestructive method. We also found that it is necessary to use extended treatment times to obtain measurable effects in the soil and made sequential measurements over a 2-yr time span. We suggest that this type of experiment could be used to investigate effects of acid rain on soil structures. This is of interest since increased levels of atmospheric CO_2 could lead to an increase in acidity of rain, thereby impacting soil structures. Second, we described measurements done on wood and root samples. It was possible to define the structures of the materials. In addition, based on attenuation coefficient measurements, it was possible to localize regions of metal content arising from either natural or anthropogenic sources. The observation of Au nanoparticles was of particular interest since it shows that CMT can be a useful technique for understanding the pathways and impacts of nanoparticles in the environment.

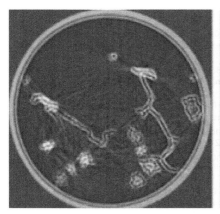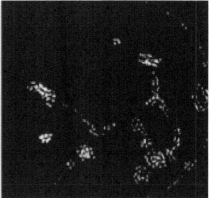

Fig. 2–13. Arbitrary section through the rice roots shown in Fig. 2–11 and 2–12. The X-ray energy was 11.5 keV. The view on the left shows all attenuation values while the view on the right excludes organic materials and shows only the high attenuation values related to Fe. The pixel size is 2 μm.

Experiments of the type described here will be moved to a new level in the next 5 to 10 yr with the ongoing upgrading of beam lines at existing synchrotrons and the construction of new synchrotron centers around the world. For example, the new NSLS II at BNL will give photon fluxes two orders of magnitude higher than are now available at NSLS I, coupled with sub-micrometer beam sizes. Time resolved experiments will become routine and times for fluorescence will be shortened so that experiments can move from a single slice mode to acquisition of true volume data.

The increased flood of data will require much innovation in the ways that the data are handled. There is a need to deliver three-dimensional-rendered data back to experimenters as they are working at the synchrotron to guide the ongoing data taking. There is also a need to merge theoretical flow and other models with the experimental data on a real-time basis. This gives the experimenter guidance on where additional experimental work would be useful input to model calculations and reality check on results of quantitative analysis.

Immersive gigapixel three-dimensional displays, such as the one at Stony Brook University (http://www.cs.stonybrook.edu/~realitydeck/index.html), will be a new help for experimenters and modelers for comparison of calculations with physical structures.

Acknowledgments

Work at Brookhaven National Laboratory was supported by the U.S. Department of Energy under Contract no. DE-AC02-98CH10886. The U.S. Department of Energy, Office of Science, and Office of Basic Energy Sciences supported use of the National Synchrotron Light Source. Other support was provided by the Department of Energy under Award no. DE-FG02-09ER64747 (SUNY Stony Brook), DE-FG02-09ER6748 (Princeton University), and KP1702030-54908 (Pacific Northwest National Laboratory). We also acknowledge the use of PRISM Imaging and Analysis Center, which is supported in part by the NSF MRSEC program through the Princeton Center for Complex Materials (grant DMR-0819860).

References

Cai, R., W.B. Lindquist, W. Um, and K.W. Jones. 2009. Tomographic analysis of reactive flow induced pore structure changes in column experiments. Adv. Water Resour. 32:1396–1403. doi:10.1016/j.advwatres.2009.06.006

Crandell, L.E., C.A. Peters, W. Um, K.W. Jones, W.B. Lindquist. 2012. Changes in the pore network structure of Hanford sediment after reaction with caustic tank wastes. J Contam Hydrol. 131:89–99.

De Nolf, W., and K. Janssens. 2010. Micro X-ray diffraction and fluorescence tomography for the study of multilayered automotive paints. Surf. Interface Anal. 42:411–418. doi:10.1002/sia.3125

Gregory, P.J., and P. Hinsinger. 1999. New Approaches to studying chemical and physical changes in the rhizosphere: An overview. Plant Soil 211:1–9. doi:10.1023/A:1004547401951

Gregory, P.J., D.J. Hutchinson, D.B. Fead, P.M. Jenneson, W.B. Gilboy, and E.J. Morton. 2003. Non-invasive imaging of roots with high-resolution X-ray micro-tomography. Plant Soil 255:351–359. doi:10.1023/A:1026179919689

Hainsworth, J.M., and L.A.G. Aylmore. 1983. The use of computer assisted tomography to determine spatial distribution of soil water content. Aust. J. Soil Res. 21:435–443. doi:10.1071/SR9830435

Hamza, M.A., S.H. Anderson, and L.A.G. Aylmore. 2001. Studies of soil water drawdowns by single radish roots at decreasing soil water content using computer-assisted tomography. Aust. J. Soil Res. 39:1387–1396. doi:10.1071/SR98057

Hansel, C.M., S. Fendorf, S. Sutton, and M. Newville. 2001. Characterization of Fe plaque and associated metals on the roots of mine-waste impacted aquatic plants. Environ. Sci. Technol. 35:3863–3868.

Limaye, A. 2012. Drishti ver. 2.0. http://anusf.anu.edu.au/Vizlab/drishti/

Lombi, E., M.D. de Jonge, E. Donner, P.M. Kopittke, D.L. Howard, R. Kirkham, C.G. Ryan, and D. Paterson. 2011. Fast X-ray Fluorescence Microtomography of Hydrated Biological Samples. PLoS ONE 6(6):e20626. doi:10.1371/journal.pone.0020626

Moyer, C. 2012. The role of iron plaques in immobilizing arsenic in the rice–root environment. M.Sc. Thesis. DE1971t6. Univ. of Delaware, Newark, DE.

National Institutes of Health. 2011. ImageJ ver. 1.46. National Institutes of Health, Bethesda, MD.

Rivers, M. 2010. GSECARS tomography processing software. University of Chicago, Chicago, IL. http://cars9.uchicago.edu/software/idl/tomography.html

Sabo-Attwood, T., J.M. Unrine, J.W. Stone, C.J. Murphy, S. Ghoshroy, D. Blom, P.M. Bertsch, and L.A. Newman. 2011. Uptake, distribution and toxicity of gold nanoparticles in tobacco (Nicotiana xanthi) seedlings. Nanotoxicology 6:353–360.

Schmidt, S., P.J. Gregory, A.G. Bengough, D.V. Grinec, and I.M. Young. 2010. Visualizing and quantifying rhizosphere processes: Root-soil contact and water uptake. 19th World Congress of Soil Science, Soil Solutions for a Changing World, 1–6 Aug. 2010, Brisbane, Australia.

Spanne, P., K.W. Jones, L. Prunty, and S.H. Anderson. 1994. Potential applications of synchrotron computed microtomography to soil science. SSSA Spec. Publ. 36. SSSA, Madison, WI. p. 43–57.

Taina, I.A., R.J. Heck, and T.R. Eliot. 2008. Application of X-ray computed tomography to soil science: A literature review. Can. J. Soil Sci. 88:1–19. doi:10.4141/CJSS06027

Tracy, S.R., J.A. Roberts, C.R. Black, A. McNeill, R. Davidson, and S.J. Mooney. 2010. The X factor: Visualizing undisturbed root architecture in soils using X-ray computed tomography. J. Exp. Bot. 61:311–313. doi:10.1093/jxb/erp386

Vaz, C.M.P., I.C. de Maria, P.O. Lasso, and M. Tuller. 2011. Evaluation of an advanced benchtop micro-computed tomography system for quantifying porosities and pore-size distributions of two brazilian oxisols. Soil Sci. Soc. Am. J. 75:832–841. doi:10.2136/sssaj2010.0245

Vincze, L., B. Vekemans, I. Szalolu, K. Janssens, R. Van Grieken, H. Feng, K.W. Jones, and F. Adams. 2002. High resolution x-ray fluorescence microtomography on single sediment particles. In: U. Bonse, editor, Proceedings of developments in x-ray tomography III, SPIE Vol. 4503. SPIE, Bellingham, WA. p. 240–248.

3

Synchrotron X-Ray Microtomography —New Means to Quantify Root Induced Changes of Rhizosphere Physical Properties

Jazmín E. Aravena, Markus Berli,* Manoj Menon, Teamrat A. Ghezzehei, Ajay K. Mandava, Emma E. Regentova, Natarajan S. Pillai, John Steude, Michael H. Young, Peter S. Nico, and Scott W. Tyler

Abstract

The rhizosphere, a thin layer of soil surrounding plant roots, plays a dynamic role in the hydrologic cycle by governing plant water and nutrient uptake. Study of rhizosphere soil structure formation due to mechanical processes has been limited by a lack of nondestructive techniques to quantify the dynamic nature of this region. In this chapter, we present recent developments in visualizing how growing roots modify their physical environment by moving soil particles, deforming aggregates and decreasing the amount of inter-aggregate pores while creating hydraulic pathways that connect neighboring soil aggregates using noninvasive, synchrotron X-ray microtomography (XMT). Image-processing tools were applied for quantifying root-induced rhizosphere alterations from XMT grayscale images as well as to transform XMT images into finite element meshes, building a bridge from nondestructive rhizosphere visualization to micromechanical and hydraulic simulations.

Abbreviations: CCD, charge-coupled-device; FCM, Fuzzy C-means; ROI, region of interest; XMT, X-ray microtomography.

J.E. Aravena, Dep. of Civil and Environmental Engineering, Univ. of Nevada, Reno, NV 89557. M. Berli and M.H. Young, Division of Hydrologic Sciences, Desert Research Institute, Las Vegas, NV 89119. M. Menon, J. Steude, and S.W. Tyler, Dep. of Geological Sciences and Engineering, Univ. of Nevada, Reno, NV 89557. T.A. Ghezzehei, School of Natural Sciences, Univ. of California, Merced, CA 95343. A.K. Mandava, E.E. Regentova, and N.S. Pillai, Dep. of Electrical and Computer Engineering, Univ. of Nevada, Las Vegas, NV 89154. P.S. Nico, Geochemistry Dep., Lawrence Berkeley National Lab., Berkeley, CA 94720; M. Menon, now at: Kroto Research Institute, Dep. of Civil and Structural Engineering, Univ. of Sheffield, S3 7HQ, UK. M.H. Young, now at: Bureau of Economic Geology, Jackson School of Geosciences, Univ. of Texas at Austin, Austin, TX 78712. *Corresponding author: Desert Research Institute, Division of Hydrologic Sciences, 755 E. Flamingo Rd., Las Vegas, NV 89052. *Corresponding author (markus.berli@dri.edu).

doi:10.2136/sssaspecpub61.c3

Soil–Water–Root Processes: Advances in Tomography and Imaging. SSSA Special Publication 61. S.H. Anderson and J.W. Hopmans, editors. © 2013. SSSA, 5585 Guilford Rd., Madison, WI 53711, USA.

Soil is a physical environment where roots, micro-organisms, and soil fauna live. It is often a difficult environment where resources (water, air, nutrients) can be scarce and patchy, with considerable variations within the soil profile. Compared with animals, plant and soil micro-organisms have only a limited ability to move toward nutrient-enriched zones. Even when abundant, availability of soil resources is often limited to organisms due to the capacity of the soil matrix to bind water and nutrients, so that roots have evolved to adapt and to influence their environment (Hinsinger et al., 2005; Richardson et al., 2009).

For almost 150 yr, researchers have recognized the role of soil structure in a broad range of vadose zone processes. The structure of the soil controls its ability to receive, store, and transmit water, to cycle carbon and nutrients, and to disperse anthropogenic and natural contaminants. It also supports root development and determines resistance to soil erosion. The scales of interest in the study of soil structure range from angstroms to kilometers. Although there is some degree of similarities across these disparate scales, as shown by the application of fractal theory (Kay, 1997), the most appropriate scale will depend on the processes under consideration. To study processes such as plant growth, root penetration, storage of water, and movement of nutrients, the most relevant scales range from microns to centimeters or meters depending on the pedon scale.

Soil Structure and Root Architecture

Microstructure is responsible for most of the soil physical properties that are necessary for the proper function of soil in agriculture and the environment (Dexter, 2004). Soils containing more than 12% clay or even pure sandy soils with some salts tend to form aggregates (Horn and Smucker, 2005). Soil aggregate formation, stabilization, and turnover occur through complex interactions of physical processes, chemical associations, and biological activity (Six et al., 2004). Many processes make soil aggregate structure dynamic. These processes include wetting and drying, freezing and thawing, root growth, plant water uptake, soil faunal activities, and tillage and other disturbances. Aggregated soils have hierarchical structure. Primary particles and silt-sized aggregates (<50 μm diameter) bind together to form micro-aggregates (50–250 μm diameter), which are generally more stable and less susceptible to disturbance (Tisdall and Oades, 1982; Dexter, 1988). These primary and secondary structures bind to form macro-aggregates (>250 μm diameter).

Aggregated soils are typically composed of textural and structural porosity. Textural porosity occurs between the primary mineral particles and is little affected by soil management. Structural porosity is composed of cracks, biopores, and other macrostructures produced by swelling and shrinkage, faunal burying activities, or tillage (Dexter, 2004). It is sensitive to management factors such as tillage, trafficking, and harvest, leading to compaction and other forms of structure deterioration. Structural porosity can be destroyed by soil compaction. This preferential loss of the largest pores changes the pore size distribution and hence the water retention characteristics.

Root architecture is the result of plant genetics, productivity, and interactions between the roots and the soil, where water and nutrients are spatio-temporally variable and scarce (Gregory, 2006). Root architecture depends on the tradeoff between minimizing energy expenditure and carbon transport, while maximizing soil exploration for water and nutrients uptake (Hargreaves et al., 2008; Javaux et al., 2008). And also it is influenced by soil structure and nutrient distribution at the plant scale (Lynch, 1995; Pierret et al., 2007). Growing roots modify their physical environment by moving soil particles, deforming aggregates, and decreasing the amount of inter-aggregate pores, while creating hydraulic pathways that connect neighboring soil aggregates. The study of intra- and inter-aggregate pore connectivities, to resolutions of a few microns, may provide better knowledge of root, microbial, and soil particle interactions, feedbacks, and rhizosphere processes (Horn and Smucker, 2005).

The Rhizosphere

The rhizosphere is the thin layer of soil that surrounds living roots; it is distinguished from the bulk soil by the root influence (Greacen et al., 1968; Dexter, 1987; Anderson et al., 1993; Whalley et al., 2005; Gregory, 2006; Bengough et al., 2010). It is directly influenced by root growth, mucilage production, exudates, and microbial activity (Feeney et al., 2006; Gregory, 2006; Watt et al., 2006). Rhizosphere characteristics are very important as they control physical, chemical, and biochemical processes that affect plant growth (Young, 1998; Whalley et al., 2005; Gregory, 2006; Hinsinger et al., 2006; Dessaux et al., 2009; Hinsinger et al., 2009). It was first described by Lorenz Hiltner in 1904 (Hiltner, 1904; Hartmann et al., 2008), and since then it has been the focus of agricultural research for many years (Anderson et al., 1993).

Most of the work related to the rhizosphere has focused on the biological and chemical processes (Young, 1998), the physical processes occurring in the rhizosphere being the least documented, despite their potential consequences for the movement of water and solutes in and out of the root (Gregory and Hinsinger, 1999; Hinsinger et al., 2005) and through the vadose zone.

Pore-scale soil physical processes are getting increasing attention due to their potential to better understand plant-soil interactions, such as water flow and nutrient transport from the soil to the root. When a root expands radially, it causes considerable structural alteration in the rhizosphere, such as localized compaction and increasing bulk density, which are well documented (Greacen et al., 1968; Dexter, 1987; Whalley et al., 2005). Rhizosphere compaction generally has been considered a disadvantage for root growth (Hamza and Anderson, 2005). However, recent studies on micro-mechanics and hydraulics of partially saturated aggregated soils (Berli et al., 2008; Carminati et al., 2008, Aravena et al., 2011) indicate that rhizosphere compaction may be beneficial for the plant by enhancing hydraulic contact between neighboring soil aggregates (which promotes water and nutrient flow to the roots) while preserving the volume and connectivity of large voids between aggregates that are essential for gas exchange (Fig. 3–1).

Ascertaining the role of rhizosphere compaction for water and gas flow is a difficult undertaking because of the scale and delicacy of the involved processes. Earlier attempts used miniaturized infiltration devices (Hallett et al., 2003; Whalley et al., 2004, 2005) and traditional imaging techniques (e.g., thin-section microscopy) to explore rhizosphere hydraulic properties. However, these techniques involve substantial soil or root disturbance or even root destruction, and

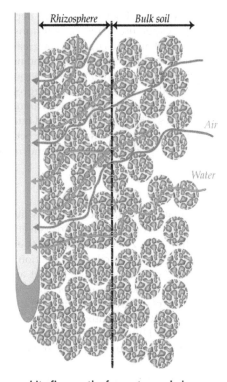

Fig. 3–1. Rhizosphere and its flow paths for water and air.

none of them allow studying living roots in situ. For many years, the understanding of soil–root–water interactions was limited by available technologies (Aylmore, 1994). The instruments typically used for soil analyses focused on understanding the soil properties at the macro-scale, disturbing the soil sample, and neglecting heterogeneities within the sampled volume. In previous decades, new methods such as neutron imaging and magnetic resonance imaging have allowed the study of water content distribution in undisturbed soils (Oswald et al., 2008; Carminati et al., 2010). However, these methods are less useful to visualize soil structure. X-ray microtomography (XMT) is a promising technique that opens exciting new ways to study soil physical properties at the pore scale, in particular soil–root–water interactions. It has been used for more than two decades as a non-invasive, nondestructive technique for studying processes of soil structure development, water movement, and solute transport in soils (Hainsworth and Aylmore, 1983; Crestana and Vaz, 1998; Taina et al., 2008). XMT can achieve resolutions of micrometers, improving the soil measurement technique itself on a spatial scale hitherto unreachable (Hopmans et al., 1994; Spanne et al., 1994). The information derived from the XMT may be essential for understanding pore-network morphology, infiltration, redistribution, preferential flow phenomena, soil–root–water interactions, and transport of nutrients and pollutants in the soil. This information is important for the progress of environmental and soil science (Spanne et al., 1994; Al-Raoush et al., 2003; Peth et al., 2008).

X-Ray Microtomography in Soil Sciences

Tomography was first developed for medical research in the 1970s (Hounsfield, 1973). Its use in soil science started approximately two decades ago as a means of quantifying the spatial distribution of soil properties such as bulk density or water content (Petrovic et al., 1982; Hainsworth and Aylmore, 1983; Anderson and Hopmans, 1994). Since then, synchrotron-based techniques have become increasingly employed in various fields of soil, plant, and biological sciences. The current available XMT systems range from benchtop scanners (Vaz et al., 2011) to synchrotron microtomographs, differing in X-ray source and energy intensity, type of detector, and resolution. Traditional medical scanners can achieve resolutions within the range of 1001000 μm, benchtop industrial scanners can reach resolutions in the range of 1 to 100 μm depending on the sample size, while the resolution of synchrotron XMT ranges between 1 and 10 μm (Wildenschild et al., 2002; Adams et al., 2005; Iassonov et al., 2009; Vaz et al., 2011). The parallel beam, produced by a coherent synchrotron radiation, allows image reconstruction with a minimum of geometrical artifacts.

A recent surge in micro-scale porous media research has been possible due to easier access to XMT facilities and an increase in computational capabilities. The advances that have been made include characterization of pore geometry and macropore spaces (Johns et al., 1993; Crestana and Vaz, 1998; Perret et al., 1999; Lindquist et al., 2000; Pierret et al., 2002; Van Geet et al., 2000; Cislerova and Votrubova, 2002; Betson et al., 2004; Al-Raoush and Willson, 2005a; Lehmann et al., 2006; Udawatta et al., 2006; Luo et al., 2008, Peth et al., 2008; Sander et al., 2008), detection and measurement of soil-liquid interfaces (Spanne et al., 1994; Tippkotter et al., 2009) linked to soil hydraulic properties (Kasteel et al., 2000; Carminati et al., 2007), pore geometry, and fluid flow in porous media (Schwartz et al., 1994; Al-Raoush and Willson, 2005b; Culligan et al., 2006; Menon et al., 2011) to measure the diffusion coefficient of heavy ions in saturated porous media (Nakashima, 2000) and to study earthworm burrow systems (Capowiez et al., 2001; Jégou et al., 2001; Pierret et al., 2002), the impact of tillage on soil properties (Olsen and Børresen, 1997; Gantzer and Anderson, 2002; Langmaack et al., 2002), the spatial distribution of roots in soils (Heeraman et al., 1997; Pierret et al., 1999), multiphase flow in porous media (Turner et al., 2004; Al-Raoush and Willson, 2005b; Wildenschild et al., 2005; Schnaar and Brusseau, 2006), measurement of fluid-fluid interfaces (Culligan et al., 2004; Brusseau et al., 2007; Costanza-Robinson et al., 2008), polluted soils containing heavy metals (Seignez et al., 2010), solute and contaminant transport (Perret et al., 2000), and characterization of reactive sites in soils (Gouze et al., 2003; Altman et al., 2005). In addition, the pore-network information of noncohesive soils, extracted from XMT images, has been used to support the development of pore-scale mathematical modeling using Lattice-Boltzmann methods (Ferreol and Rothman, 1995; Martys and Chen, 1996; Sukop et al., 2008, Menon et al., 2011). In recent years, important efforts and a large amount of resources have been devoted to improve XMT technology, microscale analysis, and fluid flow modeling (Iassonov et al., 2009; Tippkotter et al., 2009). For instance, Kutilek and Nielsen (2007) recommended a combination of hydrology and micropedology to better understand the real pore and water properties in soil.

X-Ray Microtomography to Study
Soil–Plant–Root Interactions

One area that has not been studied in great depth with tomographic techniques is the soil–plant–root interactions, as most of the efforts have focused on characterizing porous media. Aylmore (1993) presented a study of water movement around roots using XMT. Tollner et al. (1994) studied the effect of root development on soil pore area and water content in sand columns. Heeraman et al. (1997) determined the volume of roots of size equal or larger than 0.36 mm. Macedo et al. (1998) presented XMT images of roots in sand and studied soil clods. Pierret et al. (1999) analyzed the spatial distribution and rooting patterns of trees. Hamza et al. (2001) used XMT to study the drawdowns in soil water content associated with roots. Gregory et al. (2003) studied the use of XMT to obtain root architecture (root diameter and three-dimensional root axial structures). Kaestner et al. (2006) presented a method to reconstruct a root network from a three-dimensional XMT image. Hamza et al. (2007) studied the effect of increasing osmotic strength on the shrinkage and recovery of roots. Perret et al. (2007) presented quantifications of root architecture (number of lateral roots, volume, length, wall area, tortuosity, and orientation) of relatively large cores using XMT. Hargreaves et al. (2008) determined root characteristics (root ranking and root angular spread) of plants using XMT. Seignez et al. (2010) used XMT images to visualize pollutant distribution in soils with roots. Recently, Aravena et al. (2011) investigated the effect of root-induced compaction on soil hydraulic properties.

XMT images have also been used to study roots by themselves, that is, without porous media. Scheckel et al. (2007) used absorption-edge XMT to study the distribution and compartmentalization of thallium in roots. Leroux et al. (2009) presented a XMT preparation method to enhance contrast of fresh plant tissues by staining roots with heavy metals. Bogart et al. (2010) used XMT images to study root tissues within the corm.

In this chapter, we present recent developments in visualizing how growing roots modify their physical environment by moving soil particles, deforming aggregates, and decreasing the amount of inter-aggregate pores while creating hydraulic pathways that connect neighboring soil aggregates using noninvasive synchrotron XMT. Image-processing tools were applied for quantifying root-induced rhizosphere alterations from XMT grayscale images as well as to transform XMT images into finite element meshes, building a bridge from nondestructive rhizosphere visualization to micromechanical and hydraulic simulations.

Imaging Using Synchrotron-Based
X-Ray Microtomography
Principles

Synchrotron light is an electromagnetic radiation emitted when ultra-high-speed charged electrons interact with a magnetic field. The radiated energy is proportional to the fourth power of the particle speed and is inversely proportional to the square of the radius of the path. Although X-rays exist naturally in outer space, the use of synchrotron radiation generated in circular particle accelerators is relatively recent (Baldwin, 1975; Pollock, 1983). The physics research community took advantage of the first generation of synchrotron particle accelerators to study the fundamental properties of matter. Since then, the third generation of synchrotron facilities have become multidisciplinary research centers, supporting broad research fields in physics, chemistry, biology, soil science, and engineering, allowing the study of three-dimensional objects at the micrometer scale (Wildenschild et al., 2002; Lombi and Susini, 2009; Pires et al., 2010).

The interaction of X-rays with matter produces partial or total absorption of the photons from the beam. The photons lose part of their energy and some deviate from their original path (Pires et al., 2010). These processes occur for each photon of the beam, producing an exponential decay of the beam intensity. The ratio between the absorbed and the transmitted photons is known as the attenuation coefficient. According to the Beer–Lambert law, the ratio between the transmitted (I) and incident photons (I_0) depends on the mass attenuation coefficient and the path that the photons follow through the sample (Beer, 1851; Colgate, 1952):

$$\frac{I}{I_0} = \exp(-\mu x) \tag{1}$$

where μ is the mass attenuation coefficient, and x is the length of the photons path through the sample.

For heterogeneous materials such as soils, the attenuation coefficient is a function of the path length (x_i) of each component (primarily solid and water), where gaseous components such as air can be neglected:

$$\frac{I}{I_0} = \exp(-\mu_w x_w - \mu_s x_s) = \exp\left[-\left(\mu_{w,mass}\rho_w\theta + \mu_{s,mass}\rho_s\right)x\right]$$

$$\tag{2}$$

where subscripts s and w are for soil and water, respectively, θ is the volumetric water content, and ρ is the density of the material.

Image Acquisition

There are three essential components involved in the image acquisition process: X-ray source (point source or synchrotron), detector, and sample to be scanned. The characteristics of the beam (e.g., geometry of the beam, fixed or adjustable energy) vary according to the facility; some scanners use a cone beam or planar fan beam whereas a synchrotron uses a parallel beam. In this chapter, all the images were acquired using the synchrotron facility of Lawrence Berkeley National Laboratory. The beam line 8.3.2 of the Advanced Light Source facility is equipped with a synchrotron X-ray source (1.9 GeV), a superbend magnet (4.37 Tesla), and a multilayer monochromator to deliver a monochromatic beam with an adjustable energy range from 8 to 40 keV to the specimen (beam size 40 by 4.6 mm, flux density 5 by 105 hν μm^{-2}). Transmitted X-ray light is converted to visible light using a $CdWO_4$ single crystal scintillator, magnified by either a Canon or a Mitutoyo 2X lens, and imaged on a Cooke PCO 4000 CCD camera. Soil and root specimens contained in 10-mL plastic syringes with an inner diameter of 14 mm

(a)

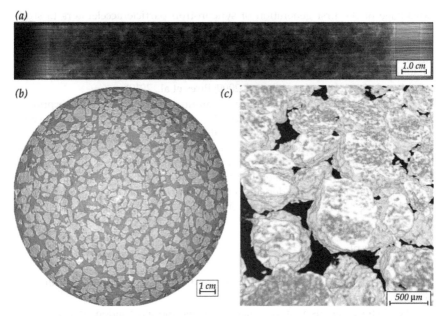

(b)   *(c)*

Fig. 3–2. (a) Two-dimensional depth-integrated map of a sample of dry soil aggregates; (b) Reconstruction of a two-dimensional projection of a sample of dry soil aggregates; (c) Partial reconstruction of a three-dimensional projection of a sample of dry soil aggregates.

were scanned with beam energy of 33 keV, and by acquiring transmission images every 0.125 degrees. The image pixel size was 4.4 μm.

The synchrotron X-ray source is directed to a monochromator to deliver a monochromatic beam, which is passed through the sample. The detector is typically a crystal scintillator panel optically linked via magnification lenses to a high-resolution charge-coupled device (CCD) camera. The scintillator converts the transmitted X-rays photons into visible light and the CCD camera converts the light into digital image data. As the X-ray beam passes through the sample, the CCD camera creates a two-dimensional depth-integrated map of the linear attenuation of the sample (Fig. 3–2a). To obtain a full three-dimensional attenuation map, the acquisition of a large number of two-dimensional projections is necessary (Fig. 3–2b and 3–2c) Therefore, after the image is collected, the sample is rotated in small increments of an angle (e.g., 0.125°), and another image is acquired. This process is repeated until images are collected between 0° and 180°. This stack of images is then reconstructed to produce a three-dimensional image of the sample.

The spatial resolution of the image varies depending on the field of view, detector and sample sizes, and magnification lenses utilized during the scan. For cone and fan beam scanners that utilize geometrical magnification, the spatial resolution depends on the relative position of the sample and the source, the detector resolution, and other factors, such as beam spot size.

Fig. 3–3. X-ray microtomography (XMT) image with the detector located near the sample (left); Phase contrast XMT image with the detector at 70 cm from the sample (right). Both images have a spatial resolution of 4.4 μm.

Phase Contrast Images

If a sufficiently coherent X-ray beam is used, "phase contrast images" with an increased sensitivity can be obtained by simply moving the detector downstream of the imaged object (Fig. 3–3). The phase contrast images display sharper and easier to detect edges, which correspond with the phase changes. Although important qualitative information is gained, this method complicates segmentation of the image and the measurement of the different phases (Werth et al., 2010).

Image Post-Processing

XMT images are not artifact free, which makes the quantitative image analysis a little more complicated. The attenuation values can experience problems including beam hardening, high-frequency noise, scattered X-rays, and poorly centered samples (Wildenschild et al., 2002; Iassonov et al., 2009). In addition, errors and distortions from the XMT reconstruction process are also common. These kinds of problems can be partially solved, for example, by carefully calibrating the instrument, centering the sample, and applying metal filters. Therefore, the analysis and measurement of soil properties using XMT images is not an easy task. Image-derived porosity may differ from directly measured porosity depending on the segmentation method used in the binarization process (Iassonov et al., 2009).

Basic Image Processing: Correction and Normalization

Once the sample has been completely scanned, the images are subject to darkfield correction and normalization processes. For darkfield corrections, the signal measured in the absence of any X-ray beam is used to correct detector offsets (Weitkamp et al., 2011). The normalization process utilizes brightfield exposures, which are the measured images of X-rays without the sample, to determine (i) incident X-ray beam intensity, (ii) beam fluctuations, (iii) non-uniformities in the incident X-ray beam, and (iv) the detector response. Other post-processing steps can include ring-noise filtering, which is intended to reduce ring artifacts produced by drifts or nonlinearities in the detector response and poor centering of the sample. These images are then reconstructed to develop macroscopic cross-

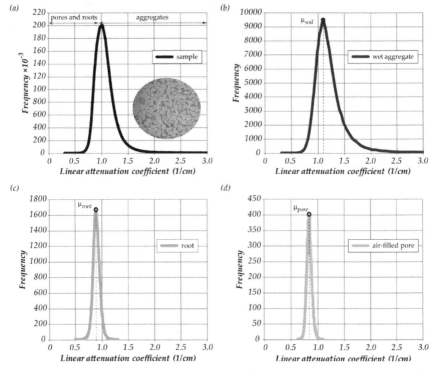

Fig. 3–4. (a) Histogram of linear attenuation coefficient of *Helianthus annuus* sample; (b) Histogram of linear attenuation coefficient of a group of wet soil aggregates; (c) Histogram of linear attenuation coefficient of a root; (d) Histogram of linear attenuation coefficient of an air-filled pore.

sections and three-dimensional visualizations of the sample. The reconstructed images show, in grayscales, the attenuation of the X-ray beam in discrete locations, called voxels. The magnitude of the attenuation depends on chemical composition, effective atomic number, and density of the material, as well as the X-ray energy. Materials with low atomic numbers absorb fewer X-rays (depicted as darker areas in the XMT image), so image contrast occurs when materials of different atomic numbers and/or densities are present within the sample. For example, Fig. 3–4 presents the histogram distribution of linear attenuation coefficients of a sample of sunflower (*Helianthus annuus* L.) in a bed of aggregates, scanned at 33 keV, and also the histogram distribution of linear attenuation coefficients for a root, an air-filled pore, and a group of wet aggregates within the sample. The histogram is unimodal so for most of the cases, attenuation coefficients above 0.9 cm^{-1} correspond with soil aggregates, while lower attenuation coefficients correspond with either pores or roots. The attenuation coefficients for pores, roots, and aggregates are 0.83, 0.89, and 1.1 cm^{-1}, respectively.

Figure 3–5 shows XMT images of *Helianthus annuus* and sweet pea (*Lathyrus odoratus* L.) roots growing in a bed of aggregates. Gray levels obtained in the XMT slices correspond to different X-ray attenuations. As explained before, the attenuation is a function of the X-ray energy and atomic number (in general proportional

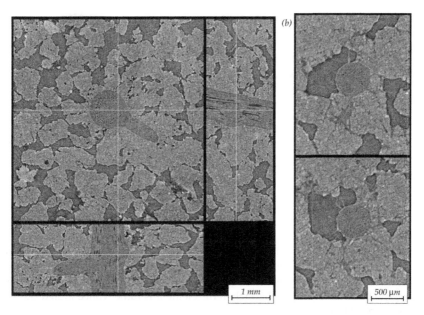

Fig. 3–5. X-ray microtomography examples of roots and soil: (a) Main and lateral root system of *Helianthus annuus* (branching) with xylem system. (b) Tap root cross-sectional area with root hair of *Lathyrus odoratus*.

to the density) of the material being imaged. Thus, denser (or less porous) materials are shown as lighter areas, whereas air- or water-filled spaces or organic materials, e.g., roots, are shown as darker spots. The roots have circular shapes, which distinguish them from voids. The orthogonal views, presented in Fig. 3–5a, clearly show the circular cross-section of the root, its cellular structure, its xylem system, and the development of the lateral root system. Figure 3–5b shows the presence of root hairs in areas where the root lost contact with the soil.

Phase Separation and Absorption Edge

Separation of fluid phases and its distributions in the porous structure can be achieved by subtracting aligned images scanned at different water contents (Aylmore, 1993; Wildenschild et al., 2002). An example is presented in Fig. 3–6 where the subtraction between a dry aggregate and a wet aggregate (after the addition of 20% w/w of water) image results in the enhancing of the areas of the sample that were affected by the water addition. It was found that 8.8% of the added water was held in the inter-aggregate pore system, while the rest of the water allowed the swelling of the aggregates. The increased volume of the aggregates was estimated in 12.5% of the sample volume and affected only the outer layer of the aggregates (Aravena et al., 2011).

The development of high-energy X-ray sources with multiple energy ranges is a unique attribute of X-ray microscopy. If the instrument allows scanning at different energies, it is possible to take advantage of the absorption edge of an element, which is an abrupt change in absorptivity or the mass absorption coefficient

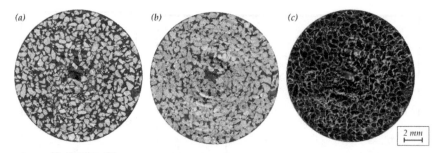

Fig. 3–6. (a) X-ray microtomography (XMT) image of the sample of dry aggregates; (b) XMT image of the sample of wet aggregates (after the addition of 20% w/w of water); (c) Area of the aggregates affected by the water addition (light blue [modified from Aravena et al., 2011]).

that occurs at the energy or wavelength of the X-ray necessary to eject an electron from an atom. In this technique, images are collected above and below the absorption edge of the element under study. Then, the spatial distribution of the element within the sample can be obtained by applying subtraction. However, this method can only be successfully applied when the element is rather concentrated (Lombi and Susini, 2009).

XMT Image Segmentation

One of the most crucial steps in XMT post-processing is image segmentation. Segmentation is the binarization process that allows obtaining morphological representation of the pore system. Segmentation influences quantitative analysis such as porosity, surface area, and pore network structure, and subsequently, modeling. During the segmentation, the XMT images are separated into distinct phases (e.g., solid particles, air- and water-filled pores and roots). Typical segmentation techniques that are applied to XMT images include

Global thresholding: This is the most common approach of segmentation and it is generally used when background and foreground are to be separated. To separate the phases of interest, a single threshold gray value is selected. One method used for the selection of the optimal threshold value for segmentation of two phases is based on the analysis of the histogram shape (Zack et al., 1977; Otsu, 1979; Velasco, 1980; Kapur et al., 1985; Tsai, 1985; Kittler and Illingworth, 1986; Perez and Gonzales, 1987; Glasbey, 1993; Li and Lee, 1993; Shanbhag, 1994; Huang and Wang, 1995; Yen et al., 1995; Prasanna et al., 2004; Prewitt and Mendelsohn, 2006). The shape of the histogram varies typically from unimodal (bell) to bimodal (double-peaked) shape, depending on the sample and its characteristics (Fig. 3–7).

For unimodal histograms, the threshold value is usually picked, or automatically calculated, either at the most frequent value (mode) or at the point that corresponds with the maximum of the first derivative of the histogram (see red dots in Fig. 3–7a). For bimodal histograms, the threshold value is usually picked in the valley between peaks.

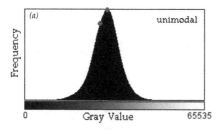 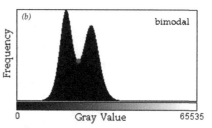

Fig. 3–7. Histogram shapes, typically found in X-ray microtomography images of soil samples.

The global thresholding method may lead to misclassification errors that are proportional to the overlapping values of the two considered classes (Oh and Lindquist, 1999; Gantzer and Anderson, 2002; Rachman et al., 2005; Iassonov et al., 2009).

Figures 3–8 and 3–9 show results of many known global thresholding techniques applied to a stack of a unimodal and bimodal XMT images, respectively. Over-segmentation and disturbing effect of ring artifacts can be observed in the segmentation of Fig. 3–8 produced by iterative, Li's, moments, and mean techniques. The triangle method fails to segment the *L. odoratus* sample, as it is designed for thresholding images with a bimodal histogram, which is not the case for this sample. Shanbhag's method fails to segment the dry clay sample (Fig. 3–9).

Locally adaptive thresholding: Includes a wide range of methods that use local image statistics to select different and dynamic intensities of thresholding for different regions in the image (Oh and Lindquist, 1999; Sheppard et al., 2004; Bloom et al., 2010). This method takes advantage of local thresholding and, at the same time, prevents over-segmentation with the global image information. Figure 3–10 presents the results of the Bloom segmentation for the *L. odoratus* sample and the dry aggregates. In this case, the method fails to correctly segment the *L. odoratus* sample.

Region growing: These segmentation techniques are designed to segment images based on the user input (selection) in the region considered to be of interest. It involves the selection of initial seed points that are used to determine if neighboring pixels should be added to a region of interest (ROI). It assumes that all pixels belonging to a particular ROI are connected and sufficiently similar (Gonzàlez and Wood, 2002; Ketcham, 2005; Schlüter et al., 2010). A drawback of this approach is the need to manually define initial seed regions, based on a certain criterion. Region growing is intended for cases when automated segmentation fails and user interaction is required. Post-processing is usually necessary.

Generic clustering: Representatives of the clustering algorithms are K-means and Fuzzy C-means (FCM). The methods divide data, using either intensity values or feature values calculated using intensities of the pixel's neighborhood, into classes or clusters according to a similarity criterion. The success of clustering depends on the selection of features and the power of the grouping technique. Each data point is assigned to a group based on a probability of belongingness that depends on the proximity to the cluster's centroid, and then it performs an iterative process by re-calculating and re-grouping data until an error of the subsequent iteration

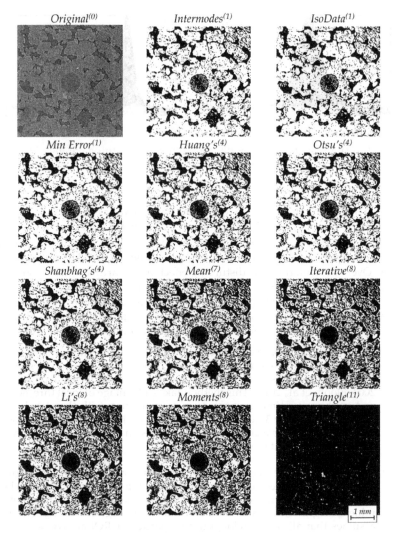

Fig. 3–8. Results of several global thresholding techniques, sorted from best (1) to worst (11) in ascending order, applied to an X-ray microtomography image of a *Lathyrus odoratus* root in bed of clay aggregates (unimodal histogram).

drops below a certain specified limit. FCM allows a feature point to be a member of more than one class, and thus it is more flexible to fuzziness nature of features. It utilizes fuzzy membership functions and rules when the boundary of an object or the classification of a pixel is not clear. Thus, a probability of being a member of a class is assigned, and then the notion of fuzzy clustering becomes relevant (Blais, 2005; Iassonov et al., 2009). For phases that cannot be clearly differentiated due to limitations of soil image tomography, image noise, and phase conditions, the FCM grouping may be a better choice; however, it is not perfect either. Figure 3–11 presents the FCM segmentation of the XMT images using two classes.

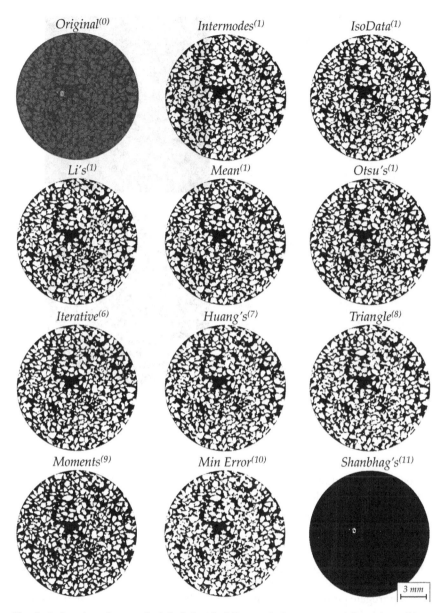

Fig. 3–9. Results of several global thresholding techniques, sorted from best (1) to worst (11) in ascending order, applied to an X-ray microtomography image of dry clay aggregates (bimodal histogram).

The choice of the most appropriate algorithm or combination of methods will depend on a number of factors, such as type of sample, image under study, image quality, and artifacts present. Segmentation remains an open research area for soil science. For high-performance methods, there is need of user interaction

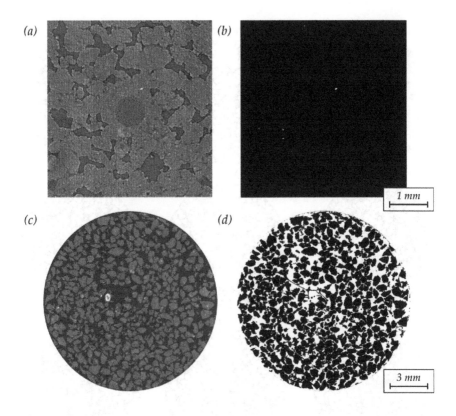

Fig. 3–10. (a) X-ray microtomography (XMT) of an *Lathyrus odoratus* root growing in a bed of aggregates; (b) Bloom segmentation of *L. odoratus* root sample; (c) XMT of dry clay aggregates; (d) Bloom segmentation of dry clay aggregates.

for either the selection of training data or setting parameters for the algorithm to obtaining the desirable result. Additional post-processing is required depending on the task. A series of post-processing steps is presented in Fig. 3–12.

Quantifying Soil Properties Using XMT Images

Once the image is segmented, the XMT data can be used to quantify soil properties such as porosity, as illustrated in Fig. 3–13 and 3–14. The radial distribution of inter-aggregate porosity was estimated from the images using concentric ellipses centered at the root (or micro-balloon) and computing the porosity of the multiple regions bounded by the concentric ellipses. Selection of the segmentation technique significantly affects the estimation of porosity. This information can be used to populate flow models, as will be described later in this chapter.

Modeling Using XMT Images

More complex analysis includes the use of segmented XMT images to create finite element meshes that can be used to simulate fluid flow and mechanical

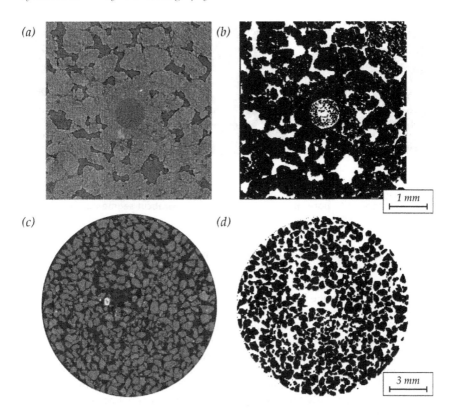

Fig. 3–11. Clustering segmentation using Fuzzy C-means (FCM) of two classes (a) X-ray microtomography (XMT) of an *Lathyrus odoratus* root growing in a bed of aggregates; (b) FCM segmentation of *L. odoratus* root sample; (c) XMT of dry clay aggregates; (d) FCM segmentation of dry clay aggregates.

deformation of the mesh and to describe the compaction of the soil induced by root growth. This process is illustrated in Fig. 3–15.

Recent studies (Aravena et al., 2011) have focused on the impact of root-induced compaction on rhizosphere hydraulic properties. These studies included both experimental work and water flow modeling within aggregates. Using numerical modeling, Aravena et al. (2011) showed that the effective hydraulic conductivity of a pair of aggregates undergoing deformation increased following a nonlinear relationship as the inter-aggregate contact area increased due to deformations. Aravena et al. (2011) also presented synchrotron-based XMT numerical models to quantify the effect of aggregate deformation on a pack of multiple aggregates (aggregated soil) around a root surrogate, providing insights into root growth and water uptake patterns; these simulations show an increase in unsaturated water flow toward the simulated root under induced deformation.

Figure 3–16 shows the evolution of soil water potential over 100 s throughout the domain for the uncompacted and compacted bed of aggregates. As predicted by the simulations of the pair of aggregates, the deformation of the bed of aggregates increased its connectivity. Therefore, the root extracted 21% more

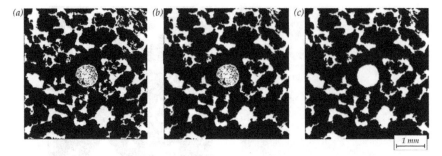

Fig. 3–12. Post-processing of Fig. 3–11a. (a) Fuzzy C-means segmentation of the X-ray microtomography image; (b) Connected component filter, to eliminate disconnected pixels, was applied to pore space and aggregates; (c) Grabcut segmentation to eliminate root area.

water from the compacted soil compared with the uncompacted soil. They found that the average connectivity for the uncompacted and compacted bed was 0.38 and 0.58, respectively, which produced an increase in effective hydraulic conductivity of 83%; the change in hydraulic parameter produced a decrease of 39% in water flow toward the root. The combined effect was estimated to increase water flow by 12%, less than the 21% observed in the simulations; they suggested that the difference was due to the steeper hydraulic gradient in the compacted case and to radial flow effects, where changes in the soil properties near the root have a larger impact on water flow than changes that occur far away from it. They also suggested that the increase in effective hydraulic conductivity cannot proceed indefinitely, as eventually the inter-aggregate porosity cannot accommodate the majority of the strain, and the aggregates themselves will begin to significantly compact. Thus, the rhizosphere compaction observed and simulated in their study may pass through an optimum for water extraction, depending on initial inter-aggregate porosity, root diameter, and soil texture.

Most recently, Berli et al. (2011) presented the first results of numerical simulations that describe both the mechanical deformation and the feedback to hydraulic behavior of the soil, as a root radially grows. By coupling these results, it is now possible to directly quantify the effect of compaction on soil hydraulic properties and plant water uptake.

Advantages and Disadvantages

A major advantage of synchrotron XMT is its coherent, monochromatic, parallel beam that allows image reconstruction with minimal artifacts possible, improving the quality of the images. The beam energy is adjustable, providing flexibility to the scanning process and allows scanning of the same sample at multiple energy levels. Its resolution is typically in the range of 1 to 10 μm. The main disadvantage of XMT is its low sensitivity (Lombi and Susini, 2009). This limits the separation of soft materials, especially when dense structures are overlapping. Phase contrast can be used to obtain images with an increased sensitivity and easier to detect edges. However, even when qualitative information is gained, this method complicates segmentation and thus the measurement of the different phases.

(a)

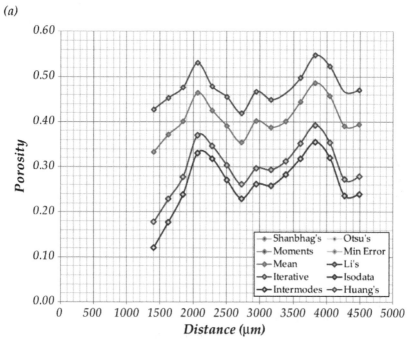

(b)

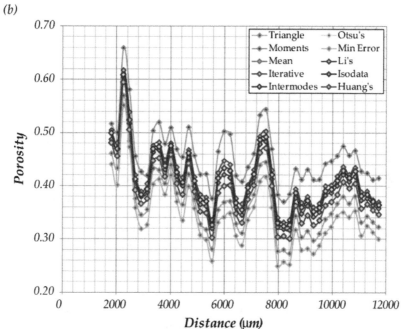

Fig. 3–13. Quantification of soil porosity from the X-ray microtomography image using global thresholding techniques (a) *Lathyrus odoratus* in a bed of aggregates; (b) Dry clay aggregates.

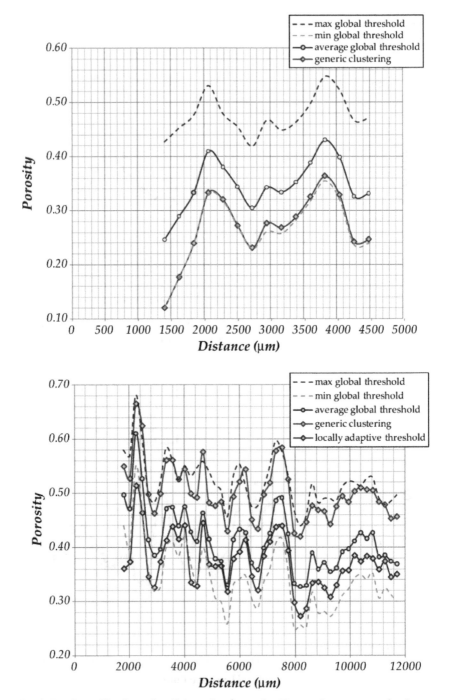

Fig. 3–14. Quantification of soil porosity from the X-ray microtomography image using different segmentation techniques (a) *Lathyrus odoratus* in a bed of aggregates; (b) Dry clay aggregates.

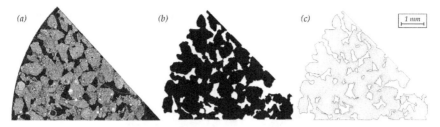

Fig. 3–15. (a) Selection of the area of interest from the X-ray microtomography image; (b) Segmentation of the image; (c) Digitalization and mesh creation.

Other disadvantages of the XMT are the size limitation of the sample and the time required to complete a three-dimensional scan. The sample size limitation is on the order of a few centimeters and is imposed by the synchrotron's beam. The time required to complete a three-dimensional scan can vary from minutes to hours; this does limit the ability of the method to characterize flowing fluids. The scanning time can be reduced at the expense of the quality of the image by decreasing the number of rotations of the sample or by decreasing the photons measuring time at each step.

Summary and Conclusions

In the past 20 yr, significant efforts and resources have been devoted to improve XMT technology and its application to soil science. This technique has allowed the study of micro-scale soil structure and soil–plant–root interactions in a non-destructive and noninvasive way. Quantification of soil properties using XMT has become more routine as more segmentation techniques have been developed and adapted to soil and root images.

We have presented the current state of the art in observing soil–plant–root interactions using XMT. We discussed techniques to improve sensitivity of XMT images, when soft materials, such as roots, are being scanned. The high resolution achieved with the synchrotron XMT technique allowed the observation of soil microstructure, root architecture, and root structure (e.g., xylem system and root hairs). A discussion of the feasibility of various segmentation techniques and examples of their use to quantify soil properties and to visualize them was provided and examples were included. Finally, we have shown the application of XMT to study root-induced compaction to quantify hydrodynamic changes in the rhizosphere. We analyzed the case of a bed of aggregates undergoing deformation due to root growth. These results have begun to open a new chapter in the understanding of biological feedbacks in soil structure and formation. XMT now indicates that soil compaction, previously thought to be detrimental to vascular plants, is now indeed beneficial. Root growth, by definition, must compact soil, and it is shown that this is, in fact, beneficial to a certain point for root water uptake. This "soil engineering" by roots is likely to explain the morphology and strategies of root growth in the diverse soil environment. In unstructured soils with limited compaction capacity, roots are likely to develop into more dense architecture to exploit all available soil water. In aggregated soils, root development may follow a different path, in which root growth

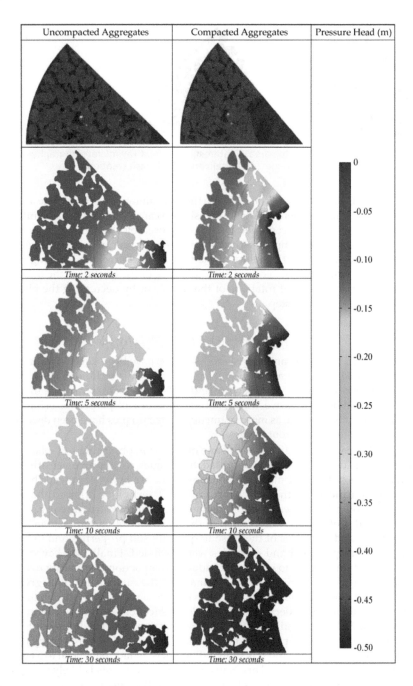

Fig. 3–16. Evolution of the pressure head distribution in a two-dimensional bed of aggregates after 2, 5, 10, and 30 s of simulation (Aravena et al., 2011).

is controlled by soil compaction. Roots capable of water uptake will continue to grow radially until compaction begins to reduce intra-aggregate porosity, at which point, root water uptake function is likely to move to other root areas.

Acknowledgments

This material is based on work supported by the National Science Foundation under Grants No. DEB-0816726 and DEB-0817073. The Advanced Light Source is supported by the Director, Office of Science, Office of Basic Energy Sciences, of the U.S. Department of Energy under Contract No. DE-AC02-05CH11231. We also wish to recognize the valuable comments and inputs from Dr. Francisco Suárez.

References

Adams, F., L. Van Vaeck, and R. Barrett. 2005. Advanced analytical techniques: Platform for nano materials science. Spectrochim. Acta, B At. Spectrosc. 60:13–26. doi:10.1016/j.sab.2004.10.003

Al-Raoush, R.I., K.E. Thompson, and C.S. Willson. 2003. Comparison of network generation techniques for unconsolidated porous media systems. Soil Sci. Soc. Am. J. 67:1687–1700. doi:10.2136/sssaj2003.1687

Al-Raoush, R.I., and C.S. Willson. 2005a. Extraction of physically realistic pore network properties from three-dimensional synchrotron x-ray microtomography images of unconsolidated porous media systems. J. Hydrol. 300:44–64. doi:10.1016/j.jhydrol.2004.05.005

Al-Raoush, R.I., and C.S. Willson. 2005b. A pore-scale investigation of a multiphase porous media system. J. Contam. Hydrol. 77:67–89. doi:10.1016/j.jconhyd.2004.12.001

Altman, S.J., M.L. Rivers, M.D. Reno, R.T. Cygan, and A.A. McLain. 2005. Characterization of adsorption sites on aggregate soil samples using synchrotron X-ray computerized microtomography. Environ. Sci. Technol. 39:2679–2685. doi:10.1021/es049103y

Anderson, S.H., and J.W. Hopmans. 1994. Tomography of soil-water-root processes. SSSA Spec. Publ. 36. SSSA, Madison, WI.

Anderson, T.A., E.A. Guthrie, and B.T. Walton. 1993. Bioremediation in the rhizosphere. Plant roots and associated microbes clean contaminated soil. Environ. Sci. Technol. 27:2630–2636. doi:10.1021/es00049a001

Aravena, J.E., M. Berli, T.A. Ghezzehei, and S.W. Tyler. 2011. Effects of root-induced compaction on rhizosphere hydraulic properties- X-ray microtomography imaging and numerical simulations. Environ. Sci. Technol. 45:425–431. doi:10.1021/es102566j

Aylmore, L.A.G. 1993. The use of computer assisted tomography in studying water movement around plant roots. Adv. Agron. 49:1–54. doi:10.1016/S0065-2113(08)60792-0

Aylmore, L.A.G. 1994. Application of computer assisted tomography to soil-plant-water studies: An overview. SSSA Spec. Publ. 36. SSSA, Madison, WI. p. 7–15.

Baldwin, G.C. 1975. Origin of synchrotron radiation. Phys. Today 28:9–17.

Beer, A. 1851. Versus der absorptions- verhaltnisse des cordietes fur rothes licht zu bestimmen. Annu. Phys. Chem. 84:37–52.

Bengough, A.G., J. Hans, M.F. Bransby, and T.A. Valentine. 2010. PIV as a method for quantifying root cell growth and particle displacement in confocal images. Microsc. Res. Tech. 73:27–36.

Berli, M., A. Carminati, T.A. Ghezzehei, and D. Or. 2008. Evolution of unsaturated hydraulic conductivity of aggregated soils due to compressive forces. Water Resour. Res. 44:W00C09. doi:10.1029/2007WR006501

Berli, M., S.A. Ruiz, J.E. Aravena, L. Bolduc, T.A. Ghezzehei, D.P. Cook, M. Menon, S.W. Tyler, and M.H. Young. 2001. Simulating rhizosphere structure alterations using finite element calculations. European Geosciences Union General Assembly, Vienna, April 2011.

Betson, M., J. Barker, P. Barnes, T. Atkinson, and A. Jupe. 2004. Porosity imaging in porous media using synchrotron tomographic techniques. Transp. Porous Media 57:203–214. doi:10.1023/B:TIPM.0000038264.33451.4a

Blais, K.E. 2005. Measurements of physical and hydraulic properties of organic soil using computed tomographic imagery. M.Sc. thesis, Simon Fraser University, Burnaby, BC.

Bloom, M., M.J. Russell, A. Kustau, S. Mandayam, and B. Sukumaran. 2010. Measurement of Porosity in Granular Particle Distributions Using Adaptive Thresholding. IEEE Trans. Instrum. Meas. 59:1192–1199. doi:10.1109/TIM.2010.2040902

Bogart, S.J., G. Spiers, and E. Cholewa. 2010. X-ray μCT imaging technique reveals corm microstructures of an arctic-boreal cotton-sedge, *Eriophorum vaginatum*. J. Struct. Biol. 171:361–371. doi:10.1016/j.jsb.2010.06.009

Brusseau, M.L., S. Peng, G. Schnaar, and A. Murao. 2007. Measuring air–water interfacial areas with X-ray microtomography and interfacial partitioning tracer tests. Environ. Sci. Technol. 41:1956–1961. doi:10.1021/es061474m

Capowiez, Y., P. Monestiez, and L. Belzunces. 2001. Burrow systems made by *Aporrectodea nocturna* and *Allolobophora chlorotica* in artificial cores: Morphological differences and effects of interspecific interactions. Appl. Soil Ecol. 16:109–120. doi:10.1016/S0929-1393(00)00110-4

Carminati, A., A. Kaestner, R. Hassanein, O. Ippisch, P. Vontobel, and H. Flühler. 2007. Infiltration through series of soil aggregates: Neutron radiography and modeling. Adv. Water Resour. 30:1168–1178. doi:10.1016/j.advwatres.2006.10.006

Carminati, A., P. Kaestner, R. Lehmann, and H. Flühler. 2008. Unsaturated water flow across soil aggregate contacts. Adv. Water Resour. 31:1221–1232. doi:10.1016/j.advwatres.2008.01.008

Carminati, A., A.B. Moradi, D. Vetterlein, P. Vontobel, E. Lehmann, U. Weller, H.J. Vogel, and S.E. Oswald. 2010. Dynamics of soil water content in the rhizosphere. Plant Soil 332:163–176. doi:10.1007/s11104-010-0283-8

Cislerova, M., and J. Votrubova. 2002. CT derived porosity distribution and flow domains. J. Hydrol. 267:186–200. doi:10.1016/S0022-1694(02)00149-X

Colgate, S.A. 1952. Gamma-ray absorption measurements. Phys. Rev. 87:592–600. doi:10.1103/PhysRev.87.592

Costanza-Robinson, M.S., K.H. Harrold, and R.M. Lieb-Lappen. 2008. X-ray microtomography determination of air-water interfacial area-water saturation relationships in sandy porous media. Environ. Sci. Technol. 42:2949–2956. doi:10.1021/es072080d

Crestana, S., and C.M.P. Vaz. 1998. Non-invasive instrumentation opportunities for characterizing soil porous systems. Soil Tillage Res. 47:19–26. doi:10.1016/S0167-1987(98)00068-3

Culligan, K.A., D. Wildenschild, B.S.B. Christensen, W.G. Gray, and M.L. Rivers. 2006. Pore-scale characteristics of multiphase flow in porous media: A synchrotron-based CMT comparison of air-water and oil-water experiments. Adv. Water Resour. 29:227–238. doi:10.1016/j.advwatres.2005.03.021

Culligan, K.A., D. Wildenschild, B.S.B. Christensen, W.G. Gray, and A.F.B. Tompson. 2004. Interfacial area measurements for unsaturated flow through a porous medium. Water Resour. Res. 40:W12413. doi:10.1029/2004WR003278

Dessaux, Y., P. Hinsinger, and P. Lemanceau. 2009. Rhizosphere: Achievements and challenges. Springer, New York.

Dexter, A.R. 1987. Compression of soil around roots. Plant Soil 97:401–406. doi:10.1007/BF02383230

Dexter, A.R. 1988. Advances in characterization of soil structure. Soil Tillage Res. 11:199–238. doi:10.1016/0167-1987(88)90002-5

Dexter, A.R. 2004. Soil physical quality part I. Theory, effects of soil texture, density, and organic matter, and effects on root growth. Geoderma 120:201–214. doi:10.1016/j.geoderma.2003.09.004

Feeney, D.S., J.W. Crawford, T. Daniell, P.D. Hallett, N. Nunan, K. Ritz, M. Rivers, and I.A. Young. 2006. Three-dimensional microorganization of the soil-root-microbe system. Microb. Ecol. 52:151–158. doi:10.1007/s00248-006-9062-8

Ferréol, B., and D.H. Rothman. 1995. Lattice–Boltzmann simulations of flow-through Fontainebleau sandstone. Transp. Porous Media 20:3–20. doi:10.1007/BF00616923

Gantzer, C.J., and S.H. Anderson. 2002. Computed tomographic measurement of macroporosity in chisel-disk and no-tillage seedbeds. Soil Tillage Res. 64:101–111. doi:10.1016/S0167-1987(01)00248-3

Glasbey, C.A. 1993. An analysis of histogram-based thresholding algorithms. Graph. Models Image Processing 55:532–557. doi:10.1006/cgip.1993.1040

González, R.C., and R.E. Wood. 2002. Digital image processing, Addison-Wesley, Boston, MA.

Gouze, P., C. Noiriel, C. Bruderer, D. Loggia, and R. Leprovost. 2003. X-ray tomography characterization of fracture surfaces during dissolution. Geophys. Res. Lett. 30(5):1267–1271. doi:10.1029/2002GL016755

Greacen, E.L., D.A. Farrell, and B. Cockroft. 1968. Proceedings of the 9th international congress of soil science. Angus and Roberton, Adelaide. p. 769–779.

Gregory, P.J. 2006. Roots, rhizosphere and soil: The route to a better understanding of soil science? Eur. J. Soil Sci. 57:2–12. doi:10.1111/j.1365-2389.2005.00778.x

Gregory, P.J., and P. Hinsinger. 1999. New approaches to studying chemical and physical changes in the rhizosphere: An overview. Plant Soil 211:1–9. doi:10.1023/A:1004547401951

Gregory, P.J., D.J. Hutchison, D.B. Read, P.M. Jenneson, W.B. Gilboy, and E.J. Morton. 2003. Non-invasive imaging of roots with high resolution X-ray micro-tomography. Plant Soil 255:351–359. doi:10.1023/A:1026179919689

Hainsworth, J.M., and L.A.G. Aylmore. 1983. The use of computer-assisted tomography to determine spatial distribution of soil water content. Aust. J. Soil Res. 21:435–443. doi:10.1071/SR9830435

Hallett, P.D., D.C. Gordon, and A.G. Bengough. 2003. Plant influence on rhizosphere hydraulic properties: Direct measurements using a miniaturized infiltrometer. New Phytol. 157:597–603. doi:10.1046/j.1469-8137.2003.00690.x

Hamza, M.A., and W.K. Anderson. 2005. Soil compaction in cropping systems: A review of the nature, causes and possible solutions. Soil Tillage Res. 82:121–145. doi:10.1016/j.still.2004.08.009

Hamza, M.A., S.H. Anderson, and L.A.G. Aylmore. 2007. Computed tomographic evaluation of osmotica on shrinkage and recovery of lupin (*Lupinus angustifolius* L.) and radish (*Raphanus sativus* L.) roots. Environ. Exp. Bot. 59:334–339. doi:10.1016/j.envexpbot.2006.04.004

Hargreaves, C.E., P.J. Gregory, and A.G. Bengough. 2008. Measuring root traits in barley (*Hordeum vulgare* ssp. *vulgare* and ssp. *spontaneum*) seedlings using gel chambers, soil sacs and X-ray microtomography. Plant Soil 316:285–297. doi:10.1007/s11104-008-9780-4

Hartmann, A., M. Rothballer, and M. Schmid. 2008. Lorenz Hiltner, a pioneer in rhizosphere microbial ecology and soil bacteriology research. Plant Soil 312:7–14. doi:10.1007/s11104-007-9514-z

Heeraman, D.A., J.W. Hopmans, and V. Clausnitzer. 1997. Three dimensional imaging of plant roots in situ with X-ray computed tomography. Plant Soil 189:167–179.

Hiltner, L. 1904. Über neuere Erfahrungen und Probleme auf dem Gebiete der Bodenbakteriologie unter besonderer Berücksichtigung der Gründüngung und Brache. Arb. DLG 98:59–78.

Hinsinger, P., A.G. Bengough, D. Vetterlein, and I.M. Young. 2009. Rhizosphere: Biophysics, biogeochemistry and ecological relevance. Plant Soil 321:117–152. doi:10.1007/s11104-008-9885-9

Hinsinger, P., G.R. Gobran, P.J. Gregory, and W.W. Wenzel. 2005. Rhizosphere geometry and heterogeneity arising from root-mediated physical and chemical processes. New Phytol. 168:293–303. doi:10.1111/j.1469-8137.2005.01512.x

Hinsinger, P., C. Plassard, and B. Jaillard. 2006. Rhizosphere: A new frontier for soil biogeochemistry. J. Geochem. Explor. 88:210–213. doi:10.1016/j.gexplo.2005.08.041

Hopmans, J.W., M. Cislerova, and T. Vogel. 1994. X-ray tomography of soil properties. SSSA Spec. Publ. 36. SSSA, Madison, WI. p. 17–28.

Horn, R., and A.J.M. Smucker. 2005. Structure formation and its consequences for gas and water transport in unsaturated arable and forest soils. Soil Tillage Res. 82:5–14. doi:10.1016/j.still.2005.01.002

Hounsfield, G.N. 1973. Computerized transverse axial scanning (tomography): Part I. Description of the system. Br. J. Radiol. 46:1016–1022. doi:10.1259/0007-1285-46-552-1016

Huang, L.K., and M.J. Wang. 1995. Image thresholding by maximizing the index of nonfuzziness of the 2-D grayscale histogram. Pattern Recognit. 28:41–51. doi:10.1016/0031-3203(94)E0043-K

Iassonov, P., T. Gebrenegus, and M. Tuller. 2009. Segmentation of X-ray computed tomography images of porous materials: A crucial step for characterization and quantitative analysis of pore structures. Water Resour. Res. 45:W09415. doi:10.1029/2009WR008087

Javaux, M., T. Schröder, J. Vanderborght, and H. Vereecken. 2008. Use of a three-dimensional detailed modeling approach for predicting root water uptake. Vadose Zone J. 7:1079–1088. doi:10.2136/vzj2007.0115

Jégou, D., Y. Capowiez, and D. Cluzeau. 2001. Interactions between earthworm species in artificial soil cores assessed through the 3D reconstruction of the burrow systems. Geoderma 102:123–137. doi:10.1016/S0016-7061(00)00107-5

Johns, R.A., J.S. Steude, L.M. Castanier, and P.V. Roberts. 1993. Nondestructive measurements of fracture aperture in crystalline rock cores using X-ray computed-tomography. J. Geophys. Res. Solid Earth 98:1889–1900. doi:10.1029/92JB02298

Kaestner, A., M. Schneebeli, and F. Graf. 2006. Visualizing three-dimensional root networks using computed tomography. Geoderma 136:459–469. doi:10.1016/j.geoderma.2006.04.009

Kapur, J.N., P.K. Sahoo, and A.K.C. Wong. 1985. A new method for gray-level picture thresholding using the entropy of the histogram. Comput. Vis. Graph. Image Process. 29:273–285. doi:10.1016/0734-189X(85)90125-2

Kasteel, R., H.J. Vogel, and K. Roth. 2000. From local hydraulic properties to effective transport in soil. Eur. J. Soil Sci. 51:81–91. doi:10.1046/j.1365-2389.2000.00282.x

Kay, B.D. 1997. Soil structure and organic carbon; a review. Soil processes and the carbon cycle. CRC Press, Boca Raton, FL. p. 169–197.

Ketcham, R.A. 2005. Three-dimensional grain fabric measurements using high-resolution X-ray computed tomography. J. Struct. Geol. 27:1217–1228. doi:10.1016/j.jsg.2005.02.006

Kittler, J., and J. Illingworth. 1986. Minimum error thresholding. Pattern Recognit. 19:41–47. doi:10.1016/0031-3203(86)90030-0

Kutilek, M., and D.R. Nielsen. 2007. Interdisciplinarity of hydropedology. Geoderma 138:252–260. doi:10.1016/j.geoderma.2006.11.015

Langmaack, M., S. Schrader, U. Rapp-Bernardt, and K. Kotzke. 2002. Soil structure rehabilitation of arable soil degraded by compaction. Geoderma 105:141–152. doi:10.1016/S0016-7061(01)00097-0

Lehmann, P., P. Wyss, A. Flish, E. Lehmann, P. Vontobel, M. Krafczyk, A. Kaestner, F. Beckmann, A. Gygi, and H. Flühler. 2006. Tomographical imaging and mathematical description of porous media used for the prediction of fluid distribution. Vadose Zone J. 5:80–97. doi:10.2136/vzj2004.0177

Leroux, O., F. Leroux, E. Bellefroid, M. Claeys, M. Couvreur, G. Borgonie, L. Van Hoorebeke, B. Masschaele, and R. Viane. 2009. A new preparation method to study fresh plant structures with X-ray computed tomography. J. Microsc. 233:1–4. doi:10.1111/j.1365-2818.2008.03088.x

Li, C.H., and C.K. Lee. 1993. Minimum cross entropy thresholding. Pattern Recognit. 26:617–625. doi:10.1016/0031-3203(93)90115-D

Lindquist, W.B., A.B. Venkatarangan, J. Dunsmuir, and T.F. Wong. 2000. Pore and throat size distributions measured from synchrotron X-ray tomographic images of Fontainebleau sandstones. J. Geophys. Res. 105(21):509–527.

Lombi, E., and J. Susini. 2009. Synchrotron-based techniques for plant and soil science: Opportunities, challenges and future perspectives. Plant Soil 320:1–35. doi:10.1007/s11104-008-9876-x

Luo, L.F., H.S. Lin, and P. Halleck. 2008. Quantifying soil structure and preferential flow in intact soil using X-ray computed tomography. Soil Sci. Soc. Am. J. 72:1058–1069. doi:10.2136/sssaj2007.0179

Lynch, J. 1995. Root architecture and plant productivity. Plant Physiol. 109:7–13.

Macedo, A., S. Crestana, and C.M.P. Vaz. 1998. X-ray microtomography to investigate thin layers of soil clod. Soil Tillage Res. 49:249–253. doi:10.1016/S0167-1987(98)00180-9

Martys, N.S., and H. Chen. 1996. Simulation of multicomponent fluids in complex three-dimensional geometries by the Lattice Boltzmann method. Phys. Rev. E Stat. Phys. Plasmas Fluids Relat. Interdiscip. Topics 53:743–750. doi:10.1103/PhysRevE.53.743

Menon, M., Q. Yuan, X. Jia, A.J. Dougill, S.R. Hoon, A.D. Thomas, and R.A. Williams. 2011. Assessment of physical and hydrological properties of biological soil crusts using X-ray microtomography and modeling. J. Hydrol. 397:47–54. doi:10.1016/j.jhydrol.2010.11.021

Nakashima, Y. 2000. The use of X-ray CT to measure diffusion coefficients of heavy ions in water-saturated porous media. Eng. Geol. 56:11–17. doi:10.1016/S0013-7952(99)00130-1

Oh, W., and W.B. Lindquist. 1999. Image thresholding by indicator kriging. IEEE Trans. Pattern Anal. Mach. Intell. 21:590–601. doi:10.1109/34.777370

Olsen, P.A., and T. Børresen. 1997. Measuring differences in soil properties in soils with different cultivation practices using computer tomography. Soil Tillage Res. 44:1–12. doi:10.1016/S0167-1987(97)00021-4

Oswald, S.E., M. Menon, A. Carminati, P. Vontobel, E. Lehmann, and R. Schulin. 2008. Quantitative imaging of infiltration, root growth, and root water uptake via neutron radiography. Vadose Zone J. 7:1035–1047. doi:10.2136/vzj2007.0156

Otsu, N. 1979. A threshold selection method from gray-level histograms. IEEE Trans. Syst. Man Cybern. 9:62–66. doi:10.1109/TSMC.1979.4310076

Perez, A., and R.C. Gonzales. 1987. An iterative thresholding algorithm for image segmentation. IEEE Trans. Pattern Anal. Mach. Intell. PAMI-9:742–751. doi:10.1109/TPAMI.1987.4767981

Perret, J.S., M.E. Al-Belushi, and M. Deadman. 2007. Non-destructive visualization and quantification of roots using computed tomography. Soil Biol. Biochem. 39:391–399. doi:10.1016/j.soilbio.2006.07.018

Perret, J., S.O. Prasher, A. Kantzas, and C. Langford. 1999. Three-dimensional quantification of macropore networks in undisturbed soil cores. Soil Sci. Soc. Am. J. 63:1530–1543. doi:10.2136/sssaj1999.6361530x

Perret, J., S.O. Prasher, A. Kantzas, and C. Langford. 2000. A two-domain approach using CAT scanning to model solute transport in soil. J. Environ. Qual. 29:995–1010. doi:10.2134/jeq2000.00472425002900030039x

Peth, S., R. Horn, F. Beckmann, T. Donath, J. Fischer, and A.J.M. Smucker. 2008. Three-dimensional quantification of intra-aggregate pore-space features using synchrotron-radiation-based microtomography. Soil Sci. Soc. Am. J. 72:897–907. doi:10.2136/sssaj2007.0130

Petrovic, A.M., J.E. Siebert, and P.E. Rieke. 1982. Soil bulk density in three dimensions by computed tomographic scanning. Soil Sci. Soc. Am. J. 46:445–450. doi:10.2136/sssaj1982.03615995004600030001x

Pierret, A., Y. Capowiez, L. Belzunces, and C.J. Moran. 2002. 3D reconstruction and quantification of macropores using X-ray computed tomography and image analysis. Geoderma 106:247–271. doi:10.1016/S0016-7061(01)00127-6

Pierret, A., C. Doussan, Y. Capowiez, F. Bastardie, and L. Pages. 2007. Root functional architecture: A framework for modeling the interplay between roots and soil. Vadose Zone J. 6:269–281. doi:10.2136/vzj2006.0067

Pierret, A., C.J. Moran, and C.E. Pankurst. 1999. Differentiation of soil properties related to the spatial association of wheat roots and soil macropores. Plant Soil 211:51–58. doi:10.1023/A:1004490800536

Pires, L.F., J.A.R. Borges, O.O.S. Bacchi, and K. Reichardt. 2010. Twenty-five years of computed tomography in soil physics: A literature review of the Brazilian contribution. Soil Tillage Res. 110:197–210. doi:10.1016/j.still.2010.07.013

Pollock, H.C. 1983. The discovery of synchrotron radiation. Am. J. Phys. 51:278–280. doi:10.1119/1.13289

Prasanna, P.K., K. Sahoo, and G. Arora. 2004. A thresholding method based on two-dimensional Renyi's entropy. Pattern Recognit. 37:1149–1161. doi:10.1016/j.patcog.2003.10.008

Prewitt, J., and M.L. Mendelsohn. 2006. The analysis of cell images. Ann. N. Y. Acad. Sci. 128:1035–1053. doi:10.1111/j.1749-6632.1965.tb11715.x

Rachman, A., S.H. Anderson, and C.J. Gantzer. 2005. Computed tomographic measurement of soil macroporosity parameters as affected by stiff-stemmed grass hedges. Soil Sci. Soc. Am. J. 69:1609–1616. doi:10.2136/sssaj2004.0312

Richardson, A.E., J.M. Barea, A.M. McNeill, and C. Prigent-Combaret. 2009. Acquisition of phosphorus and nitrogen in the rhizosphere and plant growth promotion by microorganisms. Plant Soil 321:305–339. doi:10.1007/s11104-009-9895-2

Sander, T., H.H. Gerke, and H. Rogasik. 2008. Assessment of Chinese paddy-soil structure using X-ray computer tomography. Geoderma 145:303–314. doi:10.1016/j.geoderma.2008.03.024

Scheckel, K.G., R. Hamon, L. Jassogne, M. Rivers, and E. Lombi. 2007. Synchrotron X-ray absorption-edge computed microtomography imaging of thallium compartmentalization in *Iberis intermedia*. Plant Soil 290:51–60. doi:10.1007/s11104-006-9102-7

Schlüter, S., U. Weller, and H.J.S. Vogel. 2010. Segmentation of X-ray microtomography images of soil using gradient masks. Comput. Geosci. 36:1246–1251. doi:10.1016/j.cageo.2010.02.007

Schnaar, G., and M.L. Brusseau. 2006. Characterizing pore-scale dissolution of organic immiscible liquid in natural porous media using synchrotron X-ray microtomography. Environ. Sci. Technol. 40:6622–6629. doi:10.1021/es0602851

Schwartz, L.M., F. Auzerais, J. Dunsmoir, N. Martys, D.P. Bentz, and S. Torquato. 1994. Transport and diffusion in three-dimensional composite media. Physica A 207:28–36. doi:10.1016/0378-4371(94)90351-4

Seignez, N., A. Gauthier, F. Mess, C. Brunel, M. Dubois, and J.L. Potdevin. 2010. Development of plant roots network in polluted soils: an x-ray computed microtomography investigation. Water Air Soil Pollut. 209:199–207. doi:10.1007/s11270-009-0192-8

Shanbhag, A. 1994. Utilization of information measure as a means of image thresholding. Comput. Vis. Graph. Image Process. 56:414–419.

Sheppard, A.P., R.M. Sok, and H. Averdunk. 2004. Techniques for image enhancement and segmentation of tomographic images of porous materials. Physica A 339:145–151. doi:10.1016/j.physa.2004.03.057

Six, J., H. Bossuyt, S. Degryze, and K. Denef. 2004. A history of research on the link between (micro)aggregates, soil biota, and soil organic matter dynamics. Soil Tillage Res. 79:7–31. doi:10.1016/j.still.2004.03.008

Spanne, P., K.W. Jones, L. Prunty, and S.H. Anderson. 1994. Potential application of synchrotron computed microtomography to soil science. SSSA Spec. Publ. 36. SSSA, Madison, WI. p. 43–56.

Sukop, M.C., H. Huang, C.L. Lin, M.D. Deo, K. Oh, and J.D. Miller. 2008. Distribution of multiphase fluids in porous media: Comparison between Lattice Boltzmann modeling and micro-x-ray tomography. Phys. Rev. E Stat. Nonlin. Soft Matter Phys. 77:026710. doi:10.1103/PhysRevE.77.026710

Taina, I.A., R.J. Heck, and T.R. Elliot. 2008. Application of X-ray computed tomography to soil science: A literature review. Can. J. Soil Sci. 88:1–20. doi:10.4141/CJSS06027

Tippkotter, R., T. Eickhorst, H. Taubner, B. Gredner, and G. Rademaker. 2009. Detection of soil water in macropores of undisturbed soil using microfocus X-ray tube computerized tomography (µCT). Soil Tillage Res. 105:12–20. doi:10.1016/j.still.2009.05.001

Tisdall, J.M., and J.M. Oades. 1982. Organic matter and water stable aggregates in soils. J. Soil Sci. 33:141–163. doi:10.1111/j.1365-2389.1982.tb01755.x

Tollner, E.W., C. Murphy, and E.L. Ramseur. 1994. Techniques and approaches for documenting plant root development with X-ray computed tomography. SSSA Spec. Publ. 36. SSSA, Madison, WI. p. 115–134.

Tsai, W. 1985. Moment-preserving thresholding: A new approach. Comput. Vis. Graph. Image Process. 29:377–393. doi:10.1016/0734-189X(85)90133-1

Turner, M.L., L. Knüfing, C.H. Arns, A. Sakellariou, T.J. Senden, A.P. Sheppard, R.M. Sok, A. Limaye, W.V. Pinczewski, and M.A. Knackstedt. 2004. Three-dimensional imaging of multiphase flow in porous media. Physica A 339:166–172. doi:10.1016/j.physa.2004.03.059

Udawatta, R.P., S.H. Anderson, C.J. Gantzer, and H.E. Garrett. 2006. Agroforestry and grass buffer influence on macropore characteristics: A computed tomography analysis. Soil Sci. Soc. Am. J. 70:1763–1773. doi:10.2136/sssaj2006.0307

Van Geet, M., R. Swennen, and M. Wevers. 2000. Quantitative analysis of reservoir rocks by microfocus X-ray computerized tomography. Sediment. Geol. 132:25–36. doi:10.1016/S0037-0738(99)00127-X

Vaz, C.M.P., I.C. de Maria, P.O. Lasso, and M. Tuller. 2011. Evaluation of an advanced benchtop micro-computed tomography system for quantifying porosities and pore-size distributions of two Brazilian Oxisols. Soil Sci. Soc. Am. J. 75:832–841. doi:10.2136/sssaj2010.0245

Velasco, F.R.D. 1980. Thresholding using the ISODATA clustering algorithm. IEEE Trans. Syst. Man Cybern. 10:771–774. doi:10.1109/TSMC.1980.4308400

Watt, M., W.K. Silk, and J.B. Passioura. 2006. Rates of root and organism growth, soil conditions, and temporal and spatial development of the rhizosphere. Ann. Bot. (Lond.) 97:839–855. doi:10.1093/aob/mcl028

Weitkamp, T., D. Haas, D. Wegrzynek, and A. Rack. 2011. ANKAphase: Software for single-distance phase retrieval from inline X-ray phase-contrast radiographs. J. Synchrotron Radiat. 18:617–629. doi:10.1107/S0909049511002895

Werth, C.J., C. Zhang, M.L. Brusseau, M. Oostrom, and T. Baumann. 2010. A review of non-invasive imaging methods and applications in contaminant hydrogeology research. J. Contam. Hydrol. 113:1–24. doi:10.1016/j.jconhyd.2010.01.001

Whalley, W.R., P.B. Leeds-Harrison, P.K. Leech, B. Riseley, and N.R.A. Bird. 2004. The hydraulic properties of soil at root-soil interface. Soil Sci. 169:90–99. doi:10.1097/01.ss.0000117790.98510.e6

Whalley, W.R., B. Riseley, P.B. Leeds-Harrison, N.R.A. Bird, P.K. Leech, and W.P. Adderley. 2005. Structural differences between bulk and rhizosphere soil. Eur. J. Soil Sci. 56:353–360. doi:10.1111/j.1365-2389.2004.00670.x

Wildenschild, D., J.W. Hopmans, M.L. Rivers, and A.J.R. Kent. 2005. Quantitative analysis of flow processes in a sand using synchrotron-based X-ray microtomography. Vadose Zone J. 4:112–126. doi:10.2113/4.1.112

Wildenschild, D., J.W. Hopmans, C.M.P. Vaz, M.L. Rivers, and D. Rikard. 2002. Using X-ray computed tomography in hydrology: Systems, resolutions and limitations. J. Hydrol. 267:285–297. doi:10.1016/S0022-1694(02)00157-9

Yen, J.C., F.J. Chang, and S. Chang. 1995. A new criterion for automatic multilevel thresholding. IEEE Trans. Image Process. 4:370–378. doi:10.1109/83.366472

Young, I.M. 1998. Biophysical interactions at the root–soil interface: A review. J. Agric. Sci. 130:1–7. doi:10.1017/S002185969700498X

Zack, G.W., W.E. Rogers, and S.A. Latt. 1977. Automatic measurement of sister chromatid exchange frequency. J. Histochem. Cytochem. 25:741–753. doi:10.1177/25.7.70454

4

Effects of Root-Induced Biopores on Pore Space Architecture Investigated with Industrial X-Ray Computed Tomography

Sebastian K. Pagenkemper, Stephan Peth,* Daniel Uteau Puschmann, and Rainer Horn

Abstract

Over the last decade, the application of X-ray computed tomography (CT) in soil science showed rapid improvement in terms of image quality and acquisition time. Its noninvasive nature in combination with sophisticated image analysis tools provides excellent opportunities to study soil–water–root processes at various scales. In this chapter, we show examples, demonstrating the possibilities of X-ray CT analysis for a qualitative/visual and quantitative evaluation of root-induced structure formation. Soil samples ranging in size from large monoliths (20-cm diameter, 70-cm height) to small cores (5.1-cm diameter, 4.4-cm height) were collected from a Haplic Luvisol (Hypereutic, Siltic) from loess of an experimental field trial with various preceding crops [*Medicago sativa* L. subsp. *sativa* (alfalfa), *Cichorium intybus* L. (chicory), and *Festuca arundinacea* Schreb. (fescue)]. Morphological and topological parameters, such as pore volume size distribution, biopore surface area, and properties of the network skeleton (path length, diameter, and tortuosity), were calculated and used to assess how root-induced biopores of the different preceding crops modify pore space architectures.

Abbreviations: b_c, burn number; CT, computed tomography; IK, Indicator kriging; MA, medial axis; PLD, path length density; PSF, pore solid fractal; ROI, region of interest; τ^d, interface/distance.

S.K. Pagenkemper, University Kiel, Faculty of Agricultural and Nutritional Sciences, Institute of Plant Nutrition and Soil Science, Hermann-Rodewald-Str. 2, 24118 Kiel, Germany (s.pagenkemper@soils.uni-kiel.de); S. Peth, University of Kassel, Faculty of Ecological Agriculture, Department of Soil Science, Nordbahnhofstr. 1a, 37213 Witzenhausen, Germany (peth@uni-kassel.de); D. Uteau Puschmann, University Kiel, Faculty of Agricultural and Nutritional Sciences, Institute of Plant Nutrition and Soil Science, Hermann-Rodewald-Str. 2, 24118 Kiel, Germany (d.uteau@soils.uni-kiel.de); Rainer Horn, University Kiel, Faculty of Agricultural and Nutritional Sciences, Institute of Plant Nutrition and Soil Science, Hermann-Rodewald-Str. 2, 24118 Kiel, Germany (rhorn@soils.uni-kiel.de). *Corresponding author.

doi:10.2136/sssaspecpub61.c4

Soil–Water–Root Processes: Advances in Tomography and Imaging. SSSA Special Publication 61. S.H. Anderson and J.W. Hopmans, editors. © 2013. SSSA, 5585 Guilford Rd., Madison, WI 53711, USA.

In general, the pore space can be described by an inter-connected system of both primary (i.e., textural) and secondary (i.e., structural) pores (Kutílek and Nielsen, 1994; Hajnos et al., 2006). Thus, the three-dimensional pore space is made up of a variety of different pore types, showing structural heterogeneity in size, shape, and orientation. As pores are modified due to various processes such as shrinking and swelling, freezing and thawing, mechanical deformation, chemical and mechanical weathering, as well as biological activity, morphological and topological attributes like connectivity, path length and width and tortuosity of pores, are affected. Considering the accessibility of pores by plant roots and the transport and microbial mediated release of nutrients, we should differentiate between bulk soil, intra-aggregate, and rhizosphere pore space. While roots, with the exception of root hairs, grow partially into transmission pores, in general with diameters larger than 50 μm, they also penetrate the compressible soil growing along aggregates, building a functioning root network to explore the soil for water and nutrients (Gregory, 1988).

In particular, biopores (e.g., taproot or earthworm induced pores) have a predominant influence on soil processes in subsoils, where the packing density of the bulk soil is usually high. An often cited example of the dominant effect of biopores on transport is the preferential and unstable flow in soil, which is important at the pore and also the catchment scale (Gerke et al., 2010). Influencing the transport of water (Bouma, 1981; Gerke, 2006) and oxygen (Ball, 1981; Douglas et al., 1986), large and continuous pores also accelerate the transport of dissolved nutrients and agricultural chemicals, bypassing the regions of soil matrix pores with the rapid preferential movement of water from surface to subsoil (Flury and Flühler, 1995; Ghodrati and Jury, 1992; Kodešová et al., 2010; Perret et al., 1998; Toor et al., 2005). Germann and Beven (1981) pointed out the capability of macropore rich systems to buffer heavy infiltration events, when the soil is saturated and the vertical flow velocity of soil matrix pores is exceeded or a permeability break occurs. Thus, macroporosity can generate different phenomena of preferential flow, which are described in the literature (Booltink and Bouma, 1991, 1993; Beven and Germann, 1982; Bouma, 1990; Garré et al., 2010; Jarvis, 2007; Köhne et al., 2009; White, 1985). In addition, the pore wall characteristics (e.g., organic matter composition at the surfaces) may control properties of flow pathways (Jégou et al., 2001; Leue et al., 2010; Tiunov et al., 2001). However, not only is the movement of gases, water, and solutes affected by macroporosity but also the physiology of plants may be influenced by root–soil interactions in biopores. Thus, macropores can, depending on their continuity, positively stimulate root growth as well as nutrient and water uptake (Bennie, 1991; Hillel, 1980; Gliński and Stępniewski, 1985; Sutton, 1991; Ma and Selim, 1997; Perret et al., 1999; Dreesmann, 1994). Plant roots, which are mentioned as "the hidden half" (Waisel et al., 2002), have the ability to dynamically alter the soil structure and to adapt themselves to an existing pore network in the soil. As there are characteristic types of root systems for various plant species with differences in diameter and features like number of branches, branch length, distribution and angles, particular crop sequences can generate root systems with major differences in the resulting pore architectures (Fitter, 1991). For example, plant species with a taproot system will form continuous vertical macropores (Mitchell et al., 1995), while homorhizous plants such as monocotyledons generally form dense root systems consisting mostly of lateral roots. Another enhancement is an increased root penetration supporting the following crops, provided the pores are clear of recent roots or animals (Jones et al., 2004; Mitchell et al., 1995; White and Kirkegaard, 2010; Bennie, 1991). Also the predominating fauna such as earthworms is able to produce macropores and thus increase the hydraulic conductivity (Bouma, 1991; Edwards et al., 1989; Francis and Fraser, 1998; Joschko et al., 1993).

Pore network architectures determine the ecological (micro-) environment through the interaction of physical, biogeochemical, and biological processes on various scales (Pierret et al., 2007), such as the distribution, sorption, retention, and release of soil nutrients and soil organic matter by governing gas, water, and biochemical fluxes within the pore space environment (Horn and Smucker, 2005). Thus, the opening of new interfaces in the subsoil can especially enhance intensity rates of reaction and transport processes (Boone 1988; De Wever et al., 2004; Horn and Smucker, 2005), while at the same time roots benefit due to the lower mechanical resistance (Logsdon and Linden, 1992) and the availability of oxygen and nutrients (Stewart et al., 1999). These properties can be identified as critical for a soil controlling plant growth (Angers and Caron, 1998). Kautz and Köpke (2010) argued that biopores can affect the nutrient uptake by plant roots, since they often grow along preexisting pores and re-enter the matrix soil exploring new potential nutrient sources. Following the concept of "intensity" rather than "capacity," as suggested by Horn and Kutílek (2009), the accessibility of exchange surfaces in the rhizosphere can be considered a key determinant regarding the nutrient supply to roots. Hence, the internal structure of a soil and the microscale spatial heterogeneity of functions (physical, chemical, and biological) will contribute to a better quantification of "effective" nutrient supply considering the dynamic functionality in soils over time and space. Considering this, crop performance will benefit from accelerated root growth into the subsoil (Jakobsen and Dexter, 1988; Hirth et al., 2005).

Noninvasive technologies, such as X-ray computed tomography (CT), have long been applied to investigate soil structure and especially biopores generated by plant roots (Perret et al., 1999) and earthworms (Joschko et al., 1993). Over the last decade significant technical enhancements, especially of industrial scanners, now provide an insight into soil structures at greater detail spanning a larger range of scales. Furthermore, improved and more sophisticated image analysis algorithms are now available, allowing a systematic quantitative analysis of the acquired CT data. This study aims at demonstrating the potential of X-ray CT in conjunction with quantitative image analysis tools to extract pore space characteristics (e.g., pore wall surfaces, pore volume distributions, or length, width and tortuosity of paths) that are relevant for the transport, accessibility, and uptake of water and nutrients in the subsoil. In addition, these data can be used for application in models considering root growth, root water uptake, etc., as more detailed and complex models need a larger number of input parameters (Javaux et al., 2008).

Materials and Methods
Site Description and Sampling

The experimental field site for this study is located at the experimental station Klein Altendorf of the University of Bonn (50°37'9"N lat; 6°59'29"E long) with a maritime climate (9.3°C mean annual temperature, 594 mm annual rainfall). The

Table 4–1. Description and basic properties of the reference soil profile.

Depth	Horizons	Texture				Bulk density	pH CaCl$_2$	CaCO$_3$	SOC
		Sand	Silt	Clay	Textural class				
cm		—— % ——				g cm^{-3}		g kg^{-1}	g kg^{-1}
0–27	Ap	8	77	15	SiL	1.29	6.5	<1	10.0
27–41	E/B	5	75	20	SiL	1.32	6.9	<1	4.6
41–75	Bt1	4	69	27	SiCL	1.42	6.9	<1	4.5
75–87	Bt2	4	66	30	SiCL	1.52	6.9	<1	3.9
87–115	Bt3	5	70	25	SiL	1.52	7.1	<1	2.5
115–127	Bw	5	72	23	SiL	1.46	7.3	<1	2.6
127–140+	C	8	79	13	SiL	1.47	7.4	127	/

major soil type at this site is a Haplic Luvisol (Hypereutric, Siltic) developed from loess. Table 4–1 summarizes some basic properties of the specific horizons for a reference profile. Perennial fodder crops—alfalfa, chicory and fescue—were cultivated in a field trial that was established in 2007. Before the field trial started, different crops were cultivated in varying crop rotations between 1996 and 2007: fava bean (*Vicia faba* L.), sugar beet (*Beta vulgaris* L. subsp. *vulgaris*), summer barley (*Hordeum vulgare* L.), summer wheat (*Triticum aestivum* L), triticale (× *Triticosecale* spp.), winter barley (*Hordeum vulgare* L.), and winter wheat (*Triticum aestivum* L). The main contrast is expected to be found between the allorhizous and homorhizous crops by creating a deeply penetrating system of biopore channels vs. a shallow root system mainly restricted to the topsoil, respectively. In the course of the soil sampling, we chose fescue (Fe1y) after 1 yr, and alfalfa (Af2y) and chicory (Ch2y) after 2 yr of undisturbed cultivation. Thus, fescue as a shallow rooter represents a biopore poor treatment that should not generate biopores in the subsoil. On the contrary, alfalfa and chicory are considered biopore rich treatments because they are expected to generate biopores (with larger diameters) in the subsoil by their specific root systems.

In April 2010, six undisturbed soil monoliths, 70 cm in height and 20 cm in diameter, from the three treatments with two field replicates (indicated with Roman numerals I and II) were taken from the subsoil with emphasis on investigation of structural roots. A hydraulic–mechanical device (UGT GmbH, Müncheberg, Germany) was used to push PVC pipes into the E/B and Bt horizons of the profile. The soil monoliths were taken from 45 to 115 cm depth in the soil profile. After excavation, the bottom 5 cm of each monolith was replaced with filter gravel. Subsequently, the monoliths were sealed to prevent water loss by evaporation. In addition, undisturbed soil cores (5.1 cm in height and 4.4 cm in diameter) were taken from depths of 45 to 60, 60 to 75, and 75 to 90 cm in the profile from each treatment.

Acquisition and Pre-processing of X-Ray CT Data

The monoliths were scanned at the Department of Mechanical Engineering (Technical University Dortmund, Germany) with a v|tome|x 240 (Phoenix-X-Ray, GE-Sensing and Inspection Technologies GmbH, Wunstorf, Germany).

Each monolith was scanned in five height steps (multiscans 0–12, 12–27, 27–42, 42–57, and 58–70 cm from 45 cm depth in the soil profile) at field moist condition (matric potential between −150 and −300 hPa). The scan specifications were: energy 160 keV with a current of 1000 µA, 1200 projections, and a 1-mm Cu filter to prevent effects of beam hardening (Palomo et al., 2006).

Reconstruction was done with datos|x (GE-Sensing and Technologies GmbH, Wunstorf, Germany), using a modified Feldkamp backprojection algorithm (Feldkamp et al., 1984). The achieved voxel resolution was 231.58 µm in unbinned mode. To facilitate image analysis the size of the dataset was reduced by reconstructing the projection in 2-binned mode (2^3 voxels are binned into one single voxel) resulting in a voxel edge length of 463.16 µm for the reconstructed images. However, image quality seemed to not have been significantly affected according to visual inspection of the rendered volumes, which confirmed a very good recovery of the soil structure. In the visualization of biopores section, we intended to present an example for the different pore network morphologies derived from different sample sizes currently covering a voxel resolution ranging from ~100 to ~463 µm. Therefore, we defined macropores as pores with a diameter larger than 463.16 µm. The height shifts (multiscans) were referenced by the acquisition software. Finally, the reconstructed single volume datasets for each height step were imported into VGStudio Max 2.0 (Volume Graphics GmbH), and the image stacks were rendered according to the height references into one single volume file representing the complete monolith. The volume file was exported in 8-bit gray-scale format where, according to a color depth of 2^8 (256), the minimum value 0 is black and the maximum value 255 is white, corresponding to the lowest (air) and highest (solid) X-ray attenuation, respectively. The basic principles of X-ray CT have been described in detail by various authors and will not be repeated here (Aylmore, 1993; Brunke et al., 2008; Clausnitzer and Hopmans, 2000; Ketcham and Carlson, 2001).

Before image analysis, a region of interest (ROI) for each soil monolith was determined. The regions of interest were chosen such that some physical disturbances at the top and bottom of the monoliths were excluded from the analysis. Top and bottom borders were selected at constant heights for each monolith at 12 cm (top) and 58 cm (bottom) from the monolith top corresponding to 57 and 103 cm depth in the soil profile. Subsequently the quantitative image analysis refers to the selected ROIs, and results are presented, with a few exceptions, as relative values referring to the ROI volume.

Image Analysis

Quantitative image analysis of pore structures in the samples was done with 3DMA-Rock (Lindquist et al., 2000). The principle work flow from the original tomogram to the extraction of a pore network can be described as illustrated in Fig. 4–1a and 4–1b show the soil monolith with the PVC pipe and a two-dimensional cross-section displaying the internal structure of the sample. Figure 4–1c shows the same two-dimensional cross-section after segmentation, where pores in black and soil matrix in white (including solids and pores below the resolution limit) are shown. After segmentation pore surfaces (Fig. 4–1d) are calculated, which are then used for pore network (medial axis) reconstruction (Fig. 4–1e). Visualization of pore surfaces and pore networks was done with Geomview 1.9.4 (http://www.geomview.org/, accessed 4 Aug. 2011).

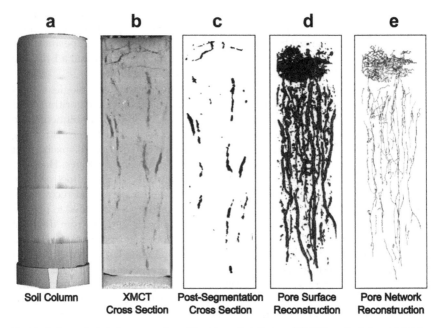

a   **b**   **c**   **d**   **e**

Soil Column   XMCT   Post-Segmentation   Pore Surface   Pore Network
Cross Section   Cross Section   Reconstruction   Reconstruction

Fig. 4–1. Steps in image processing: (a) scanned soil monolith (70-cm height and 20-cm diameter); (b) two-dimensional cross-section showing the internal structure in gray scale; (c) two-dimensional cross-section after segmentation where black and white refer to pores and soil matrix including solids and unresolved pore space, respectively; (d) reconstructed pore surfaces; (e) medial axis representing the pore network.

Image Segmentation

Peth et al. (2008) mentioned that the quality of segmentation is a key determinant for the quality of pore statistics derived from CT images. However, most approaches in quantitative image analysis require distinguishing between pore and solid voxels. Concerning soils, this is a real challenge, as a major part of the pore space is "hidden" below the resolution limit, preventing the use of benchmark measures such as total porosity as a control for the accuracy of selected segmentation thresholds. There are various studies dealing with the problem of image segmentation in soils (Iassonov et al., 2009; Baveye et al., 2010; Wang et al., 2011). One conclusion that can be drawn from them is that a simple straightforward best segmentation method is currently not available, and the selection of an appropriate procedure depends very much on image quality, resolution, frequency distribution of attenuation values, and finally the focus of the study. However, local adaptive thresholding schemes, where the spatial correlations of attenuation coefficients are utilized to determine object edges, seem to surpass global thresholding (Iassonov et al., 2009; Wang et al., 2011). This appears to be a logical consequence of the fact that image noise, partial volume effects, and other image disturbances (artifacts) are not uniformly distributed within tomograms. In regions where the pore size approaches the size of a few voxels, there is a higher uncertainty in phase classification than within larger pore volumes due to partial volume effects increasing at solid void interfaces (Oh and Lindquist, 1999).

3DMA provides a local adaptive thresholding algorithm which is based on the selection of two threshold values (T_0 and T_1) in between which voxel classification is performed using indicator kriging (Oh and Lindquist, 1999). According to Iassonov et al. (2009), indicator kriging can show very good results in segmentation quality, but with the requirement of supervision by a skilled operator, because the initialization of this algorithm depends on the method upon which the two highly sensitive threshold values are chosen. Wang et al. (2011) stated that indicator kriging provides good segmentation results, especially when bimodal frequency distributions of attenuation coefficients (expressed as gray-scale values) are available. In a bimodal gray value histogram, two peaks occur where the one corresponding to lower X-ray attenuation values (in our case darker gray-scale values) is considered the pore space peak, while the one with higher X-ray attenuation values (in our case brighter gray-scale values) is considered the soil matrix peak and consists of solid components and nonresolved pores. Between the two peaks and with increasing distance from the peak centers, the decision whether a pixel belongs to the soil matrix or the resolved pore space becomes increasingly uncertain. For segmentation, three sections in a gray-scale histogram were considered, which are separated by two threshold values: (i) between the lowest gray value (0 = black) and a lower threshold value T_0, (ii) between T_0 and an upper threshold value T_1 and (iii) between T_1 and the highest gray value (255 = white).

In this study, the two threshold values T_0 and T_1 were selected based on the peaks of the frequency distribution of the gray values. First, we assumed that the gray-scale values of the soil matrix peak are represented reasonably well by a Gaussian distribution. After fitting a Gauss function to the soil matrix peak we used the standard deviation (σ) to determine T_1 (= mean − 2σ). As the number of pore voxels is much smaller than of soil matrix voxels suppressing the pore peak, we used the gray value distribution of a smaller ROI around single biopores to determine T_0 (= center of the pore peak). The three-dimensional ROI for threshold estimation was sufficiently large to represent macropores and average gray values of the soil matrix. We selected biopores where no lateral gradients in gray values were observed, which may have been related to compaction effects resulting from earthworm or root pressure.

In the section between the lowest gray value and T_0, voxels were considered to certainly belong to the pore space while in the section between T_1 and the highest gray value they were considered to certainly belong to the soil matrix, respectively. Such voxels were classified in a first sweep of the segmentation algorithm. In the section between both thresholds (T_0 and T_1), voxels were considered to be associated with a high risk for misclassification.

Volume Thresholding and Pore Surface Reconstruction

After segmentation, we applied a cleanup algorithm to remove isolated void and solid phase (grain) voxels. Isolated grain voxels are considered segmentation artifacts; however, although usually low in number, they significantly distort medial axis construction and were therefore removed completely. As we were interested mainly in the biopore channels, we also removed disconnected pore voxel clusters up to a certain size, based on a volume threshold. This allowed us, similarly to what was demonstrated by Capowiez et al. (2001), to isolate the main percolating biopore network. The selection of the volume threshold was based on a frequency distribution of disconnected pore volumes. We assumed that the

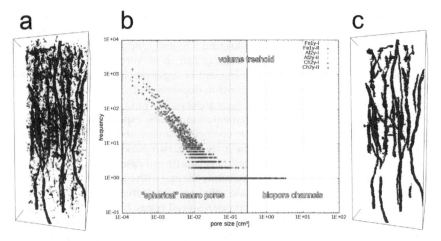

Fig. 4–2. Graphical illustration of the applied volume thresholding method. (a) All resolved pore volumes consisting of biopore channels and spherical macropores; (b) frequency distribution of disconnected pore volumes of all analyzed soil monoliths and the selected volume threshold; (c) isolated network of biopore channels.

probability to detect large biopore channels with exactly the same size (number of voxels) is very low for large pores but increases with a decrease in pore size. Based on the frequency distributions of disconnected pore volumes for all treatments and replicates (Fig. 4–2), we determined a single volume threshold value of 0.3 cm³ (Fig. 4–2b) that was subsequently applied for isolating the major biopore network for all monoliths (Fig. 4–2c).

Medial Axis–Pore Network Reconstruction

The reconstruction of pore surfaces can be used to visualize the spatial distribution of biopores. Moreover, a network skeleton can be reconstructed as a simplified representation of the pore space geometry, which subsequently can be analyzed for descriptive components, such as width, length, and tortuosity of pore space channels, etc. To achieve this, an algorithm for the construction of a medial axis (MA) has been used (assuming 26-connectivity where MA-voxels are connected via faces, edges and corners) to extract the skeleton for the void phase (Lindquist et al., 1995).

Four types of network paths were distinguished: (i) paths connected with both ends to a cluster (branch-branch), (ii) paths connected with only one end to a cluster (branch-leaf), (iii) paths with no connection to a cluster (leaf-leaf), and (iv) paths where both ends are connected in a closed loop to the same point (needle-eye).

Needle-eye paths, leaf-leaf paths, and branch-leaf paths up to a certain size as well as isolated MA-voxels were removed using a trimming routine. Both needle-eye paths and leaf-leaf paths do not have a connection to the "percolating" pore network and are considered of minor relevancy for transport processes, and therefore all were removed. Branch-leaf paths, on the other hand, often are generated due to surface noise (roughness of the pore wall) within pore channels not necessarily representing paths entering the bulk soil from a biopore (Peth, 2010). To remove such paths, which are likely associated with surface roughness, a length threshold was used. The threshold length was selected locally according to the maximum burn number (b_c) of the connected MA cluster (Lindquist et al., 1995). The burn number represents the distance in voxels from the center to the pore wall. Branch-leaf paths were removed when their length $l_{bl} < b_c + 1$ retaining all longer "dead-end" paths, which we considered to be part of the network (such paths are likely associated with lateral roots or cracks).

Network Parameters

Pore network parameters such as path length, pore diameter, and tortuosity of connected paths were extracted from the medial axis. Pore channel length is calculated by measuring the distance along the medial axis between vertices of connected path clusters. As a result, the length of all curved pore channels connecting at least two nodal pores is determined. Local pore channel width is calculated from the specific burn number associated with each MA voxel. The burn number can physically be interpreted as the largest ball centered at an MA voxel which just fits the void space and thus determines the radius (in units of number of voxels) of the pore channel along its curved skeleton (Lindquist et al., 1995). The vertical distance tortuosity or interface tortuosity (τ^d) in 3DMA is characterized as the ratio between curved path length and the Euclidean distance of the path connecting two parallel planes and is expressed as a number ≥ 1, where 1 denotes a straight path and increasing τ^d indicates increasing tortuosity (Lindquist et al., 1996). We calculated vertical distance tortuosities for the major connected paths extracted from the biopore network. For all the possible connecting paths between the voxels in the top and bottom plane, both the Euclidean and actual distances were calculated. In addition to the vertical distance tortuosity, we calculated the local path tortuosity at a smaller scale for branch-branch (τ^{b-b}) and branch-leaf (τ^{b-l}) paths within the pore network. Here the path length and the Euclidean distance between the start and the end voxel of the considered path cluster are used to determine the tortuosity.

Results
Visualization of Biopores

The most obvious advantage of noninvasive techniques such as X-4ay CT is that the dynamics of the pore space and the interaction of soil structure with various soil processes (transport, root growth, etc.) can be investigated. This is illustrated for a biopore, approximately 1 cm in diameter, which is blocked by aggregates that may be associated with earthworm infillings (Fig. 4–3). This is a good example of how biological activity can rearrange the pore space, hence significantly influencing pore functioning, gas and water flow for example. However, while most studies concentrate on the effect of pore space architectures on physical soil functions, biological functions such as root growth are also clearly influenced. Figure 4–3 shows a section from 55 to 72 cm depth containing a biopore that is invaded by a small root with approximately 0.1-cm diameter. In this case, the root preferred to grow into the biopore where it apparently elongated readily 20 cm downward without adhering to the pore wall until it reached the bottom of the biopore. In the upper part, it is noticeable that the root follows a more or less vertical growth first and then breaks through into the biopore (Fig. 4–3). In the lower

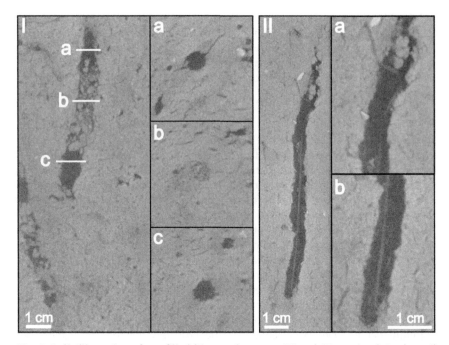

Fig. 4–3. (I) Illustration of a refilled biopore between 55 and 67 cm depth in the soil profile in a vertical (left) and three horizontal (right) cross-sections. (II) Another biopore with a recolonizing plant root. The upper (a) and lower (b) section of the biopore is shown with magnification at the right side.

section, it seems that the root branches out, while one branch penetrates back into the biopore wall (Fig. 4–3).

In Fig. 4–4 we demonstrate three-dimensional visualizations for the six soil monoliths before (Fig. 4–4a) and after (Fig. 4–4b) volume thresholding, which was employed to isolate the network of biopore channels. It is shown that most of the spherical macropores could be removed with this simple procedure, while all biopore channels were preserved. In the subsequent analysis, this helps to distinguish between pores associated with root channels or earthworm burrows from other, more spherical type macropores, which could be either remnants of very old refilled root channels or faunal casts. Functionally, the spherical macropores are considered to behave differently than biopore channels so that a separation of the two will allow a separate quantification of pore characteristics (e.g., pore wall, surface area, tortuosity, number of lateral branching channels, etc.).

Scale Issues

Soils show a variety of pore size classes where pores at different scales determine different functions. Thus, pores should be comprehensively analyzed over multiple scales, which we have done by investigating soil samples of different sizes from the same treatments. The samples range from larger soil monoliths (20-cm diameter, 70-cm height), which we consider as pedon scale, over soil cores (5.1-cm diameter,

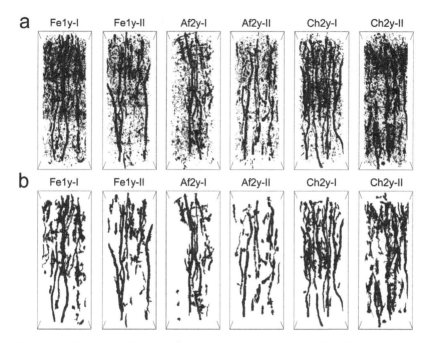

Fig. 4–4. Volume rendering of pore surfaces for all soil monoliths (a) before (all macropores) and (b) after volume thresholding (isolated biopore channels). The monoliths are 70 cm in height and 20 cm in diameter.

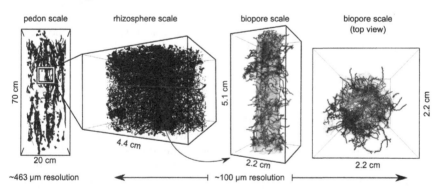

Fig. 4–5. Pore surface and medial axis on different scales ranging from a large soil monolith (pedon scale) over a smaller core including one large biopore channel (rhizosphere scale) down to the single biopore scale showing lateral channels penetrating the rhizosphere.

4.4-cm height) resolving the transition from the rhizosphere to the bulk soil, finally down to the single biopore and aggregate scale (Fig. 4–5). Using this approach, we were able to quantify pore network characteristics across different scales which can be analyzed in view of specific processes, such as lateral oxygen and nutrient diffusion between biopores and rhizosphere or transport of water and gas through the soil profile via the interconnected set of biopores. Implementation of the derived

Table 4–2. Porosity, surface/volume ratio, and pore wall surface area for the resolved pore space before (macroporosity) and after (bioporosity) volume thresholding.

		Fe1y-I	Fe1y-II	Af2y-I	Af2y-II	Ch2y-I	Ch2y-II
Porosity (%)	all macropores	2.05	1.47	1.28	1.08	1.82	2.22
	biopore channels	1.43	1.21	1.01	0.82	1.46	1.68
Surface to volume ratio (cm^2 cm^{-3})	all macropores	14.0	12.6	12.6	10.4	12.2	15.2
	biopore channels	9.8	8.6	9.4	8.7	9.8	11.2
Pore wall surface (cm^2)	All macropores	3340	2317	1768	1389	2722	4001
	biopore channels	1638	1292	1050	877	1748	2228

network parameters representing different scales into modeling approaches and comparisons with localized and continuum scale physical measurements (micro sensors and flow experiments) will have to be discussed in future analysis.

Pore Network Quantification

Pore Surface Area and Volumes

Macroporosity (resolved pores > ~463 μm) was highest for chicory-II (2.22%) and fescue-I (2.05%), while alfalfa had the lowest porosity in both replicates (1.28 and 1.08%). Both alfalfa replicates contained a slightly lower total biopore volume (1.01 and 0.82%) compared to chicory (1.46 and 1.68%) and fescue (1.43 and 1.21%). For all samples, between 70 and 80% of the total resolved macroporosity was associated with biopore channels. Differences for the porosity of biopores and macropores between were as high as between treatments (Table 4–2). The pore surface of biopore channels, representing the potential reaction interface for roots growing in such pores, was the highest for the chicory treatment (1748 and 2228 cm^2), with 50 to 65% of the total pore surface found in the biopore channels. Alfalfa showed a smaller difference between macro- and bioporosity as well as available pore surface of about 40% after volume thresholding. The surface/volume ratio was in the order of 10:1 for all treatments and decreased after volume thresholding.

No major differences between treatments were observed for the derived disconnected pore volume distributions for the smaller pore sizes (<2E−03 cm^3) indicated by a very similar logarithmic trend (Fig. 4–6). However, higher numbers of larger pores (continuous biopore channels) were found for alfalfa compared to fescue, while chicory ranged between the two with a high deviation between replicates. For larger pore sizes (>1E−02 cm^3) the frequency distribution deviated from the logarithmic trend, with a high variance of medium and large pore sizes of a few pores (low relative frequency), extending over two orders of magnitude (between circa 1E−02 and 1E+00 cm^3).

For a better itemization, Fig. 4–7 illustrates the cumulative relative frequency distribution of the biopore fraction only subdivided into pore size classes of <1 cm^3, 1–10 cm^3, and >10 cm^3. For all six treatments, the bioporosity consisted of 79 to 83% of pores smaller than 1 cm^3, where the maximum porosity for this class was found with 0.4% for fescue (replicate I). Between 17 and 21% of the bioporosity was found for pore classes between 1 and 10 cm^3 and only 1 to 3% of the pore

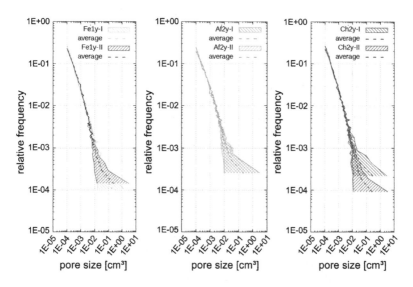

Fig. 4–6. Relative frequency distribution of disconnected pore volumes for all treatments and replicates (indicated by I and II) before volume thresholding.

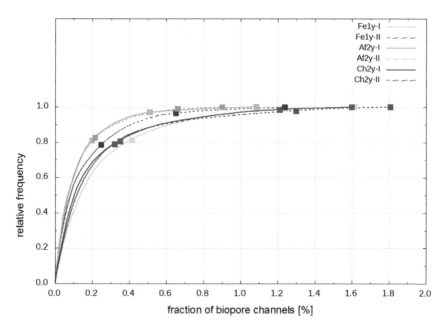

Fig. 4–7. Cumulative biopore fraction for the six treatments with reference to three pore size classes illustrated as squares: First square on the line (from left) marks the class <1, second square marks the class 1–10 and the last square marks the class >10 cm³).

sizes had a volume greater than 10 cm³. Chicory showed the highest fraction of the larger biopores channels for both replicates compared to alfalfa and fescue.

Pore Network Skeleton (Medial Axis)

The size distributions shown in the previous section give us an indication of connected and disconnected pore volumes but do not preserve any information on the geometrical characteristics of the biopore network. Parameters on the architecture of the pore space, which are essential for transport functions for example, can be derived from the medial axis. Figure 4–8 illustrates the pore skeleton (medial axis) for the investigated treatments. The associated colors reflect the pore width at that specific location, with red-yellow-orange describing the presence of very thin to medium pore channels and green-blue-purple describing medium to very thick pore channels, respectively. It is obvious that the visual impression denotes features that have been shown in the visualization of biopores section, but allows further statistical network analysis.

The flow path width, which is an important parameter for flow processes in soils, is shown in Fig. 4–9. Up to a path width of 0.6 cm, all treatments and replicates have shown only slight differences in the pore diameter frequencies. For wider paths (>0.6-cm diameter) the curves clearly deviated; however, variation was as large between replicates as between treatments. Fescue, alfalfa, and chicory had the same maximum path width (~0.93 cm), however, with the highest frequency for chicory followed by alfalfa and then fescue.

Similar to the distribution of pore sizes (Fig. 4–6) all treatments showed higher frequencies for smaller path lengths but with minor differences between the treatments for path lengths <1 cm (Fig. 4–10). For larger path lengths (>1 cm) differences between treatments and replicates were more pronounced, where chicory had the longest paths (>20 cm) followed by alfalfa and fescue, which is in line with the highest bioporosity and the total path lengths found for chicory.

Total path lengths have been calculated providing some information of the overall length of biopore networks. Chicory was found to have in average a 1.5 to 1.8 times longer network than fescue and alfalfa where the ranking of all six

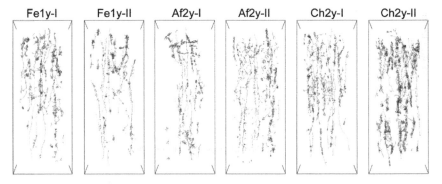

Fig. 4–8. Volume rendering of the computed pore skeleton (medial axis) for all soil monoliths after volume thresholding. The monoliths are 70 cm in height and 20 cm in diameter. Red colors represent very narrow channels while orange, yellow, green, blue, and purple colors indicate progressively larger local channel diameters.

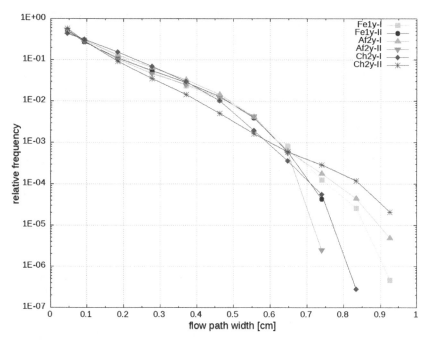

Fig. 4–9. Relative frequency distribution of flow path widths for all investigated soil monoliths after volume thresholding.

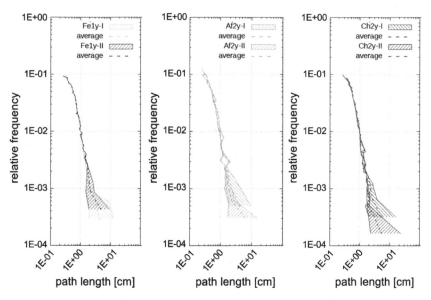

Fig. 4–10. Relative frequency distribution of path lengths for all soil monoliths after volume thresholding and medial axis trimming.

Table 4–3. Total/branch–branch path length and total/branch-branch path frequency densities for all soil monoliths.

	Fe1y-I	Fe1y-II	Af2y-I	Af2y-II	Ch2y-I	Ch2y-II
Total path length density (cm cm^{-3})	0.15	0.10	0.10	0.09	0.14	0.24
Total path frequency density (cm^{-3})	0.30	0.19	0.19	0.17	0.25	0.51
Branch-branch path length density (cm cm^{-3})	0.09	0.07	0.06	0.09	0.09	0.15
Branch-branch path frequency density (cm^{-3})	0.16	0.11	0.10	0.11	0.15	0.31

Table 4–4. Mean values for the vertical distance tortuosity (τ^d) and number of continuous paths for which tortuosity was calculated.

	Fe1y-I	Fe1y-II	Af2y-I	Af2y-II	Ch2y-I	Ch2y-II
τ^d	2.4	1.8	2.0	3.2	2.3	4.3
Number of continuous paths	6	2	8	12	8	7

monoliths was Ch2y-I (2586 cm), Fe1y-I (1601 cm), Ch2y-II (1511 cm), Af2y-II (1319 cm), Fe1y-II (1132 cm), and Af2y-I (942 cm). Accordingly, the total path length density and frequency density showed the same order (Table 4–3). Additionally the total branch-branch path length density and branch-branch path frequency density indicate the interconnectivity of the pore system. Chicory-II had the highest number for branch-branch paths (0.31 cm cm^{-3}) indicating that the biopores have numerous intersecting lateral channels (Fig. 4–9). In contrast, alfalfa and fescue-II had in average less of such secondary lateral side channels.

Tortuosity is an important parameter for diffusion processes. We have calculated tortuosity values on a larger scale distance tortuosity (τ^d) for the soil monoliths (Table 4–4). The calculated mean vertical distance tortuosities for the soil monoliths ranged from 1.8 to 4.3; however, variation was higher between replicates than between treatments.

Rhizosphere Pore Networks

The reconstruction of the pore surface and network of smaller soil cores containing biopore channels are useful to study the pore space characteristics at the rhizosphere scale. A large biopore with lateral channels is shown for a sample of the chicory treatment (75 cm depth) in Fig. 4–11. We illustrated the pore network in three perspectives, including a top view to demonstrate the connectivity of the paths that are connected to a large vertical biopore channel and laterally penetrate the soil matrix. Compared to the larger scale soil monoliths, here significantly smaller pores with a connection to the biopore were resolved (at ~100 µm resolution).

As an example, we calculated the same statistical parameters for the particular sample shown in Fig. 4–11 to compare them with data of the large scale chicory monoliths (Fig. 4–12). The pore size distribution reveals that, by resolving pore volumes <1E–04 cm^3 for the smaller soil core, which corresponds to the lower detection limit of pore volume sizes for the large scale, we are able to extend the characteristic log–log relationship for the frequency distribution toward two orders of magnitude smaller pore volumes. The log–log relationship for pores <1E-04 cm^3 suggests that the size of the smaller cores is just large enough to cover

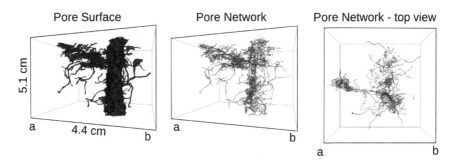

Fig. 4–11. Extracted soil core sample from a chicory treatment at 75 cm depth showing the pore surface and pore network connected to a large biopore channel from different perspectives. Resolution of the scan was ~100 μm.

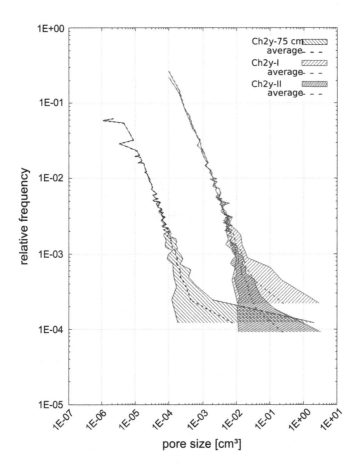

Fig. 4–12. Relative frequency distribution of disconnected pore volumes for both chicory treatments after volume thresholding and a small scale soil core from the same treatment at 75 cm depth referring to the interconnected pore network shown in Fig. 4–13.

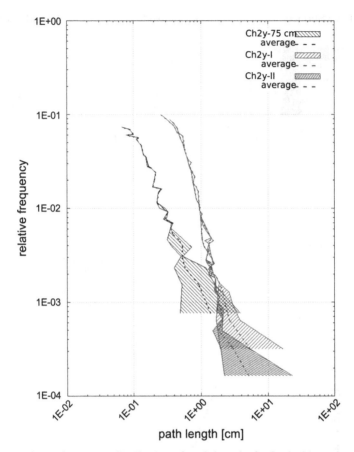

Fig. 4–13. Relative frequency distribution of path lengths for both chicory treatments after volume thresholding and a small scale soil core from the same treatment at 75 cm depth referring to the interconnected pore network shown in Fig. 4–13.

a representative number of pores close to the detection limit of the larger monoliths, which could be useful, for example, for scaling hydraulic functions that potentially could be derived from the frequency distributions.

The small scale pore network shown in Fig. 4–12 has a total path length of 283 cm with a path length density (PLD) of 4.04 cm cm^{-3}. Compared to the smaller scale, a much lower path length density was found for the larger monoliths of the chicory treatments (1511 cm; PLD = 0.14 and 2586 cm; PLD = 0.24), indicating a concentration of lateral paths close to biopores. In addition, more paths were detected around the biopore in smaller pores (at the smaller scale) adding up to the total length because of a higher resolution. The frequency distribution of path lengths shows that, in general, shorter paths are found on the smaller scale but also a number of longer paths were observed (Fig. 4–13) which are associated with the biopore and may represent channels created by branching roots (Fig. 4–12). While the distribution for the large scale also reveals paths >1E+01 cm, the longest path at the small scale is found between 2 and 3 cm, but with much higher frequencies.

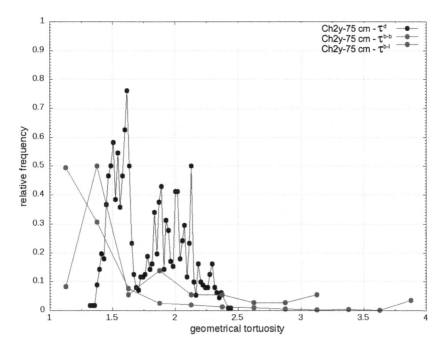

Fig. 4–14. Relative frequency distribution of vertical distance tortuosity (τ^d), branch-branch tortuosity ($\tau^{b\text{-}b}$), and branch-leaf tortuosity ($\tau^{b\text{-}l}$) for the small scale chicory soil core after medial axis trimming.

Compared to the large scale chicory treatment, where we found a tortuosity ranging between 2.3 and 4.3 (Table 4–4), τ^d at the small scale ranged between 1.25 and 2.3 with a peak between 1.5 and 1.75 (Fig. 4–14), which represents the vertical biopore and laterally connected pores (Fig. 4–12). Branch-branch path tortuosities, reflecting in this case the restriction on diffusion processes from the biopore to the bulk soil (95.2% of the pore space in Fig. 4–11 is connected to the biopore), showed a higher frequency for smaller tortuosity values.

Discussion

Noninvasive analysis of soils with X-ray micro-tomography has proven its capacity for many key aspects in scientific research. In particular, the extension from two- to three-dimensional datasets has become indispensable for a characterization of morphological and topological properties of soils. Pierret et al. (2002) argued that two-dimensional imaging cannot be satisfactorily used for the observation and description of complex soil structures, which are inherently three-dimensional. Despite recent advances in CT imaging, the analysis of soil pore systems is still limited by the finite resolution, which is in the order of 1:1000 of the sample diameter.

The purpose of the study was to investigate the effect of root growth on biopore systems and structure dynamics associated with pre-crops differing in root architectures (alfalfa, chicory, and fescue). Plant roots can alter soil structure on various scales. Therefore, we used two different sample sizes to cover structural features from the profile down to the rhizosphere level. Large soil monoliths represented the major biopore channels very well, which could be separated from macropores that are disconnected from this main network by a simple volume thresholding approach. Quantitative image analysis provided network characteristics and surface properties on a soil profile level (pedon scale) of this higher hierarchy biopore system. Smaller soil cores, allowing a fourfold higher resolution, provided a good representation of the biopores at the rhizosphere scale, where lateral channels penetrating the bulk soil from the biopore could be analyzed.

Image Segmentation and Scaling

Indicator kriging (IK) as a local thresholding scheme proved to work very well for the segmentation of biopores. Thresholds were estimated based on the gray-scale distribution by selecting regions of interest around biopores and only needed slight calibration to provide the visually best recovery of the structure. Although thresholding methods still need to be further improved in the future, the localized thresholding method applied in this study provided much better results for structural pores compared to "simple thresholding" and Otsu's method, which partly resulted in remnant voxels within biopores due to noise. This effect would have greatly obscured the analysis of the pore skeleton. On the dispense that some small but "real" pore voxels within the soil matrix are also removed by IK, the major benefit is that isolated grain phases due to segmentation artifacts within biopores are successfully avoided. Since it is inevitable to "loose" pores close to and below the resolution limit, we consider the conversion of some very small pores after thresholding with IK to be of minor importance as we were interested in the larger biopores. Nevertheless, smaller pores, which certainly may be important considering transport processes between bulk soil and rhizosphere, were resolved by analyzing the pore space on smaller core samples with an increased resolution. This multiscale approach was demonstrated to successfully close the resolution gap (see the "Pore Network Quantification" section), which is important for scaling physical functions (e.g., hydraulic conductivity). Scaling could be realized by using quantitative measures of the pore network structure, for example, employing multiscale pore solid fractal (PSF) models (Perrier and Bird, 2003) or pore network models (Delerue and Perrier, 2002). With the approach shown here, we were able to resolve the structural pore space >50 μm. However, the resolution limit is currently extended in ongoing work down to the single aggregate level, where pores <10 μm can be detected (Peth et al., 2008). The wide span of pore sizes analyzed in this multiscale approach allows us to recover the complete range of structural pores, which are considered most relevant for physical soil functions related to gas and water transport (Cresswell et al., 1992).

Characterization of Bioporosity

In our study, we implemented a criterion for extracting biopore channels based on volume thresholding, which conforms to the definition for a biopore as a continuous large pore that is formed by soil fauna and by roots of previous crops (Stirzaker et al., 1996). We isolated the biopore channels from those which we referred to as spherical macropores. Such macropores, which reached volumes of up to 0.3 cm³ in this study, could be residues of old refilled biopore channels or cavities. In particular, endogeic earthworm species have a higher proportion of burrow backfilling compared to anecic species (Capowiez et al., 2011), which is the dominating species at our site (field observation). However, since these pores are not significantly contributing to rapid vertical translocation of water and gas or root growth, they were separated from the continuous biopores. Capowiez et al. (2011) used a similar volume based threshold for the definition of earthworm burrows (0.5 cm^3) and their calculated burrow porosities were close to what we found. Pierret et al. (2002) found bioporosities of large tubular pores between 0.65 and 0.97% for one third of the volume we used. Capowiez et al. (2011) found biopore length densities (in this case earthworm burrows) between 0.09 and 0.23 cm cm^{-3} confirming our results (Table 4–3).

Bastardie et al. (2005) mentioned that there is a lack of information for determining species related differences of burrow topology in field samples. Capowiez et al. (2001) in turn detected differences in branching and continuity of biopores; however, differentiation based on diameter and total length was not statistically significant, while our findings showed that the biopore channels vary in maximum pore diameter and distributions of diameters >0.7 cm (with the largest biopores at the chicory treatment). Since we are not yet able to distinguish between biopores originating from plant roots and earthworm activity, we analyzed biopores irrespective of their origin. Although it is known that plants can create radial pressures of up to 2 MPa resulting in cylindrical compression zones (Goss, 1991; Dexter, 2004), this may not be an unambiguous indicator for biopore genesis since also earthworms do change the bulk density around burrows (Schrader et al., 2007). However, the detection of lateral branching channels (e.g., Fig. 4–11) suggests that biopores are colonized by roots which develop secondary laterals. Furthermore, earthworms are likely to produce pore systems with a different angularity and branching than roots, resulting in different tortuosities. Therefore, the tortuosity of a biopore network may be a useful parameter to distinguish between burrows and root channels (Langmaack et al., 2002; Bastardie et al., 2002). Compared to the branching rate (number of branching points per path length) determined by Bastardie et al. (2003) and Capowiez et al. (2001), which for earthworms was between 1 and 3 m^{-1} and 20–50 m^{-1}, respectively, we found much higher branching rates in the order of 1 cm^{-1} for our biopore network. This seems to be in a similar range of what could be estimated from visualizing secondary lateral roots by endoscopy measurements (Kautz and Köpke, 2010) suggesting that most of our biopores have been colonized by roots.

The earthworm burrows investigated by Capowiez et al. (2011) are more tortuous compared to our findings (Fig. 4–5), which could be the result of initially root generated biopores in our study. However, this point would require further systematic investigation which is currently under progress. It is also reported that biopore channels are recolonized by plant roots (Hirth et al., 2005; Bengough, 2003). White and Kirkegaard (2010) studied wheat root distributions in subsoil and found that roots prefer to grow in cracks and macropores with a greater diameter than the root itself. Moreover, Böhm and Köpke (1977) and Watt et al. (2006) found that subsoil roots in general prefer to grow into biogenetically or pedogenetically created macropores or in their direct neighborhood (Pierret et al., 1999), which was confirmed in our study. With X-ray CT, we were able to detect plant roots which recolonized biopores. Plants with allorhiz and homorhiz root systems likely develop different biopore systems. For example, the root system of alfalfa is described by long, almost straight taproots with large diameters (Mitchell et al., 1995). Mitchell et al. (1995) and Meek et al. (1989) found that alfalfa could develop its enhancements for water flow properties during a longer production cycle. At first, nondecayed root remains decreased transport but with an initiating decomposition of remaining organic matter an increase can be found. Such dynamic processes, in addition to the dynamic changes in soil structure, could also be investigated by X-ray CT as a noninvasive technique, which allows for analyzing the change in pore space by multiple scans. As mentioned in Javaux et al. (2008), this would fulfill the need for more detailed models, as consequently more detailed information about spatial characteristics is needed.

Modeling Approaches

Quantification of morphological and topological parameters as shown in this study can be useful for various practical model applications in soil science. The parameters derived from the medial axis such as length and width of paths of the pore network, for example, could be used as input for pore network models by which effective hydraulic parameters of the imaged soil structures and the complex three-dimensional phase distribution of gas and fluids in an unsaturated porous media can be simulated (e.g., Lindquist, 2001; Delerue and Perrier, 2002; Raeesi and Piri, 2009). Monga et al. (2009) described a model for simulation of biological activity and organic matter decomposition based on three-dimensional geometrical data of soils. Such models are computationally expensive, demanding simplification of the complex soil structure with geometrical primitives like chains of connected balls and cylinders recovering the properties of the three-dimensional pore network. Ngom et al. (2011) have recently shown that Delaunay triangulation provided a reasonable recovery of pore network characteristics when compared to a voxel based approach which was considered as a reference for the complex three-dimensional structure. In doing so, they were able to enhance the simulation of organic matter decomposition without having to rely on voxel based spatial discretization which would require laborious calculations.

As mentioned in Gerke et al. (2010) observations and quantification of preferential flow is essential for the conceptual development, validation, and parametrization of models. Concepts of two pore domains are used in dual-continuum and dual-porosity models (e.g., Gerke et al., 2009; Vogel et al., 2000, 2010), which can be supported with our data (e.g., pore size distribution, macropore surface area). In particular, earthworms can create coatings at biopore walls, which are then stated to have varying biochemical and physical properties (Jégou et al., 2000). With respect to the free surface films at macropore walls shown in Nimmo (2010), these conditions could be considered and pore surface parameters extracted from the pore networks could be implemented. Quantifications of pore network geometries can also be useful for three-dimensional modeling of root growth and nutrient uptake (Somma et al., 1998), since a previous pore network may be recolonized by succeeding roots. In addition to the estimation of porosities of the domains and following the intensity–capacity concept proposed by Horn and Kutílek (2009), the tortuosity of pores in particular, which we illustrated as vertical distance tortuosity (pedon scale) and tortuosity of branch-branch paths (rhizosphere and biopore scale), could provide a measure to estimate the intensity of flow (advection and diffusion) through pores at various scales. Such information is for example needed for a consideration of structure dependent gas diffusion in models which commonly employ tortuosity as a constraining parameter (Liu et al., 2006; Moldrup et al., 2000).

Conclusions and Perspectives

With application of X-ray CT, we illustrated the effect of roots on soil structural changes at different scales and the interaction between structure and biology (e.g., root growth into biopores). Using quantitative image analysis, pore surface, and network parameters (pore size and surface, path length, width, and tortuosity) were extracted from the tomograms to document differences between biopore systems. Such parameters could be very useful for modeling transport processes and biogeochemical cycling, especially nutrient acquisition by plant roots, and the influence of structural changes on soil processes. It is recommended to investigate different sample sizes and thus resolve scale dependent soil architectures. This multiscale approach allows connecting pedon and rhizosphere scale processes to obtain a more complete picture of soil structure–root interactions.

As the investigated soils comprised an older macropore system, differences between the three treatments were not as pronounced as would have been expected when the preceeding crops were planted in a homogenized soil. Within ongoing research, more replicates will be analyzed to better distinguish between root effects and the inherent variability of the older macropore system. However, we could demonstrate that the employed image analysis algorithms were suitable to quantify differences in the biopore network characteristics between the six soil columns. Future investigations, combining noninvasive techniques with physical measurement of related transport and exchange parameters, which are currently under progress, will facilitate the assessment of relationships between morphological characteristics of biopore systems and their functioning in soils from the rhizosphere to the pedon scale. This will further our understanding of soil–water–root relations and help to develop better mechanistic prediction models.

Acknowledgments

This study was supported by the German Research Foundation (Deutsche Forschungsgemeinschaft–DFG) within the framework of the research unit DFG-FOR 1320, speaker: Ulrich Köpke, Bonn'.

References

Angers, D.A., and J. Caron. 1998. Plant-induced changes in soil structure: Processes and feedbacks. Biogeochemistry 42:55–72. doi:10.1023/A:1005944025343

Aylmore, L.A.G. 1993. Use of computer-assisted tomography in studying water movement around plant roots. In: D.L. Sparks, editor, Advances in Agronomy Vol. 49. Academic Press, London. p. 2–50.

Ball, B.C. 1981. Pore characteristics of soils from two cultivation experiments as shown by gas diffusivities and permeabilities and air-filled porosities. J. Soil Sci. 32:483–498. doi:10.1111/j.1365-2389.1981.tb01724.x

Bastardie, F.B., M.C. Cannavacciuolo, Y.C. Capowiez, J.R. de Dreuzy, A.B. Bellido, and D.C. Cluzeau. 2002. A new simulation for modelling the topology of earthworm burrow systems and their effects on macropore flow in experimental soils. Biol. Fertil. Soils 36:161–169. doi:10.1007/s00374-002-0514-0

Bastardie, F., Y. Capowiez, J.R. de Dreuzy, and D. Cluzeau. 2003. X-ray tomographic and hydraulic characterization of burrowing by three earthworm species in repacked soil cores. Appl. Soil Ecol. 24:3–16. doi:10.1016/S0929-1393(03)00071-4

Bastardie, F., Y. Capowiez, and D. Cluzeau. 2005. 3D characterisation of earthworm burrow systems in natural soil cores collected from a 12-year-old pasture. Appl. Soil Ecol. 30:34–46. doi:10.1016/j.apsoil.2005.01.001

Baveye, P.C., M. Laba, W. Otten, L. Bouckaert, P. Dello Sterpaio, R.R. Goswami, D. Grinev, A. Houston, Y. Hu, J. Liu, S. Mooney, R. Pajor, S. Sleutel, A. Tarquis, W. Wang, Q. Wei, and M. Sezgin. 2010. Observer-dependent variability of the thresholding step in the quantitative analysis of soil images and X-ray microtomography data. Geoderma 157:51–63. doi:10.1016/j.geoderma.2010.03.015

Bengough, A.G. 2003. Root growth and function in relation to soil structure, composition and strength. In: H. De Kroon and E.W.J Visser, editors, Root ecology. Springer, Berlin. p. 151–171.

Bennie, A.T.P. 1991. Growth and mechanical impedance. In: Y. Waisel et al., editors, Plant roots: The hidden half. Marcel Dekker, New York. p. 393–416.

Beven, K., and P. Germann. 1982. Macropores and water flow in soils. Water Resour. Res. 18:1311–1325. doi:10.1029/WR018i005p01311

Böhm, W. and U. Köpke. 1977. Comparative root investigations with two profile wall methods. J. Agron. Crop Sci. (1949–1985). 144: 297–303.

Boone, F.R. 1988. Weather and other environmental factors influencing crop responses to tillage and traffic. Soil Tillage Res. 11:283–324. doi:10.1016/0167-1987(88)90004-9

Booltink, H.W.G., and J. Bouma. 1991. Physical and morphological characterization of bypass flow in a well structured clay soil. Soil Sci. Soc. Am. J. 55:1249–1254. doi:10.2136/sssaj1991.03615995005500050009x

Booltink, H.W.G., and J. Bouma. 1993. Sensitivity analysis on processes affecting bypass flow. Hydrol. Processes 7:33–43.

Bouma, J. 1981. Soil morphology and preferential flow along macropores. Agr. Water Manage. 3:235–250. doi:10.1016/0378-3774(81)90009-3

Bouma, J. 1990. Using morphometric expressions for macropores to improve soil physical analyses of field soils. Geoderma 46:3–11. doi:10.1016/0016-7061(90)90003-R

Bouma, J. 1991. Influence of soil macroporosity on environmental quality. In: L.S. Donald, editor, Advances in Agronomy Vol. 46. Academic Press, London. p. 1–37.

Brunke, O., K. Brockdorf, S. Drews, B. Müller, T. Donath, J. Herzen, and F. Beckmann. 2008. Comparison between x-ray tube-based and synchrotron radiation-based mCT. In: S.R. Stock, editor, Developments in X-Ray Tomography VI. San Diego, CA, USA. 2008. Proc. SPIE 7078, 70780U (2008). http://dx.doi.org/10.1117/12.794789

Capowiez, Y., P. Monestiez, and L. Belzunces. 2001. Burrow systems made by Aporrectodea nocturna and Allolobophora chlorotica in artificial cores: Morphological differences and effects of interspecific interactions. Appl. Soil Ecol. 16:109–120. doi:10.1016/S0929-1393(00)00110-4

Capowiez, Y., S.P. Sammartino, and E. Michel. 2011. Using X-ray tomography to quantify earthworm bioturbation non-destructively in repacked soil cores. Geoderma 162:124–131. doi:10.1016/j.geoderma.2011.01.011

Clausnitzer, V., and J.W. Hopmans. 2000. Pore-scale measurements of solute breakthrough using microfocus X-ray computed tomography. Water Resour. Res. 36:2067–2079. doi:10.1029/2000WR900076

Cresswell, H., D. Smiles, and J. Williams. 1992. Soil structure, soil hydraulic properties and the soil water balance. Aust. J. Soil Res. 30:265–283. doi:10.1071/SR9920265

Delerue, J.F., and E. Perrier. 2002. DXSoil, a library for 3D image analysis in soil science. Comput. Geosci. 28:1041–1050. doi:10.1016/S0098-3004(02)00020-1

De Wever, H., D.T. Strong, and R. Merckx. 2004. A system for studying the dynamics of gaseous emissions in response to changes in soil matric potential. Soil Sci. Soc. Am. J. 68:1242–1248. doi:10.2136/sssaj2004.1242

Dexter, A.R. 2004. Soil physical quality: Part I. Theory, effects of soil texture, density, and organic matter, and effects on root growth. Geoderma 120:201–214. doi:10.1016/j.geoderma.2003.09.004

Douglas, J.T., M.G. Jarvis, K.R. Howse, and M.J. Goss. 1986. Structure of a silty soil in relation to management. J. Soil Sci. 37:137–151. doi:10.1111/j.1365-2389.1986.tb00014.x

Dreesmann, S. 1994. Pflanzenbauliche Untersuchungen zu Rotklee- und Luzernegras-Grünbrachen in der modifizierten Fruchtfolge Zuckerrüben– Winterweizen–Wintergerste. Ph.D. diss., Rheinische Friedrich-Wilhelms-Universität, Bonn, Germany.

Edwards, W.M., M.J. Shipitalo, L.B. Owens, and L.D. Norton. 1989. Water and nitrate movement in earthworm burrows within long-term no-till cornfields. J. Soil Water Conserv. 44:240–243.

Feldkamp, L.A., L.C. Davis, and J.W. Kress. 1984. Practical cone-beam algorithm. J. Opt. Soc. Am. 1:612–619. doi:10.1364/JOSAA.1.000612

Fitter, A.H. 1991. Characteristics and functions of root systems. In: Y. Waisel et al., editors, Marcel Dekker, New York. p. 3–24.

Flury, M., and H. Flühler. 1995. Tracer characteristics of Brilliant Blue. Soil Sci. Soc. Am. J. 59:22–27. doi:10.2136/sssaj1995.03615995005900010003x

Francis, G.S., and P.M. Fraser. 1998. The effects of three earthworm species on soil macroporosity and hydraulic conductivity. Appl. Soil Ecol. 10:11–19. doi:10.1016/S0929-1393(98)00045-6

Garré, S., J. Koestel, T. Günther, M. Javaux, J. Vanderborght, and H. Vereecken. 2010. Comparison of heterogeneous transport processes observed with electrical resistivity tomography in two soils. Vadose Zone J. 9:336–349. doi:10.2136/vzj2009.0086

Gerke, H.H. 2006. Preferential flow descriptions for structured soils. J. Plant Nutr. Soil Sci. 169:382–400. doi:10.1002/jpln.200521955

Gerke, H.H., A. Badorreck, and M. Einecke. 2009. Single- and dual-porosity modelling of flow in reclaimed mine soil cores with embedded lignitic fragments. J. Contam. Hydrol. 104:90–106. doi:10.1016/j.jconhyd.2008.10.009

Gerke, H.H., P. Germann, and J. Nieber. 2010. Preferential and unstable flow: From the pore to the catchment scale. Vadose Zone J. 9:207–212. doi:10.2136/vzj2010.0059

Germann, P., and K. Beven. 1981. Water flow in soil macropores I. An experimental approach. J. Soil Sci. 32:1–13. doi:10.1111/j.1365-2389.1981.tb01681.x

Ghodrati, M., and W.A. Jury. 1992. A field study of the effects of soil structure and irrigation method on preferential flow of pesticides in unsaturated soil. J. Contam. Hydrol. 11:101–125. doi:10.1016/0169-7722(92)90036-E

Gliński, J., and W. Stępniewski. 1985. Soil aeration and its role for plants. CRC Press, Boca Raton, FL.

Goss, M.J. 1991. Consequences of the activity of roots on soil. In: D. Atkinson, editor, Plant root growth: An ecological perspective. Blackwell Scientific Publications, London. p. 171–186.

Gregory, P.J. 1988. Growth and functioning of plant roots. In: A. Wild, editor, Soil conditions & plant growth. Longman Scientific & Technical, Harlow. p. 113–167.

Hajnos, M., J. Lipiec, R. Świeboda, Z. Sokolowska, and B. Witkowska-Walczak. 2006. Complete characterization of pore size distribution of tilled and orchard soil using water retention curve, mercury porosimetry, nitrogen adsorption, and water desorption methods. Geoderma 135:307–314. doi:10.1016/j.geoderma.2006.01.010

Hillel, D. 1980. Fundamentals of soil physics. Academic Press, New York.

Hirth, J.R., B.M. McKenzie, and J.M. Tisdall. 2005. Ability of seedling roots of Lolium perenne L. to penetrate soil from artificial biopores is modified by soil bulk density, biopore angle and biopore relief. Plant Soil 272:327–336. doi:10.1007/s11104-004-5764-1

Horn, R., and A. Smucker. 2005. Structure formation and its consequences for gas and water transport in unsaturated arable and forest soils. Soil Tillage Res. 82:5–14. doi:10.1016/j.still.2005.01.002

Horn, R., and M. Kutílek. 2009. The intensity-capacity concept–How far is it possible to predict intensity values with capacity parameters. Soil Tillage Res. 103:1–3. doi:10.1016/j.still.2008.10.007

Iassonov, P., T. Gebrenegus, and M. Tuller. 2009. Segmentation of X-ray computed tomography images of porous materials: A crucial step for characterization and quantitative analysis of pore structures. Water Resour. Res. 45: W09415.

Jakobsen, B.E., and A.R. Dexter. 1988. Influence of biopores on root growth, water uptake and grain yield of wheat (Triticum aestivum) based on predictions from a computer model. Biol. Fertil. Soils 6:315–321. doi:10.1007/BF00261020

Jarvis, N.J. 2007. A review of non-equilibrium water flow and solute transport in soil macropores: Principles, controlling factors and consequences for water quality. Eur. J. Soil Sci. 58:523–546. doi:10.1111/j.1365-2389.2007.00915.x

Javaux, M., T. Schröder, J. Vanderborght, and H. Vereecken. 2008. Use of a three-dimensional detailed modeling approach for predicting root water uptake. Vadose Zone J. 7:1079–1088. doi:10.2136/vzj2007.0115

Jégou, D., D. Cluzeau, V. Hallaire, J. Balesdent, and P. Tréhen. 2000. Burrowing activity of the earthworms *Lumbricus terrestris* and *Aporrectodea giardi* and consequences on C transfers in soil. Eur. J. Soil Biol. 36:27–34. doi:10.1016/S1164-5563(00)01046-3

Jégou, D., S. Schrader, H. Diestel, and D. Cluzeau. 2001. Morphological, physical and biochemical characteristics of burrow walls formed by earthworms. Appl. Soil Ecol. 17:165–174. doi:10.1016/S0929-1393(00)00136-0

Jones, D.L., A. Hodge, and Y. Kuzyakov. 2004. Plant and mycorrhizal regulation of rhizodeposition. New Phytol. 163:459–480. doi:10.1111/j.1469-8137.2004.01130.x

Joschko, M., P.C. Müller, K. Kotzke, W. Döhring, and O. Larink. 1993. Earthworm burrow system development assessed by means of X-ray computed tomography. Geoderma 56:209–221. doi:10.1016/0016-7061(93)90111-W

Kautz, T., and U. Köpke. 2010. In situ endoscopy: New insights to root growth in biopores. Plant Biosyst. 144:440–442. doi:10.1080/11263501003726185

Ketcham, R., and W.D. Carlson. 2001. Acquisition, optimization and interpretation of X-ray computed tomographic imagery: Applications to the geosciences. Comput. Geosci. 27:381–400. doi:10.1016/S0098-3004(00)00116-3

Kodešová, R., J. Šimůnek, A. Nikodem, and V. Jirků. 2010. Estimation of the Dual-Permeability Model Parameters Using Tension Dsik Infiltrometer and Guelph Permeameter. Vadose Zone J. 9:213–225. doi:10.2136/vzj2009.0069

Köhne, J.M., S. Köhne, and J. Šimůnek. 2009. A review of model applications for structured soils: A) Waterflow and tracer transport. J. Contam. Hydrol. 104:4–35. doi:10.1016/j.jconhyd.2008.10.002

Kutílek, M., and D. Nielsen. 1994. Soil hydrology. Catena, Reiskirchen, Germany.

Langmaack, M., S. Schrader, U. Rapp-Bernhardt, and K. Kotzke. 2002. Soil structure rehabilitation of arable soil degraded by compaction. Geoderma 105:141–152. doi:10.1016/S0016-7061(01)00097-0

Leue, M., R.H. Ellerbrock, and H.H. Gerke. 2010. DRIFT mapping of organic matter composition at intact soil aggregate surfaces. Vadose Zone J. 9:317–324. doi:10.2136/vzj2009.0101

Lindquist, W.B. 2001. Network flow model studies and 3D pore structure. Contemp. Math. 295:355–366. doi:10.1090/conm/295/05026

Lindquist, W.B., S.M. Lee, D.A. Coker, K.W. Jones, and P. Spanne. 1995. Medial axis analysis of three dimensional tomographic images of drill core samples. SUNY-Stony Brook Tech. Rep., SUNYSB-AMS-95-01, Stony Brook, New York.

Lindquist, W.B., S.M. Lee, D.A. Coker, K.W. Jones, and P. Spanne. 1996. Medial axis analysis of void structure in three-dimensional tomographic images of porous media. J. Geophys. Res. 101:8297. doi:10.1029/95JB03039

Lindquist, W.B., A. Venkatarangan, J. Dunsmuir, and T.F. Wong. 2000. Pore and throat size distributions measured from sychrotron X-ray tomographic images of Fontainebleau sandstones. J. Geophys. Res. 105:21509–21528. doi:10.1029/2000JB900208

Liu, G., B. Li, K. Hu, and M.T. van Genuchten. 2006. Simulating the gas diffusion coefficient in macropore network images: Influence of soil pore morphology. Soil Sci. Soc. Am. J. 70:1252–1261. doi:10.2136/sssaj2005.0199

Logsdon, S.D.L., and D.R. Linden. 1992. Interactions of earthworms with soil physical conditions influencing plant growth. Soil Sci. 154:330–337. doi:10.1097/00010694-199210000-00009

Ma, L., and H.M. Selim. 1997. Physical non-equilibrium modeling approaches to solute transport in soils. Adv. Agron. 58:95–153.

Meek, B.D., E.A. Rechel, and L.M. Carter. 1989. Changes in infiltration under alfalfa as influenced by time and wheel traffic. Soil Sci. Soc. Am. J. 53:238–241. doi:10.2136/sssaj1989.03615995005300010042x

Mitchell, A.R., T.R. Ellsworth, and B.D. Meek. 1995. Effect of root systems on preferential flow in swelling soil. Commun. Soil Sci. Plant Anal. 26:2655–2666. doi:10.1080/00103629509369475

Moldrup, P., T. Olesen, J. Gamst, P. Schjønning, T. Yamaguchi, and D.E. Rolston. 2000. Predicting the gas diffusion coefficient in undisturbed soil from soil water characteristics. Soil Sci. Soc. Am. J. 64:94–100. doi:10.2136/sssaj2000.64194x

Monga, O., M. Bousso, P. Garnier, and V. Pot. 2009. Using pore space 3D geometrical modelling to simulate biological activity: Impact of soil structure. Comput. Geosci. 35:1789–1801. doi:10.1016/j.cageo.2009.02.007

Ngom, N.F., P. Garnier, O. Monga, and S. Peth. 2011. Extraction of three-dimensional soil pore space from microtomography images using a geometrical approach. Geoderma 163:127–134. doi:10.1016/j.geoderma.2011.04.013

Nimmo, J.R. 2010. Theory for source-responsive and free-surface film modeling of unsaturated flow. Vadose Zone J. 9:295–306. doi:10.2136/vzj2009.0085

Oh, W., and W.B. Lindquist. 1999. Image thresholding by indicator kriging. IEEE Trans. Pattern Anal. 21:590–602. doi:10.1109/34.777370

Palomo, J., K. Subramanyan, and M. Hans. 2006. Influence of mA settings and a copper filter in CBCT image resolution. Int. J. CARS 1:391–393.

Perret, J., S.O. Prasher, A. Kantzas, and C. Langford. 1998. Characterization of macropore morphology in a sandy loam soil using X-ray computer assisted tomography and geostatistical analysis. Can. Water Resour. J. 23:143–165. doi:10.4296/cwrj2302143

Perret, J., S.O. Prasher, A. Kantzas, and C. Langford. 1999. Three-dimensional quantification of macropore networks in undisturbed soil cores. Soil Sci. Soc. Am. J. 63:1530–1543. doi:10.2136/sssaj1999.6361530x

Perrier, E.M.A., and N. Bird. 2003. The PSF model of soil structure: A multiscale approach, In: Y. Pachepsky et al., editors, Scaling methods in soil physics, CRC Press, London. p. 1–18.

Peth, S. 2010. Applications of Microtomography in Soils and Sediments. In: B. Singh and M. Gräfe, editors, Synchrotron-Based techniques in soils and sediments Vol. 34. Elsevier, Heidelberg. p. 73–101.

Peth, S., R. Horn, F. Beckmann, T. Donath, J. Fischer, and A.J.M. Smucker. 2008. Three-dimensional quantification of intra-aggregate pore-space features using synchrotron-radiation-based microtomography. Soil Sci. Soc. Am. J. 72:897–907. doi:10.2136/sssaj2007.0130

Pierret, A., C.J. Moran, and C.E. Pankhurst. 1999. Differentiation of soil properties related to spatial association of wheat roots and soil macropores. Plant Soil 211:51–58. doi:10.1023/A:1004490800536

Pierret, A., Y. Capowiez, L. Belzunces, and C.J. Moran. 2002. 3D reconstruction and quantification of macropores using X-ray computed tomography and image analysis. Geoderma 106:247–271. doi:10.1016/S0016-7061(01)00127-6

Pierret, A., C. Doussan, Y. Capowiez, F. Bastardie, and L. Pagès. 2007. Root functional architecture: A framework for modeling the interplay between roots and soil. Vadose Zone J. 6:269–281. doi:10.2136/vzj2006.0067

Raeesi, B., and M. Piri. 2009. The effects of wettablitiy and trapping on relationships between interfacial area, capillary pressure and saturation in porous media: A pore-scale network modeling approach. J. Hydrol. 376:337–352. doi:10.1016/j.jhydrol.2009.07.060

Schrader, S., H. Rogasik, I. Onasch, and D. Jégou. 2007. Assessment of soil structural differentiation around earthworm burrows by means of X-ray computed tomography and scanning electron microscopy. Geoderma 137:378–387. doi:10.1016/j.geoderma.2006.08.030

Somma, F., J.W. Hopmans, and V. Clausnitzer. 1998. Transient three-dimensional modeling of soil water and solute transport with simultaneous root growth, root water and nutrient uptake. Plant Soil 202:281–293. doi:10.1023/A:1004378602378

Stewart, J.B., C.J. Moran, and J.T. Wood. 1999. Macropore sheath: Quantification of plant root and soil macropore association. Plant Soil 211:59–67. doi:10.1023/A:1004405422847

Stirzaker, R.J., J.B. Passioura, and Y. Wilms. 1996. Soil structure and plant growth: Impact of bulk density and biopores. Plant Soil 185:151–162. doi:10.1007/BF02257571

Sutton, R.F. 1991. Soil properties and root development in forest trees: A review. Inf. rep. O–X–413. Can. For. Serv., Sault Ste. Marie, ON.

Tiunov, A.V., M. Bonkowski, J. Alphei, and S. Scheu. 2001. Microflora, protozoa and nematoda in *Lumbricus terrestris* burrow walls: A laboratory experiment. Pedobiologia 45:46–60. doi:10.1078/0031-4056-00067

Toor, G.S., L.M. Condron, B.J. Cade-Menun, H.J. Di, and K.C. Cameron. 2005. Preferential phosphorus leaching from an irrigated grassland soil. Eur. J. Soil Sci. 56:155–167. doi:10.1111/j.1365-2389.2004.00656.x

Vogel, T., H.H. Gerke, R. Zhang, and M.T. Van Genuchten. 2000. Modeling flow and transport in a two-dimensional dual-permeability system with spatially variable hydraulic properties. J. Hydrol. 238:78–89. doi:10.1016/S0022-1694(00)00327-9

Vogel, T., J. Brezina, M. Dohnal, and J. Dusek. 2010. Physical and Numerical Coupling in Dual-Continuum Modeling of Preferential Flow. Vadose Zone J. 9:260–267. doi:10.2136/vzj2009.0091

Waisel, Y., A. Eshel, T. Beeckman, and U. Kafkafi. 2002. Plant roots: The hidden half. Marcel Dekker, New York.

Wang, W., A.N. Kravchenko, A.J.M. Smucker, and M.L. Rivers. 2011. Comparison of image segmentation methods in simulated 2D and 3D microtomographic images of soil aggregates. Geoderma 162:231–241. doi:10.1016/j.geoderma.2011.01.006

Watt, M., W.K. Silk, and J.B. Passioura. 2006. Rates of root and organism growth, soil conditions, and temporal and spatial development of the rhizosphere. Ann. Bot. (Lond.) 97:839–855. doi:10.1093/aob/mcl028

White, R.E. 1985. The influence of macropores on the transport of dissolved and suspended matter through soil. Adv. Soil Sci. 3:95–120. doi:10.1007/978-1-4612-5090-6_3

White, R.G., and J.A. Kirkegaard. 2010. The distribution and abundance of wheat roots in a dense, structured subsoil– implications for water uptake. Plant Cell Environ. 33:133–148. doi:10.1111/j.1365-3040.2009.02059.x

5

Computed Tomographic Evaluation of Earth Materials with Varying Resolutions

Ranjith P. Udawatta,* Stephen H. Anderson, Clark J. Gantzer, and Shmuel Assouline

Abstract

Evaluation of earth materials with computed tomography (CT) provides data on geometrical pore parameters and spatial variability within the structure for two and three dimensions that conventional methods do not. These data can be used to improve flow and energy-transfer models, advance techniques for storage of water, and develop tools to characterize structure of materials. This paper compares geometrical pore parameters of two- and three-dimensional measures of soils and explains benefits and limitations of these measures relative to sample size, image resolution, and image analysis. CT of soil samples scanned at 190-, 74-, and 9.6-μm resolutions were using a medical scanner, high-resolution scanner, and synchrotron microtomography. Available image analysis software was used to discriminate among soil materials from natural causes or management effects. Images at high resolution provide information on path tortuosity, pore connectivity, and pore geometry in three-dimensional scale, while low-resolution scans provide data on pores number, two-dimensional features, and pore area. Low- and high-resolution scans both show the same trend pore number although high-resolution scans showed greater pores. Samples scanned at a 74 and 9.6 μm discriminated samples by pore connectivity parameters. The 9.6-μm scans showed pores with high coordination numbers had lower probability values. The 9.6-μm scans underestimated total pore length and volume compared to those detected by the 74-μm scans. Comparisons indicated that 190 μm can be used to differentiate effects for a larger number of samples and high resolution may be used to obtain added information on geometrical parameters for preselected samples. Imaging scale may be based on selecting which parameter(s) is of interest. Combination of smaller-sized samples along with larger samples may provide information on more holistic treatment, reaction, and process effects.

Abbreviations: 3-DMA, Three-Dimesional Dimensional Medial Axis software; AG, agroforestry buffer; CN, coordination number; Co, characteristic coordination number constant; CT, computed tomography; e.c.d., equivalent cylindrical diameter; GB, grass buffer; GSECARS, GeoSoilEnviroCARS sector at the Argonne Advanced Photon Source; PL, path length; PLo, characteristic path length constant; RC, row crop; UMHS, University of Missouri Hospital and Clinic's Siemens Somatom Plus 4 Volume Zoom X-ray CT Scanner; UTCT, University of Texas-Austin High Resolution X-ray Computed Tomography facility.

R.P. Udawatta, 203 ABNR Bldg., Center for Agroforestry, University of Missouri, Columbia, MO 65211. R.P. Udawatta, S.H. Anderson, and C.J. Gantzer, 302 ABNR Bldg., Dep. of Soil, Environ. and Atm. Science, University of Missouri, Columbia, MO 65211. S. Assouline, Dep. of Environmental Physics and Irrigation, Agricultural Research Organization, Volcani Center, Bet-Dagan, Israel. *Corresponding author (UdawattaR@missouri.edu).

doi:10.2136/sssaspecpub61.c5

Soil–Water–Root Processes: Advances in Tomography and Imaging. SSSA Special Publication 61. S.H. Anderson and J.W. Hopmans, editors. © 2013. SSSA, 5585 Guilford Rd., Madison, WI 53711, USA.

X-ray CT has received increasing attention in recent years in soil and earth sciences. In soil science, CT procedures have been used to examine differences in porosity (Anderson et al., 1988; Rachman et al., 2005; Udawatta et al., 2006, 2008; Udawatta and Anderson, 2008), solute movement (Anderson et al., 2003), pore continuity (Grevers and de Jong, 1994), fractal dimension of porosity (Rasiah and Aylmore, 1998; Gantzer and Anderson, 2002), and plant root development (Tollner et al., 1995). In recent studies, Rachman et al. (2005), Udawatta et al. (2006, 2008) and Udawatta and Anderson (2008) used X-ray CT to differentiate soils under row crop, no-till, and perennial buffers using macropore parameters. They observed strong correlations between macropores estimated using water retention and CT procedures. Macropore characteristics such as shape, size, orientation, and size distribution affect the rate, flow, and retention of water (Rasiah and Aylmore, 1998). Although medical CT provides additional information on geometrical pore parameters as compared to traditional porosity estimation methods based on water retention, these procedures lack information on the spatial continuity of pores and most of the time measurements are based on observations in two-dimensions (Gantzer and Anderson, 2002; Mooney, 2002).

Soil scientists are now attempting to look inside the soil system to find better methods for predicting water and gas movement, to assess the effects of management on soil pore parameters and microbial habitats, and to evaluate treatment effects on soil and biological functions. It is known that microstructure governs the flow of resources through the pore space of soil media and creates spatial and temporal differences in the media (Young and Crawford, 2004; Zhang et al., 2005). Research suggests that understanding geometrical pore parameters is critically important for issues related to movement of microfauna, water, solute, and gases as well as soil and biological functions.

Tomographic image slices obtained with X-ray, γ-radiation, or nuclear magnetic resonance energy can be combined using image analysis software and then the interior structural features of samples can be evaluated to understand the nature and spatial configuration of components at micrometer-scale resolution (Asseng et al., 2000; Taina et al., 2008). A process called "volume rendering" combines adjacent cross-sectional images to analyze volumes rendered in three-dimensional space (Mooney, 2002; Akin and Kovscek, 2003; Carlson et al., 2003) to characterize pore parameters that are not available and undetectable by other methods (Akin and Kovscek, 2003). In the absence of volume rendered three-dimesional volumes, connectivity and tortuosity, two important hydrologic pore parameters, are difficult or impossible to quantify. In addition, three-dimensional volumes enable quantification of pore geometry and examination of dynamic soil processes (Pierret et al., 2002; Mooney, 2002). Images acquired at sub-μm resolution with synchrotron facilities and laser confocal microscopy (Lindquist et al., 2000) have greatly improved micro-structural investigations as the size of the smallest pore and neck that can be measured corresponding with a single voxel (Ioannidis and Chatzis, 2000). The main drawback of high resolution is the extremely small sample size (<1 mm).

Literature shows that pore coordination number, pore size, throat-size distribution, pore body-to pore-throat size, pore body-to pore aspect ratio, pore continuity, and tortuosity are geometrical pore parameters that can be used to explain flow (Ioannidis and Chatzis, 1993; Tollner et al., 1995; Ioannidis and Chatzis,

2000; Lindquist et al., 2000; Fox et al., 2004). For example, Fontainebleau sandstone material with porosities ranging from 7.5 to 22% imaged at 5.7-μm resolution were used to compare tortuosity, pore connectivity, pore-channel length, and throat and nodal parameters (Lindquist et al., 2000). The study distinguished sandstone material by comparing differences in coordination numbers of nodal pores and channel, pore, and throat parameters. In another study, Ioannidis and Chatzis (2000) showed that coordination number distributions, and pore and neck size distributions correlated with porosity. Al-Raoush (2002) evaluated different size materials and stated that coordination number was larger for smaller diameter particles and smaller for larger diameter particles. Udawatta et al. (2008) examined soil at 4.59×10^5 μm^3 voxel size to differentiate management effects on pore connectivity, tortuosity, and pore parameters of agricultural soils. Lee et al. (2008) showed that cumulative porosity was significantly reduced by untreated soils as compared to PAM (anionic polyacrylamide) treated soils in a rainfall-infiltration study. These procedures used images at μm resolution to accurately describe changes within the media. To explain contaminant movement in aquifers and obtain model parameters associated with fluid and gas movement as well as to improve our understanding of infiltration, we need quantitative information on soil structure (Ioannidis and Chatzis, 2000). However, the literature lacks information on the random nature of pore parameters, three-dimensional volume data and direct measurements of pore parameters which restricts development of relationships between water flow and porosity parameters (Lindquist et al., 2000; Mooney, 2002). Only a few studies have evaluated soil parameters at varying resolutions and scales ranging from millimeter to micrometer to explain differences in infiltration, soil hydraulic properties, porosity, and bulk density.

There is a need to obtain information on pore geometries to explain the effects of soil management on geometrical pore parameters and to evaluate relative advantages of varying imaging resolution and analysis procedures. Such knowledge will help determine the most appropriate technique for a particular study to compare treatment and management effects on earth materials. The intent of this paper was to compare geometrical soil pore parameters obtained by three image acquisition techniques and two image analysis procedures.

Materials and Methods
Experimental Sites and Soil Samples

Two groups of soil samples were used in this study (Table 5–1). The first group of samples collected from experimental watersheds in Missouri with three treatments was scanned at two resolutions and images were processed by Image-J and 3-DMA software as described in the next section. Results of these two studies will be used for method comparisons (Udawatta et al., 2006, 2008). The second group of soils was obtained from Israel and was subjected to two compaction levels. Soil cores were scanned at the Advanced Photon Source Synchrotron facility in Argonne, IL (GSECARS, GeoSoilEnviroCARS sector at the Argonne Advanced Photon Source) and images were analyzed with 3-DMA software.

The First Group of Soils

Putnam silt loam (fine, smectitic, mesic Vertic Albaqualf) soils were sampled from row crop (RC), grass buffer (GB), and agroforestry buffer (AG; tree+grass) treatments from an experimental watershed study at the Greenley Memorial Research Center near Novelty, MO (40° 01′ N, 92° 11′ W; Udawatta et al., 2006). Glacial till and wind-blown Peorian loess are the soil parent materials (Unklesbay and Vineyard, 1992). During a soybean [*Glycine max* (L.) Merr.] crop year, soils from the surface 0- to 100-mm depth were sampled in intact soil cores with 76.2 mm diameter, 76.2 mm length, and with two replications in 3.2-mm thick wall plastic cylinders in June 2003. Soils were sampled 200 mm from the base of pin oak trees for the AG treatment. For the GB treatment, samples were taken midway between two trees (1.5 m from trees) in the grass area of the buffer. For the RC treatment, two samples were taken midway between buffers (18 m from buffers). Soil cores were trimmed, sealed with caps, placed in plastic bags, transported to the laboratory, and stored at 4°C before measurements. Additional details on management, parent materials, soils, experimental design, sampling, and climatic data can be found elsewhere (Udawatta et al., 2002, 2006). Soil cores were saturated and placed on a glass-bead tension table and equilibrated to −3.5 kPa for 24-h to prepare for scanning. This procedure enhanced image contrast between pores and soil solids by removing water from pores >86 μm equivalent cylindrical diameter (e.c.d). Two additional standard cores were prepared with fine sand to ensure uniformity of X-ray CT analysis among samples. These cores contained imbedded objects of air-filled and water-filled sealed aluminum tubes, and copper-wire for use as X-ray phantoms. The copper wire was 0.55 mm in diameter, and air and water phantoms were 1.6 mm outside diameter and 1.1 mm inside diameter.

The Second Group of Soils

Hamra soil is a loamy sand (Typic Rhodoxeralf) collected from the 0–100-mm depth of an experimental field at Bet-Dagan, Central Israel (32° 12′N and 35° 25′ E). The soil contained 87% sand, 2% silt, and 11% clay (mainly smectite). Air-dry soil was sieved through 2.0 and 0.5 mm mesh sieves to separate into two aggregate size classes: >2mm and <0.5 mm. Soil was packed in 5-mm long by 5 mm diameter aluminum cores with 1.0-mm wall thickness. Soil cores were compacted with a small press and small pistons to obtain pre-determined bulk density values of 1.51 and 1.72 Mg m^{-3}. The selected two values represent bulk densities commonly found with these soils and site conditions. The open end of the soil core was cov-

Table 5–1. Summary of sample information, scanning and image analysis procedures.

Sample	Location	Core dimension	Scanner	Voxel Size	Image analysis
1	Knox County, MO	76-mm long and 76-mm diameter	University of Missouri Hospital and Clinic's Siemens Somatom Plus 4 Volume Zoom X-ray CT scanner	1.8×10^7 μm^3	Image-J
			University of Texas-Austin High Resolution X-ray Computed Tomography facility	4.59×10^5 μm^3	3-DMA
2	Bet-Dagan, Central Israel	5-mm long and 5-mm diameter	GeoSoilEnviroCARS sector at the Argonne Advanced Photon Source	8.8×10^2 μm^3	3-DMA

ered with aluminum plates and sealed with tape to secure soils inside the core. Samples were stored at room temperature before scanning.

Image Acquisition

The First Group of Soils

University of Missouri Hospital and Clinic's Siemens Somatom Plus 4 Volume Zoom X-ray CT scanner (UMHS) was used for computed tomographic image acquisition. Details on image acquisition, system parameters, and processing can be found in Udawatta et al. (2006) and Udawatta and Anderson (2008). In brief, the scan system parameters were: 125 kV, 400 mAs, and 1.5 s scan time to provide detailed and low noise projections, the field of view, i.e., the cross-section dimension, was 100 mm with 512- by 512-mm picture elements, pixel size was 190 µm by 190 µm, the volume element (voxel) size was 1.805×10^7 µm^3, and the X-ray beam width or "slice" thickness was 500 µm. Five scans were taken at 15, 26, 37, 48, and 59 mm from the top of the horizontally placed soil cores.

The same Putnam soil cores were imaged at the University of Texas–Austin High Resolution X-ray Computed Tomography facility (UTCT, 2007). Details on image acquisition, system parameters, and processing can be found in Udawatta et al. (2008). Scanner parameters were: 180 kV and 0.222 mA, spot size of about 50 µm; sample rotation was 1600 views and an acquisition time of 133 ms per view (Ketcham and Carlson, 2001; UTCT, 2007; R.A. Ketcham, personal communication, 2007). CT field of view (cross-section dimension) was 79.3 mm (1024- by 1024-mm, pixel size was 74 by 74 µm in size, the slice thickness was 84 µm, and voxel size was 4.59×10^5 µm^3.

Images were converted to digital data and these are referred to as sinograms. Adjacent scans were stacked to render three-dimensional images. During reconstruction, sinograms were converted to two-dimensional slice images. This resulted in CT numbers for raw intensity data in the sinogram. Finally, three-dimensional images were obtained by stacking adjacent scans, a process called volume rendering. To maximize contrast between solid and void phases a CT range of 0 to 65535 (2^{16}) was used at 16-bit scale.

The Second Group of Soils

Air-dry Hamra soil cores were transported to the GeoSoilEnviroCARS (GSECARS) sector at the Argonne Advanced Photon Source for image acquisition at the X-ray computed microtomography facility. Synchrotron radiation allows collection of µm spatial resolution data with reasonable acquisition time. Soil cores were imaged at a 9.6 µm resolution. The bending magnet beam line 13-BM-D, which provides a parallel beam of high-brilliance radiation with a vertical beam size of about 5 mm, was used for scanning. Scans were conducted at 7–70 keV energy range with flux (photons sec^{-1}) of 1×10^9 @ 10keV. Resolution was ($\Delta E/E$) 10^{-4} and beam size focus was 10 µm by 30 µm. The transmitted X-rays are converted to visible light with a synthetic garnet (YAG) scintillator. The visible light from the scintillator was imaged with a 5X Mitutoyo microscope objective onto a high-speed 12-bit CCD camera (Princeton Instruments Pentamax), with 1317 by 1035 pixels, each 6.7 by 6.7 µm in size. Beamline control and data acquisition were completed using Windows and Linux workstations running EPICS with VME, SPEC, IDL, marCCD, mar345, and

Princeton Instruments Winview and WinSpec. Specific synchrotron tomographic procedures and additional details can be found in Kinney and Nichols (1992).

The data processing step consisted of three main steps: preprocessing, sinogram creation, and reconstruction. Since the image data have a constant digitization offset value (~50 counts), this value was subtracted from each pixel. The second step was to remove zingers; these are bright pixels caused by scattered X-rays striking the CCD chip. The third step of the preprocessing was completed to normalize each data frame to the field image and to correct for drift.

Tomographic reconstruction was completed using filtered back projection with the IDL programming language, Riemann function written by Rivers (1998). The raw data used for tomographic reconstruction were 12-bit images and a total of 360 such images were collected as the sample was rotated twice from 0 to 180° in 0.5° steps. The data were piped to massive parallel SGI computers to view real time data before image acquisition was completed.

Image Analysis

The First Group of Soils

The number of pores, pore area, pore perimeter, circularity, and porosity of a 50-mm square region of each scan were obtained by Image-J version 1.27 (Rasband, 2002). These data were evaluated using a two-dimensional analysis due to lower resolution and infrequent spatial scans. Additional details on image analysis can be found in Udawatta et al. (2006) and Udawatta and Anderson (2008). Thresholding was conducted to separate air-filled pore areas and the other regions within a scan using an intensity value of 40 from the water phantom (38–42, mean = 40).

The same soils imaged at the UTCT were analyzed using the 3Dimentional Medial Axis (3-DMA) computer software to examine differences in geometrical pore characteristics among the treatments (Lindquist and Venkatarangan, 1999). Image analysis was conducted as explained in Lindquist et al. (2005) and Udawatta et al. (2008). The image analysis procedure consists of the following six major steps and these can be accomplished by a number of algorithms imbedded in the image analysis software: segmentation of image, extraction and modification of the medial axis of pore paths, throat construction using the medial axis, pore surface construction, assembly of pore throat network, and geometrical characterization of pore throat network.

Although each core contained >1200 scans, only the middle 800 scans were analyzed to reduce the edge effect. Calibrated CT-values obtained from standard cores were: air (21 ± 2.6), water (30 ± 1.0), sand (38 ± 1.0), and copper (220 ± 9.3). Simple thresholding and indicator kriging were used to differentiate the two populations. In the indicator kriging, the maximum likelihood estimate of the population is used to differentiate voxels having intermediate intensities between T0 and T1 (Oh and Lindquist, 1999).

The medial axis of the pore space was extracted and modified by the medial axis algorithm of Lee et al. (1994), an erosion based algorithm. The length of a path (the distance between the centers of any two adjacent nodal pores along the mid line of the connecting path) was determined by the distance measure algorithm (Lindquist, 2002). Subsequently, a throat finding algorithm was used to locate pore throats (Venkatarangan, 2000). Throat (minimal area cross-sectional surfaces of pore space or pore neck; Kwiecien et al., 1990) surface areas were determined by the throat finding algorithm as triangulated interfaces. Marching cube algorithm was used to determine pore surfaces (Bloomenthal, 1988; Lorensen and Cline, 1987). Throat surfaces separated nodal pores. The number of nodal pore voxels within a nodal pore gives the pore volumes. If a nodal pore voxel is cut by throat surfaces, half of the throat volume voxels are assigned to each of the two nodal pores. Since throats were identified and nodal pores were isolated, pore space can be divided into a network of pores separated by throat surfaces.

The number of curve segments meeting at a vertex was measured to obtain coordination numbers (CN). Coordination numbers between 3 and 6 were used to develop exponential distribution relationships between coordination numbers and probability-density values to determine characteristic coordination number constants (Co) for each sample. A similar method was used to determine characteristic path length constants (PLo), fitting an exponential distribution of path length (PL) and probability density.

Dijkstra's algorithm (Cormen et al., 1990) with a gamma distribution for tortuosity probability distribution embedded in the 3-DMA software determined path tortuosity (Lindquist et al., 1996). Tortuosity of each pore and, average tortuosity and cumulative tortuosity were compared among treatments and analysis methods.

The software generated an assembly of pore networks and geometrical characteristics of pore networks for: effective radii, pore volume, coordination numbers, path lengths, throat area, and path tortuosity with their corresponding probability density relationships.

The Second Group of Soils

Pore characteristics were analyzed at 8.8×10^2 μm^3 voxel size with 3-DMA software as described earlier. Images were cropped into 370 by 370 by 430 μm rectangular blocks to remove artifacts. Spatial distributions for nodal pore volume, coordination numbers, pore path length, throat areas, and tortuosity were obtained for 5.89×10^7 μm^3 volumes.

Comparison of Geometrical Pore Characteristics

Two thresholding procedures were used to segment an image into two phases and the thresholding procedures vary by the image acquisition method. Images from UMHS were segmented into solid and air phases by a simple thresholding procedure as previously described. This method does not allow separation of partially filled voxels into the appropriate phase based on the proportion of the voxel filled by the respective phase. In contrast, samples scanned at the UTCT and GSECARS were segmented into two phases by simple thresholding and kriging methods. While simple thresholding allowed segmentation of completely filled voxels, partially filled voxels were segmented by the kriging method.

Parameters obtained by the two image analysis procedures were different. Since some soil pores are smaller than 190-μm resolution of the UMHS CT scanner and images were acquired at 10-mm intervals, these cannot be used to obtain pore parameters related to pore connectivity. The high resolution images acquired at the UTCT and GSECARS allowed examination of pore continuity parameters such as throats, connectivity, and tortuosity, while UMHS images were used to obtain two-dimensional pore characteristics. Therefore, comparisons were made to explain differences and similarities for parameters based on the scanning and image processing techniques.

Results and Discussion
Number of Pores

Image-J analysis of Putnam silt loam soil scans from the UMHS showed differences in number of pores, porosity, area of the largest pore, and circularity among the three treatments (Table 5–2). These scans are 10 mm apart within a soil core. The tree buffer treatment had 2.4 and 4.7 times more pores than the crop treatment and the differences were significant ($p < 0.05$). Averaged across all five depths, the tree buffer, grass buffer, and crop treatments had 142, 60, and 30 pores on a 2500 mm^2 scan area, respectively. The same Putnam soil cores scanned at the UTCT had 431, 298, and 41 pores in the respective treatments. Although these numbers can be compared relative to each other, they are obtained using different techniques. The main difference between the two estimation methods is two-dimensional and three-dimensional estimated pores by two different scanning and image analysis procedures. At the UMHS, individual scans were taken and processed, while at UTCT, the entire core was evaluated. Differences between the two sets of numbers are probably due to differences in imaging and image analysis. The crop treatment had smaller number of pores (30 and 41) by both methods; this could be due to less roots and biological activities as compared to perennial vegetative buffers. In contrast, buffers had more pores (by several magnitudes) than the crop treatment by both scanning methods. The active vegetation promotes biological activities including roots and associated soil communities which may have contributed to greater number of pores. However, the resolution of the UMHS did not identify pores smaller than 190 μm in diameter, while the UTCT procedure detected as small as 75 μm. Approximately 288 and 238 pores in tree buffer and grass buffer treatments were not detected by the UMHS method. The evaluation shows that low resolution analysis provides information to discriminate treatments before investing in high resolution image acquisition and image processing. Studies requiring additional information such as exact pore numbers may consider high tech evaluations (high resolution and image processing) after initial evaluations with a medical scanner.

Hamra soil cores examined using synchrotron microtomography indicated differences in number of pores in standardized soil cores; these soil cores were smaller than those used for the prior analyses. The two treatments, low and medium density, scanned at the synchrotron on average had 50 and 55 pores, respectively, in 5.89×10^7 $μm^3$ volumes. Between the two size classes, the 2.0-mm and 0.5-mm diameter aggregates had 50 and 54 pores, respectively. Although the smaller aggregate class had a distinct pattern of more pores than the 2.0 mm aggregates, the difference was not significant.

Pore Volume

Since pore volume could not be estimated by the UMHS scanner, only the pore volumes for the UTCT scanner will be discussed. Pore space structure is important in creating the ability to transport water (Pachepsky et al., 1996). For the Putnam silt loam soil analyzed in this study, the UTCT method persistently showed larger average pore volume for buffer treatment compared to the row

Table 5–2. Average number of pores by management treatments within the soil group one (Putnam silt loam) scanned by University of Missouri Hospital and Clinic's Siemens Somatom Plus 4 Volume Zoom X-ray CT Scanner (UMHS) and University of Texas–Austin High Resolution X-ray Computed Tomography facility (UTCT). UMHS and UTCT images were analyzed by ImageJ and 3-Dimensional Medial Axis (3-DMA) software, respectively.

Treatment	Number of pores	
	UMHS	UTCT
Row crop	30	41
Grass Buffer	60	298
Tree Buffer	142	431

Table 5–3. Average pore volume (μm^3) by treatments within Putnam soil scanned at University of Texas–Austin High Resolution X-ray Computed Tomography facility (UTCT). All images were analyzed using 3 Dimensional Medial Axis (3-DMA) software.

Scanner	Treatment	Pore Volume
		μm^3
UTCT	Row crop	1.56×10^9
	Grass buffer	4.73×10^9
	Tree buffer	4.53×10^9

crop treatment. The values ranged between a minimum of 0.85×10^9 μm^3 for a row crop sample and a maximum of 5.3×10^9 μm^3 for the tree buffers. The average size of a pore was 1.56×10^9, 4.73×10^9, and 4.53×10^9 μm^3 for row crop, grass, and tree buffers, respectively (Table 5–3). Similar results have been reported by a microtomography study where perennial grass land had larger pores than crop land (Peth et al., 2008).

Synchrotron microtomography methods can estimate pore volume as well using smaller soil cores or samples. The Hamra soil was evaluated with this technique. The average pore size of the synchrotron technique for this soil ranged between 5.8×10^5 μm^3 and 7.7×10^5 μm^3 with a mean of 6.9×10^5 μm^3 (Table 5–4). The high density samples had smaller average pores (6.6×10^5 μm^3) and low density had larger (7.1×10^5 μm^3) values irrespective of aggregate size. There was no difference between the 0.5 and 2.0 mm aggregate classes of loamy sand soils on average pore size (6.8×10^5 μm^3 and 6.9×10^5 μm^3, respectively).

The mean pore size for the UTCT technique was 5227 times larger than the synchrotron technique (although these estimates are for different soils). However, direct comparisons are not warranted due to differences in soil types. Studying pore size distribution and pore volume, Vogel et al. (2010) stated that relations were better for larger and medium pores and was weaker for smallest pores where image resolution approached the pore size. Another factor that needs consideration is the aggregate size. Uniform aggregates were used in the synchrotron study as compared to the UTCT. The difference in pore size may have been caused by aggregate size, compaction, disturbance, and resolution between the images.

Table 5–4. Average pore volume (μm^3) by treatments within Hamra soil scanned at Geo-SoilEnviroCARS sector at the Argonne Advanced Photon Source (GSECARS). All images were analyzed using 3 Dimensional Medial Axis (3-DMA) software.

Scanner	Treatment	Pore volume
		μm^3
GSECARS	High density	6.65×10^5
	Low density	7.07×10^5
	2 mm diameter	6.92×10^5
	0.5 mm diameter	6.80×10^5

Coordination Numbers

The coordination number (CN) is used to characterize pore topology and it refers to the number of paths meeting at one node. The value is greater when more paths meet and less when fewer paths meet at a node. Since the parameter is based on pore connectivity, the UMHS data cannot be compared with the high resolution imaging techniques to explain differences and similarities. In the high resolution procedures, there is no coordination number-1 pore while coordination number-2 pores are part of a pore channel. Pore coordination numbers greater than 10 occur where pore size and channel length approach the voxel resolution (Lindquist et al., 2000; Seright et al., 2001).

The distribution of coordination numbers (CN) with probability was similar for the treatments with the UTCT (Fig. 5–1). These CN values agree with other published studies on soil and earth materials (Lindquist et al., 2000; Udawatta et al., 2008). Higher coordination numbers possessed lower probability values and the distributions of CNs within the lower probability range appear to be nonlinear. Samples imaged at the UTCT indicated a scattered distribution for CN values between 10 and 20.

For synchrotron samples, the data show that samples scanned at a higher resolution provide additional information for CNs greater than 10 (Fig. 5–2). All samples imaged at the synchrotron indicated a linear range for the CN values in the 0 to 20 range. Since CN value 10 approaches scanner resolution in the UTCT procedure, only CN values 3 to 10 were used to estimate the coordination number constant [CN_o; $P (CN) \approx 10^{-CN/CN_o}$; Fig. 5–2].

The CN_o varied within samples among treatments for a scanning technique. Characteristic coordination number constant values varied between 3.29 and 5.15 for samples imaged at the UTCT. The CN_o values were 4.70 and 3.34 for row crop and tree treatments in the UTCT imaged samples. The treatment effect was significant and soils from the perennial vegetation had lower CN_o than the row crop treatment.

The range for CN_o was 4.79 to 5.29 for samples imaged at the synchrotron. Among the synchrotron imaged samples, the average CN_o values for 0.5- and 2-mm diameter classes were similar (5.06 and 5.07, respectively). The high bulk density samples also showed similar average CN_o values. In the synchrotron samples, low density (5.13) samples on average had a higher CN_o than the high density (4.99) samples. However, the difference was not significant.

The CN and CN_o are indications of pore connectivity and both imaging resolutions provided similar CN_o values. However, the distributions of CN were different between the imaging techniques. The samples scanned at the UTCT had a

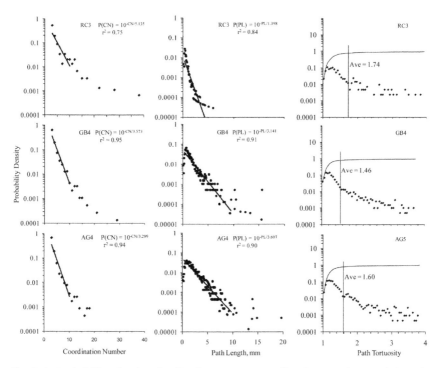

Fig. 5–1. Probability density distributions versus coordination number, path length, and tortuosity for selected Putnam soil samples imaged at University of Texas–Austin High Resolution X-ray Computed Tomography facility. Coordination number (CN) is number of curve segments meeting at the vertex and Co is the characteristic coordination number constant which is the value in each equation. Path length (PL) is number of paths and PLo is the characteristic path length constant which is the value in each equation. Average tortuosity is expressed with a vertical line and the cumulative probability density distribution is expressed with a curve. RC, GB, and AG refer to row-crop, grass buffer, and agroforestry, respectively; and the last digit denotes the replication.

lower mean CN_0 (3.89) than samples imaged at the synchrotron (5.06) but the difference was not significant. More linear distributions were exhibited by the synchrotron samples at lower probabilities than the UTCT method. This may indicate that samples imaged at a much higher resolution provide additional information such as CN, CN_0 and the patterns on pore connectivity (Fig. 5–1). Although results can be compared, conclusions cannot be made as the soils were different in the imaging techniques. The question is the usefulness of such additional information as liquid flow may be restricted in these tiny pores, but gas transport may occur. Costs to investigators for UTCT were much higher as compared to the synchrotron facility.

Path Length

Path lengths were between 0 and 20 mm for grass and tree buffers in the UTCT samples (Fig. 5–1). The row crop treatment exhibited shorter path lengths and values ranged between 0 and 6 mm. Similar to the resolution (195 by 195 μm resolution) of the UMHS technique with 2-mm-thick slices, path lengths peaked at 40

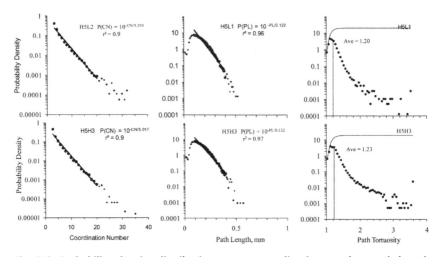

Fig. 5–2. Probability density distributions versus coordination number, path length, and tortuosity for selected Hamra soil samples imaged at GeoSoilEnviroCARS sector at the Argonne Advanced Photon Source. Coordination number (CN) is number of curve segments meeting at the vertex and Co is the characteristic coordination number constant which is the value in each equation. Path length (PL) is number of paths and PLo is the characteristic path length constant which is the value in each equation. Average tortuosity is shown with a vertical line and the cumulative probability density distribution is expressed with a curve. H5 denote 0.5 mm diameter; L and H denote low and high density.

mm for Chicot sandy loam (Perret et al., 1999). The r^2 values between path length and probability were 0.84 to 0.91 for the UTCT.

All synchrotron samples indicated path lengths less than 0.6 mm (Fig. 5–2). Comparison of results from the current study and Perret et al. (1999) indicate that path length is a function of aggregate size, bulk density, sample size, and image resolution. The r^2 values between path length and probability were higher for the synchrotron method and ranged from 0.96 to 0.98 while UTCT had a lower range (0.84–0.91).

The distribution patterns of path length versus probability were similar for both UTCT and synchrotron procedures using different soils. Short path lengths had higher probabilities and the relationship was linear. However, as path length increases, probability decreased and the relationship deviated from linearity. These results agree with other published results (Lindquist et al., 2000). Although the general relationships between path length and probability were similar, the UTCT procedure had longer path lengths compared to the synchrotron method.

The relationship between path length and probability exhibited an exponential distribution [P (PL) $\approx 10^{-PL/PLo}$; Fig. 2]. In this analysis, PL between 1 and 10 mm were used to compare differences. The PLo values for the UTCT method were between 1.398 and 3.607. The smaller PLo indicates rapid decline of path lengths in the samples than the greater values. The synchrotron method yielded significantly lower values (0.097 to 0.132) for PLo.

The sample size was also different between the two methods. Large soil cores of 76 mm diameter and 76 mm long were used in the UTCT study, while

5-mm-diameter, 5-mm-long cores were used in the synchrotron study. Sieved soil was pressed and packed to obtain the desired bulk density and thereby pore space and pore volumes were compressed. The types of soil material were different; Putnam silt loam soil in the UTCT and loamy sand (Typic Rhodoxeralf) in the synchrotron study. Sample size, soil material, and imaging technique may have contributed to differences in estimated PLo values. The results show that analysis method and the sample size should be considered when designing a study. If undisturbed soil cores from buffers were analyzed at the synchrotron, path length differences between crop and buffer soils would not have been detected. On the other hand, effects of aggregate size and density results might have been different for a UTCT analysis of loamy sand soils. Results also imply that soil geometrical properties that vary at a μm scale must be analyzed at a much high resolution similar to a synchrotron.

Path Tortuosity

As path tortuosity increases from 1 to 4, probability of a path decreased (Fig. 5–1). Tortuosity values ranged between 1.46 and 1.74 for Putnam silt loam soil analyzed at UTCT. The Hamra soil had tortuosity values between 1.20 and 1.23 for synchrotron analyzed samples. These values are comparable with other studies in which tortuosity ranged between 1.23 and 1.82 (Lindquist et al., 2000; Luo et al., 2010). The undisturbed samples imaged at the UTCT had significantly larger mean tortuosity values (1.60). Soils from the crop areas (1.73) had significantly greater values than the buffer (1.55) areas imaged at UTCT.

The mean tortuosity values of Hamra soil analyzed with synchrotron were 1.21. The mean differences were not significant between 2.0 and 0.5 mm aggregates and two density values of synchrotron samples. Shorter pores (maximum 0.6 mm) in the synchrotron analysis also indicate that small pores may be not as tortuous as longer pores in the UTCT analysis. On average, soil cores imaged at the UTCT had 32% more tortuous path than the Hamra soils analyzed at the synchrotron.

Conclusions

This study evaluated relative advantages and disadvantages of two methods for imaging earth materials. Additional data with a high resolution synchrotron scanner were also evaluated. Imaging and image analysis techniques indicated similar relationships for number of pores by treatment type. However, geometrical pore parameters such as connectivity and tortuosity resulting from volume rendering of adjacent scans cannot be compared with two dimensional images of low resolution scanners. Results indicate that a low resolution scanner may be used to identify treatment differences to select samples for high resolution imaging. The cost can be reduced by imaging more samples at low resolution and fewer samples at high resolution. Additional information that high resolution provides on spatial variability of pore parameters at three dimensions may help model development and treatment evaluation.

In the current study, path lengths and average pore volumes were smaller for synchrotron samples as compared to UTCT samples. This implies that certain parameters that are volume-dependent cannot be acquired by computed microtomography of small samples. Results also indicate that soils are not rigid and mechanical interferences can alter geometrical pore parameters. The criteria of representative sample volume will not be met by these studies and therefore sample selection must be completed after a thorough evaluation of systems, treatments, and properties of interest.

Results also showed that three-dimensional image analysis could be used to differentiate pore parameters for better understanding geometrical and quantitative differences in structure among samples and treatments than is impossible with traditional methods. Such information could be used to improve model parameters associated with solute and gas movement through earth materials. Future research goals should determine the most appropriate imaging method that meets study objectives.

References

Akin, S., and A.R. Kovscek. 2003. Computed tomography in petroleum engineering research. In: F. Mess, et al., editors, Application of X-ray computed tomography in the geosciences. The Geological Society, London, UK. p. 23–38.

Al-Raoush, R.I. 2002. Extraction of physically-realistic pore network properties from three-dimensional synchrotron microtomography images of unconsolidated porous media. Ph.D. diss. Louisiana State University, Baton Rouge, LA. 173 p.

Anderson, S.H., C.J. Gantzer, J.M. Boone, and R.J. Tully. 1988. Rapid nondestructive bulk density and soil-water content determination by computed tomography. Soil Sci. Soc. Am. J. 52:35–40. doi:10.2136/sssaj1988.03615995005200010006x

Anderson, S.H., H. Wang, R.L. Peyton, and C.J. Gantzer. 2003. Estimation of porosity and hydraulic conductivity from X-ray CT-measured solute breakthrough. In F. Mess, et al., editors, Application of X-ray Computed tomography in the geosciences. The Geological Society, London. UK. p. 135–150.

Asseng, S., L.A.G. Aylmore, J.S. MacFall, J.W. Hopmans, and P.J. Gregory. 2000. Computer-assisted tomography and magnetic resonance imaging. In A.L. Smit, et al., editors, Techniques for studying roots. Springer, Berlin. p. 343–363.

Bloomenthal, J. 1988. Polygonization of implicit surfaces. Comput. Aided Geom. Des. 5:341–355. doi:10.1016/0167-8396(88)90013-1

Carlson, W.D., T. Rowe, R.A. Ketcham, and M.W. Colbert. 2003. Application of high resolution X-ray computed tomography in petrology, meteoritics, and palaeontology. In: F. Mess, et al., editors, Application of X-ray computed tomography in the geosciences. The Geological Society, London, UK. p. 7–22.

Cormen, T.H., C.E. Leiserson, and R.L. Rivest. 1990. Introduction to algoriths. MIT Press, Cambridge, MA.

Fox, G.A., R.M. Malone, G.J. Sabbagh, and K. Rojas. 2004. Interrelationship of macropores and subsurface drainage for conservative tracer and pesticide transport. J. Environ. Qual. 33:2281–2289. doi:10.2134/jeq2004.2281

Gantzer, C.J., and S.H. Anderson. 2002. Computed tomographic measurement of macroporosity in chisel-disk and no-tillage seedbeds. Soil Tillage Res. 64:101–111. doi:10.1016/S0167-1987(01)00248-3

Grevers, M.C.J., and E. de Jong. 1994. Evaluation of soil-pore continuity using geostatistical analysis of macroporosity in serial sections obtained by computed tomography scanning. In: F. Mess, et al., editors, Application of X-ray computed tomography in the geosciences. The Geological Society, London, UK. p. 73–86.

Ioannidis, M.A., and I. Chatzis. 1993. Network modeling of pore structure and transport properties of porous media. Chem. Eng. Sci. 45:951–972.

Ioannidis, M.A., and I. Chatzis. 2000. On the geometry and topology of 3D stochastic porous media. J. Colloid Interface Sci. 229:323–334. doi:10.1006/jcis.2000.7055

Ketcham, R.A., and W.D. Carlson. 2001. Acquisition, optimization and interpretation of X-ray computed tomographic imagery: Applications to the geosciences. Comput. Geosci. 27:381–400. doi:10.1016/S0098-3004(00)00116-3

Kinney, J.H., and M.C. Nichols. 1992. X-ray tomographic microscopy (XTM) using synchrotron radiation. Annu. Rev. Mater. Sci. 22:121–152. doi:10.1146/annurev.ms.22.080192.001005

Kwiecien, M.J., I.F. Macdonald, and F.A.L. Dullien. 1990. Three dimensional reconstruction of porous media from serial section data. J. Microsc. 159:343–359. doi:10.1111/j.1365-2818.1990. tb03039.x

Lee, S.S., C.J. Gantzer, A.L. Thompson, S.H. Anderson, and R.A. Ketcham. 2008. Using high resolution computed tomography analysis to characterize soil surface seal. Soil Sci. Soc. Am. J. 72:1478–1485. doi:10.2136/sssaj2007.0421

Lee, T.C., R.L. Kashyap, and C.N. Chu. 1994. Building skeleton models via 3-D medial surface/axis thinning algorithms. CGVIP: Graph, Model Image Process. 56: 462–478.

Lindquist, W.B. 2002. Quantitative analysis of three dimensional X-ray tomographic images. In: U. Bonse, editor, Development in X-ray tomography. Proceedings of SPIE 4503, SPIE Bellingham, WA.

Lindquist, W.B., S.M. Lee, D.A. Coker, K.W. Jones, and P. Spanne. 1996. Medial axis analysis of three dimensional tomographic images of drill core samples. J. Geophys. Res. 101B:8296–8310.

Lindquist, W.B., S.M. Lee, W. Oh, A.B. Venkatarangan, H. Shin, and M. Prodanovic. 2005. 3DMA-Rock A software package for automated analysis of rock pore structure in 3-D computed microtomography images. http://www.ams.sunysb.edu/~lindquis/3dma/3dma_rock/3dma_rock.html (Accessed November 2011).

Lindquist, W.B., and A.B. Venkatarangan. 1999. Investigating 3D geometry of porous media from high resolution images. Phys. Chem. Earth 25:593–599.

Lindquist, W.B., A.B. Venkatarangan, J.H. Dunsmuir, and T.F. Wong. 2000. Pore and throat size distributions measured from synchrotron X-ray tomographic images of Fontainebleau sandstones. J. Geophys. Res. 105:21509–21528. doi:10.1029/2000JB900208

Lorensen, W.E., and H.E. Cline. 1987. Marching cubes: A high resolution 3D surface construction. ACM Comput. Graph. 21:163–169. doi:10.1145/37402.37422

Luo, L., H. Lin, and S. Li. 2010. Quantification of 3-D soil macropore networks in different soil types and land uses using computed tomography. J. Hydrol. 393:53–64. doi:10.1016/j.jhydrol.2010.03.031

Mooney, S.J. 2002. Three-dimensional visualization and quantification of soil macroporosity and water flow patterns using computed tomography. Soil Use Manage. 18:142–151. doi:10.1111/j.1475-2743.2002.tb00232.x

Oh, W., and W.B. Lindquist. 1999. Image thresholding by indicator kriging. IEEE Trans. Pattern Anal. Mach. Intell. 21:590–602. doi:10.1109/34.777370

Pachepsky, Y., V. Yakovchenko, M.C. Rabenhorst, C. Pooley, and L.J. Sikora. 1996. Fractal parameters of pore surfaces as derived from micromorphological data: Effect of long-term management practices. Geoderma 74:305–319. doi:10.1016/S0016-7061(96)00073-0

Perret, J., S.O. Prasher, A. Kantzas, and C. Langford. 1999. Three-dimensional quantification of macropore networks in undisturbed soil cores. Soil Sci. Soc. Am. J. 63:1530–1543. doi:10.2136/sssaj1999.6361530x

Pierret, A., Y. Capowiez, L. Belzunces, and C.J. Moran. 2002. 3D reconstruction and quantification of macropores using X-ray computed tomography and image analysis. Geoderma 106:247–271. doi:10.1016/S0016-7061(01)00127-6

Peth, S., R. Horn, F. Beckmann, T. Donath, J. Fisher, and A.J.M. Smucker. 2008. Three-dimensional quantification of intra-aggregate pore-space features using synchrotron-radiation-based microtomography. Soil Sci. Soc. Am. J. 72:897–907. doi:10.2136/sssaj2007.0130

Rachman, A., S.H. Anderson, and C.J. Gantzer. 2005. Computed-tomographic measurement of soil macroporosity parameters as affected by stiff-stemmed grass hedges. Soil Sci. Soc. Am. J. 69:1609–1616. doi:10.2136/sssaj2004.0312

Rasband, W. 2002. ImageJ: Image processing and analysis in Java. Available at Research Services Branch. National Institute of Health, Bethesda, MD. http://rsb.info.nih.gov/ij/index.html (accessed August 2010).

Rasiah, V., and L.A.G. Aylmore. 1998. The topology of pore structure in cracking clay soil: I. The estimation of numerical density. J. Soil Sci. 39:303–314.

Rivers, M.L. 1998. Tutorial introduction to X-ray computed microtomography data processing. http://www-fp.mcs.anl.gov/xray-cmt/rivers/tutorial.html (accessed November 2011).

Seright, R.S., J. Liang, W.B. Lindquist, and J.H. Dunsmuir. 2001. Characterizing disproportionate permeability reduction using synchrotron X-ray computed tomography. Society of Petroleum Engineers Annual Technical Meeting. Sep 3–Oct 3, 2001. New Orleans, LA. SPE 71508.

Taina, I.A., R.J. Heck, and T.R. Elliot. 2008. Application of X-ray computed tomography to soil science: A literature review. Can. J. Soil Sci. 88:1–20. doi:10.4141/CJSS06027

Tollner, E.W., D.E. Radcliffe, L.T. West, and P.F. Hendrix. 1995. Predicting hydraulic transport parameters from X-ray CT analysis. Paper No. 95–1764. ASAE, St. Joseph, MI.

Udawatta, R.P., and S.H. Anderson. 2008. CT-measured pore characteristics of surface and subsurface soils as influenced by agroforestry and grass buffers. Geoderma 145:381–389. doi:10.1016/j.geoderma.2008.04.004

Udawatta, R.P., S.H. Anderson, C.J. Gantzer, and H.E. Garrett. 2006. Agroforestry and grass buffer influence on macropore characteristics: A computed tomography analysis. Soil Sci. Soc. Am. J. 70:1763–1773. doi:10.2136/sssaj2006.0307

Udawatta, R.P., C.J. Gantzer, S.H. Anderson, and H.E. Garrett. 2008. Agroforestry and grass buffer effects on high resolution X-ray CT-measured pore characteristics. Soil Sci. Soc. Am. J. 72:295–304. doi:10.2136/sssaj2007.0057

Udawatta, R.P., J.J. Krstansky, G.S. Henderson, and H.E. Garrett. 2002. Agroforestry practices, runoff, and nutrient loss: A paired watershed comparison. J. Environ. Qual. 31:1214–1225. doi:10.2134/jeq2002.1214

Unklesbay, A.G., and J.D. Vineyard. 1992. Missouri Geology. Univ. of Missouri Press, Columbia, MO.

UTCT. 2007. High-Resolution X-ray Computed Tomography Facility at the University of Texas at Austin: Lab Capabilities and Instrumentation. University of Texas, Austin, TX. http://www.ctlab.geo.utexas.edu/labcap/index.php (accessed and verified November 2011).

Venkatarangan, A.B. 2000. Geometric and statistical analysis of porous media. Ph.D. Diss., State Univ. of New York, Stony Brook, NY.

Vogel, H.-J., U. Weller, and S. Schlüter. 2010. Quantification of soil structure based on Minkowski functions. Comput. Geosci. 36:1236–1245. doi:10.1016/j.cageo.2010.03.007

Young, I.M., and J.W. Crawford. 2004. Interactions and self-organization in the soil-microbe complex. Science 304:1634–1637. doi:10.1126/science.1097394

Zhang, X., L.K. Leeka, A.G. Bengough, J.W. Crawford, and I.M. Young. 2005. Determination of soil hydraulic conductivity with the lattice Boltzmann method and soil thin-section techniques. J. Hydrol. 306:59–70. doi:10.1016/j.jhydrol.2004.08.039

6

Applications of Neutron Imaging in Soil–Water–Root Systems

Ahmad B. Moradi,* Sascha E. Oswald, Manoj Menon, Andrea Carminati, Eberhard Lehmann, and Jan W. Hopmans

Abstract

Neutron imaging provides an exceptional noninvasive tool for studying root architecture and water distribution in root–soil systems in situ and without interfering with the root-zone processes. This unique property of neutron imaging results from the high sensitivity of neutrons to hydrogen nuclei. Since water and roots are the main hydrogen-bearing materials in soil systems, they both can be visualized. The neutron attenuation coefficient of water is an order of magnitude larger than that of most soils; therefore soil water content distributions can be quantified from neutron images with high precision. Depending on whether neutron radiography or tomography is performed, a two-dimensional or three-dimensional image of the root–soil system is obtained. Neutron radiography can be performed in an order of seconds; therefore, quasi-real time dynamics of water in the system can be obtained. Neutron tomography, however, typically requires a few hours. Spatial resolution in neutron images is usually in the range of 100 to 300 µm, but for small samples can be finer. Rapid advancements in imaging and processing tools have helped both visualization and quantification, and along with increased availability of neutron beams, have made this method viable in the recent years. These improvements can be expected to continue leading to increasing numbers of studies with this method. At present, the method is limited to laboratory scale and access to neutron imaging facilities. In this chapter, we provide a comprehensive overview on the methodological approach and application of neutron imaging in soil science, particularly on root growth, root water uptake, and soil hydraulic properties.

Abbreviations: CCD, charge-coupled device; CT, computed tomography, FOV, field of view; MRI, magnetic resonance imaging; NT, neutron radiography and tomography; QNI, quantitative neutron imaging.

A.B. Moradi and J.W. Hopmans, Dep. of Land, Air and Water Resources, Univ. of California, Davis, CA 95616. S.E. Oswald, Institute of Earth and Environmental Science, Univ. of Potsdam, Germany. M. Menon, Kroto Research Institute, Dep. of Civil and Structural Engineering, Univ. of Sheffield, UK. A. Carminati, Institute of Soil Science, Georg-August-Universität, Göttingen, Germany. E. Lehmann, Paul Scherrer Institut, Villigen, Switzerland. *Corresponding author (amoradi@ucdavis.edu).

doi:10.2136/sssaspecpub61.c6

Soil–Water–Root Processes: Advances in Tomography and Imaging. SSSA Special Publication 61. S.H. Anderson and J.W. Hopmans, editors. © 2013. SSSA, 5585 Guilford Rd., Madison, WI 53711, USA.

"Progress in science depends on new techniques, new discoveries and new ideas, probably in that order" (Brenner, 2002).

Various imaging methods have recently been developed and successfully tested in soil and other porous media including neutron radiography and tomography (NT), X-ray computed tomography (CT), and magnetic resonance imaging (MRI). Thermal neutron radiography and tomography are conceptually similar to using X-rays as presented in earlier chapters but measure the attenuation of thermal neutrons instead of photons. Compared with NT, X-ray CT is insensitive to water and poor at decoupling differences in soil bulk density from differences in water content. In contrast, neutron imaging works especially well for substances that contain a high volumetric proportion of H atoms as small changes in volumetric water content (θ) can clearly be quantified, while soil minerals are nearly transparent. Ferromagnetic and paramagnetic materials in soil hinder MRI and result in poor signal-to-noise ratio (Hall et al., 1997; Carlson, 2006). Thus, MRI is typically limited to carefully selected media such as pre-treated sand and soil, agar, and glass beads.

Although neutron attenuation techniques have been used routinely in engineering, relatively little is known about their application to soils. Early exceptions are the works by Solymar et al. (2003) and Deinert et al. (2004), demonstrating NT application to estimate spatial distribution of water content in rock and soil, respectively. The potential of neutron imaging for root investigations was realized as early as the 1970s. Willatt et al. (1978) used thermal neutron radiography to monitor the root growth of soybean and corn in soil in aluminum boxes. He combined an indium neutron collector with Kodak film to capture neutron images and presented the first neutron image of a soil–root system (Fig. 6–1). Already, a great deal of qualitative information could be obtained from these first neutron images. Later, the advances in digital imaging in late 1990s opened up the possibility of quantification in the images as well as improved spatial resolution of the images. Another step in neutron imaging methods followed soon afterward, in that the simplifications associated with the recording, handling, and treatment of digital data enabled neutron tomography, in addition to radiography, to yield a three-dimensional visualization and analysis of objects. These developments made neutron imaging a highly attractive tool to explore water dynamics in the root zone of plants, and it was found to be superior over other methods for studying roots and water flow in soil. More recent studies that have demonstrated the potential application of neutron radiography to determine root water depletion zones and ion uptake mechanisms were presented by Korosi et al. (1999) and Furukawa et al. (1999). Especially groundbreaking was the study by Hassanein (2006b), presenting various correction methods for thermal neutron tomography applications to account for backscattering and beam hardening.

In this chapter, we first introduce the basics of neutron imaging, including neutron radiography and tomography, and explain the principles behind the neutron quantifications in the images. We then bring examples on neutron application in soil–plant systems highlighting the capabilities of neutron imaging for soil and plant studies. Finally, we discuss the strengths and shortcomings of neutron imaging with a look to future improvements and benefits for soil and plant scientists.

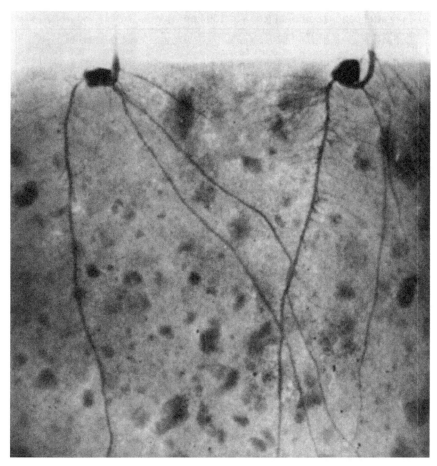

Fig. 6–1. One of the first neutron images of soil and roots (Willatt et al., 1978) showing two corn root systems 9 d after planting in a loamy sand (after Willatt et al., 1978).

Methodological Background
Comparison of Neutron with X-Ray Imaging

Neutrons are uncharged particles that usually form the nuclei of atoms, together with protons. If they are not bound to nuclei, they decay with a life time of about 15 min, as opposed to free protons, which are stable and constitute a hydrogen atom (^{1}H). Due to not carrying a charge, they do not experience electromagnetic forces and do not interact with electrons. This bears the potential to obtain information on matter which is of a different character compared with X-rays. When neutrons interact with matter, they only interact with the nuclei of atoms, while X-rays predominantly interact with the electron shell. When neutrons or X-rays pass through a material, the material can eliminate a part of this incoming radiation. This is called attenuation and its extent is described by an attenuation coefficient. The X-ray attenuation coefficient of elements increases as the atomic number of the

a) X-ray attenuation coefficient (cm⁻¹) at 100 keV

1a	2a	3b	4b	5b	6b	7b	8			1b	2b	3a	4a	5a	6a	7a	0
H 0.02																	He 0.02
Li 0.06	Be 0.22											B 0.28	C 0.27	N 0.11	O 0.16	F 0.14	Ne 0.17
Na 0.13	Mg 0.24											Al 0.38	Si 0.33	P 0.25	S 0.30	Cl 0.23	Ar 0.20
K 0.14	Ca 0.26	Sc 0.48	Ti 0.73	V 1.04	Cr 1.29	Mn 1.32	Fe 1.57	Co 1.78	Ni 1.96	Cu 1.97	Zn 1.64	Ga 1.42	Ge 1.33	As 1.50	Se 1.23	Br 0.90	Kr 0.73
Rb 0.47	Sr 0.86	Y 1.61	Zr 2.47	Nb 3.43	Mo 4.29	Tc 5.06	Ru 5.71	Rh 6.08	Pd 6.13	Ag 5.67	Cd 4.84	In 4.31	Sn 3.98	Sb 4.28	Te 4.06	I 3.45	Xe 2.53
Cs 1.42	Ba 2.73	La 5.04	Hf 19.70	Ta 25.47	W 30.49	Re 34.47	Os 37.92	Ir 39.01	Pt 38.61	Au 35.94	Hg 25.88	Tl 23.23	Pb 22.81	Bi 20.28	Po 20.22	At	Rn 9.77
Fr	Ra 11.80	Ac 24.47	Rf	Ha													

	Ce	Pr	Nd	Pm	Sm	Eu	Gd	Tb	Dy	Ho	Er	Tm	Yb	Lu
Lanthanides	5.79	6.23	6.46	7.33	7.68	5.66	8.69	9.46	10.17	10.91	11.70	12.49	9.32	14.07
	Th	Pa	U	Np	Pu	Am	Cm	Bk	Vf	Es	Fm	Md	No	Lr
*Actinides	28.95	39.65	49.08											x-ray

b) Thermal neutron attenuation coefficient (cm⁻¹)

1a	2a	3b	4b	5b	6b	7b	8			1b	2b	3a	4a	5a	6a	7a	0
H 3.44																	He 0.02
Li 3.30	Be 0.79											B 101.60	C 0.56	N 0.43	O 0.17	F 0.20	Ne 0.10
Na 0.09	Mg 0.15											Al 0.10	Si 0.11	P 0.12	S 0.06	Cl 1.33	Ar 0.03
K 0.06	Ca 0.08	Sc 2.00	Ti 0.60	V 0.72	Cr 0.54	Mn 1.21	Fe 1.19	Co 3.92	Ni 2.05	Cu 1.07	Zn 0.35	Ga 0.49	Ge 0.47	As 0.67	Se 0.73	Br 0.24	Kr 0.61
Rb 0.08	Sr 0.14	Y 0.27	Zr 0.29	Nb 0.40	Mo 0.52	Tc 1.76	Ru 0.58	Rh 10.88	Pd 0.78	Ag 4.04	Cd 115.11	In 7.58	Sn 0.21	Sb 0.30	Te 0.25	I 0.23	Xe 0.43
Cs 0.29	Ba 0.07	La 0.52	Hf 4.99	Ta 1.49	W 1.47	Re 6.85	Os 2.24	Ir 30.46	Pt 1.46	Au 6.23	Hg 16.21	Tl 0.47	Pb 0.38	Bi 0.27	Po	At	Rn
Fr	Ra 0.34	Ac	Rf	Ha													

	Ce	Pr	Nd	Pm	Sm	Eu	Gd	Tb	Dy	Ho	Er	Tm	Yb	Lu
*Lanthanides	0.14	0.41	1.87	5.72	171.47	94.58	1479.04	0.93	32.42	2.25	5.48	3.53	1.40	2.75
	Th	Pa	U	Np	Pu	Am	Cm	Bk	Cf	Es	Fm	Md	No	Lr
**Actinides	0.59	8.46	0.82	9.80	50.20	2.86								neut.

Fig. 6–2. X-ray (a) and thermal neutron (b) attenuation coefficients (cm⁻¹) of elements in the periodic table.

elements increases (Fig. 6–2), with hydrogen having the lowest and uranium the largest attenuation coefficient. In contrast to X-ray, the neutron attenuation coefficient is not directly correlated with the atomic number, and only a few elements, such as gadolinium, cadmium, boron, and hydrogen, strongly attenuate and scatter neutrons. Also, neutron attenuation varies for isotopes of elements. In the following, isotopic effects will not be discussed, with exception of water and heavy water, and elements will be considered in their natural isotopic composition.

Hydrogen plays a unique role and is the most common element in the soil–plant system, with high molar density values, both as part of molecular structures or as pore water. Because of hydrogen's relatively high neutron attenuation, H-rich organic materials and water can create relatively opaque features in neutron radiographs, while many soil components and structural materials such as silicon, calcium, and aluminum are nearly transparent. Figure 6–3 shows

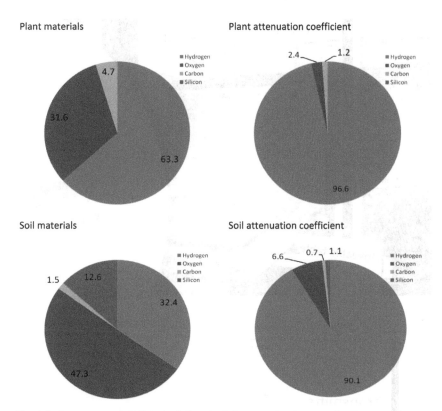

Fig. 6–3. Percentage abundance of chemical elements in plants and soil system (Mason and Moore, 1982; Salisbury and Ross, 1992; Pansu and Gautheyrou, 2006) along with their relative contribution to the neutron attenuation coefficients of the two systems. The plants and soil assumed to have water contents of 0.8 and 0.2.

abundance percentages of the dominant chemical elements in the plant–soil system, along with their relative neutron attenuation coefficients. In both plant and soil, hydrogen is responsible for more than 90% of the neutron attenuation. This is why soil water and plant roots can be visualized in great detail using neutron imaging. While X-ray imaging can observe root structures as being more transparent than mineral soil, neutron attenuation methods can easily distinguish water from sediments or air. Clearly, neutron and X-ray imaging provide complementary information in soil–water–plant studies. While neutrons are sensitive to water and organic materials, information on soil structure and density can be better obtained from X-ray imaging.

Principles of Neutron Imaging

Neutrons passing through a sample can interact with the atomic nuclei by two primary processes: absorption and scattering (Fig. 6–4). By absorption, the neutron is captured by the atomic nucleus while emitting a γ-ray, α-particle, or a proton. If the neutron is not captured by the nucleus, but is knocked out of the neutron beam direction, the process is called scattering. From a macroscopic

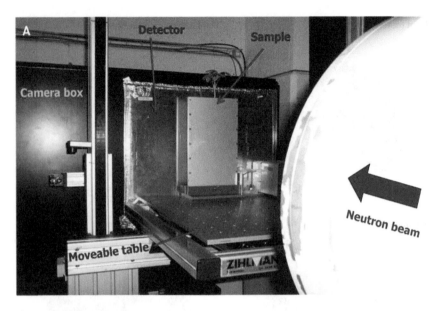

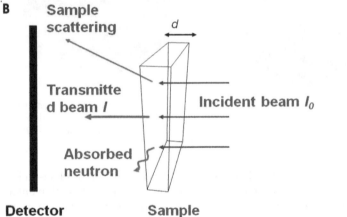

Fig. 6–4. A lupin plant sample at the neutron imaging setup at NEUTRA station, Paul Scherrer Institute, Villigen, Switzerland (a), and schematic view of the interactions between neutrons and the sample with notation of Eq. [1] (b).

perspective, the attenuation of an incoming neutron beam through a sample material is described by the Beer–Lambert law of attenuation known from light absorption along a straight line:

$$I = I_0 \cdot e^{-\Sigma_{abs} \cdot d} \tag{1}$$

where I is the detected attenuated neutron flux ($cm^{-2}\ s^{-1}$) from an incident neutron flux, I_0, passing through material of thickness d (cm) with a neutron absorption coefficient of Σ_{abs} (cm^{-1}).[1] The neutron absorption coefficient is a bulk physical property of the material. It is also a function of the energy level of the incoming neutrons and can be calculated from the interaction probability of the neutrons with matter.

Attenuation also results from scattering and in the case of neutrons is often the dominant process. Accounting for this, the attenuation coefficient, Σ, includes both processes. However, scattered neutrons are not necessarily eliminated from the neutron beam and can reach the detector plane by their scattering at a smaller angle or by repeated scattering as they pass through the imaged material. In addition, neutrons that have been removed from the sample by scattering can recur in the beam by back-scattering from materials surrounding the neutron beam. For example, objects close to the experimental setup, such as the camera box or neutron beam shielding can back-scatter into the imaged neutron beam. Therefore the Beer–Lambert law does not strictly apply and requires adaptation, or:

$$I = I_0 \cdot e^{-\Sigma_{abs} \cdot d} + \varphi \cdot I_0 \cdot e^{-\Sigma_{scat} \cdot d} + \beta \cdot I_0 \qquad [2]$$

The coefficients φ and β have values between 0 and 1 and represent the fraction of neutrons that reach the detector from repeated scattering or back-scattering, respectively. If φ and β are small, Eq. [2] reduces to its well-known Beer–Lambert form:

$$I = I_0 \cdot e^{-\Sigma \cdot d} \quad \text{with} \quad \Sigma = \Sigma_{abs} + \Sigma_{scat}$$
$$or \quad I' = I / I_0 = e^{-\Sigma \cdot d} \qquad [3]$$

The distinction between absorbed and scattered neutrons is critical for quantitative neutron radiography and tomography analysis (Moradi et al., 2009b), as the additional scattering components in Eq. [2] increase image intensity and lead to an underestimation of neutron attenuation. Additional deviations from the exponential law of attenuation are due to beam hardening and the energy dependency of detector efficiency (Hassanein, 2006b).

Neutron Interactions with the Soil–Water–Plant System

For samples that are made up of different materials, the total neutron attenuation coefficient, Σ, is the sum of individual material attenuation coefficients. For example, in a system that consists of a sample container, soil minerals, roots, and water (Fig. 6–5), the measured Σ is the sum of the attenuation coefficients of the container, soil material, and water, assuming that the neutron attenuation of air can be neglected. In that case, Eq. [3] leads to:

$$-\ln(I')(x,y) = \Sigma_{soil} \cdot d_{soil}(x,y) + \Sigma_{container} \cdot d_{container}(x,y) + \Sigma_{water} \cdot d_{water}(x,y)$$
$$= \Sigma_{total} \cdot d_{total}(x,y) \qquad [4]$$

where (x,y) are the coordinates of the two-dimensional detector perpendicular to the neutron beam direction and d_{mat} is the "thickness" of the individual material in the direction of the neutron beam, defined as the cumulative length of a

---

[1] Σ not to be confused with a summation sign.

a)

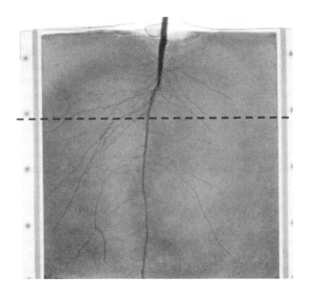

b)

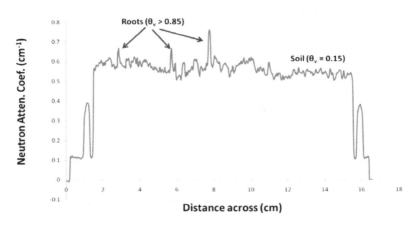

Fig. 6–5. A neutron radiograph of a lupin plant growing in an aluminum slab filled with a sandy soil (a), and the neutron attenuation coefficient extracted from the radiograph along the dotted line in the radiograph (b). Note the contrast between roots and the surrounding soil with relatively high water content.

neutron path through the individual material. In the case of soil, this is usually the inner width of the soil-filled container. In the case of water, it is the amount of water in the soil pores (and physio-chemically bound in the soil material) that is expressed as a water thickness. By rearranging the equation and knowing the attenuation coefficients of dry soil, water, and the container and the thickness of soil and container walls, the volumetric water content (θ_v) can be calculated as:

$$\theta_v(x,y) = \frac{d_{water}(x,y)}{d_{soil}(x,y)}$$

$$= -\frac{\ln(I'(x,y))}{\Sigma_{water} \cdot d_{soil}(x,y)} - \frac{\Sigma_{soil}}{\Sigma_{water}} - \frac{\Sigma_{container}}{\Sigma_{water}} \cdot \frac{d_{container}(x,y)}{d_{soil}(x,y)} \qquad [5]$$

In the pixels where root is present, the calculated water content from Eq. [5] gives the sum of water residing in soil and in the roots along the beam direction. However, pixels containing no roots give the soil water content.

Almost all of the containers that have been used for soil and plant studies using neutron imaging are made of aluminum, as it has a very low neutron attenuation coefficient of only 0.1 cm^{-1}, thus making it effectively transparent to neutrons. In that case, the volumetric water content can be approximated by:

$$\theta_v(x,y) \approx -\frac{\ln(I'(x,y))}{\Sigma_{water} \cdot d_{soil}(x,y)} - \frac{\Sigma_{soil}}{\Sigma_{water}} \qquad [6]$$

Neutron Imaging Configuration and Resolution

Unlike X-ray or MRI systems, neutron imaging systems are not part of a stand-alone laboratory but require a large-scale dedicated facility. The availability of neutron imaging is comparable to synchrotron imaging. A neutron imaging facility consists of a neutron source, collimator object holder, and a two-dimensional neutron detector, which is connected to a charge-coupled device (CCD) camera for the light produced by neutron capture (Fig. 6–6).

High-energy neutrons are either produced by nuclear fission in a research nuclear reactor or by direction of high-energy protons onto a massive target, by which neutrons and other forms of radioactivity are produced by the collision of protons with nuclei of target material (neutron spallation). In both cases, neutrons are generated with high energies and a distribution of directions. The high energy neutrons are slowed down to lower energies (thermal or cold neutrons) by a moderator.

The collimator is an especially designed tube that collects and sorts the thermal-equilibrated neutrons in the moderator tank and guides the thermal neutron beam toward the object to be imaged (Lehmann et al., 1996). It is built to yield a high-intensity neutron beam with the desired narrow energy range (thermal or cold) and with a close to parallel direction in the imaging field of view (FOV). The collimator geometry is expressed by the collimation ratio L/D, where L is the collimator length and D is the diameter of the inlet aperture of the collimator. In radiography, a sample with a known thickness is usually placed in the neutron beam ahead of the detector. This could also be done at different positions along the collimator, providing different fields of view with different neutron intensities.

The neutron detection assembly consists of a scintillator and a CCD camera. The scintillator first captures the neutrons that have passed through the sample and then converts the neutrons into light. The light is then recorded by the CCD camera resulting in a raw transmission image. In tomography, a large number

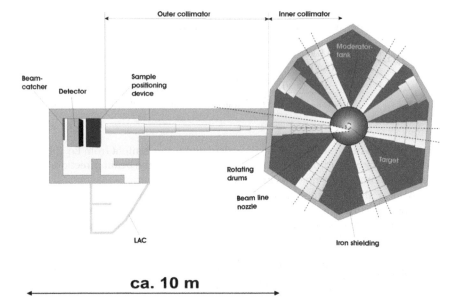

ca. 10 m

Fig. 6–6. Schematic view of the thermal neutron imaging facility (NEUTRA) at the Paul Scherrer Institut, Villigen, Switzerland.

of radiographs are taken at different angles by rotating the imaged sample step-wise on a turning table by 180 or 360°. As the sample is rotated, it must remain inside the FOV to avoid truncation artifacts. After full rotation, all individual projections are assembled for subsequent reconstruction into a three-dimensional image (Strobl et al., 2009). The exposure time for each radiograph for soil and plant materials depend on beam line properties, sample thickness, and sample materials, but typically ranges between 2 and 30 s, so that a complete scan can last up to a few hours.

For radiographies of samples larger than the FOV, either only a part of the sample can be selected for imaging, or multiple images can be taken with some overlap between images. This is usually done with a moveable table that can move the sample in every direction in three-dimensional space at high spatial resolution. Such a positioning system is also useful when different-sized samples are imaged repeatedly, for example, for taking longer term time-series of root growth.

Spatial resolution in neutron images depends on sample size and FOV, pixel size, scintillator thickness, the collimator geometry, and the imaging positions in the collimator. The spatial resolution improves as the sample size, pixel size, scintillator thickness, and the collimator ratio L/D decrease. It usually ranges between 50 and 200 μm in most of the studies reported. Although this spatial resolution is slightly finer than in MRI studies, it is at least an order of magnitude coarser than the spatial resolutions achieved with X-ray CT machines. Typical neutron imaging parameters are listed in Table 6–1.

Table 6–1. Typical imaging parameters for neutron radiography and tomography at Neutron imaging stations at Paul Scherrer Institute, Villigen, Switzerland.

Neutron	Camera array (pixels)	Pixel size (µm)	Effective resolution (µm)	Exposure time
Radiography	2048 × 2048	13–170	30 … 100	~ 1–30 s
Tomography	2048 × 2048	13–100	< 56	~ 2–5 h

The exposure time determines which gray level values are recorded by the CCD camera, since the light intensity is cumulatively measured during the exposure. The longer the selected exposure time, the higher the obtainable signal to noise ratio. However, some limitations apply. First, in neutron images white spots appear on variable, arbitrary locations. These are called "zingers" and result from X-rays or γ-rays, produced by neutrons hitting sample or other nearby materials that directly interact with the CCD chip thereby exciting random locations of the detector. Since the number of zingers is proportional to exposure time, one rather prefers to take several subsequent images with medium exposure than one image with long exposure. Second, longer exposure time also implies longer sample exposure to ionizing radiation. So far, the radiation dose does not seem to be a major stress component for plant–soil systems, but this is difficult to quantify and may change when using higher beam intensities and increasing number of tomographs. In addition, neutron exposure causes sample activation and radioactivity, which decays in a relatively short time, but increases with the duration of exposure to the neutron beam. Finally, the exposure time should be adjusted to yield pixel values inside the gray level range chosen (e.g., a maximum of 65,536 for a 16-bit image). If gray values reach this upper limit, the neutron image would be overexposed and a shorter exposure time should be chosen.

Image Processing

Once the images are acquired, the next step is image processing. The transmission neutron radiographs first need to be corrected for beam variation and camera noise using a flat field correction. The corrected image (I') is calculated as:

$$I'(x,y) = f_{OB} \cdot \frac{I_{raw}(x,y) - I_{dark}(x,y)}{I_{OB}(x,y) - I_{dark}(x,y)} \qquad [7]$$

where $I_{raw}(x,y)$ is the raw image, $I_{dark}(x,y)$ is the image with neither beam nor sample, $I_{OB}(x,y)$ is the open beam image containing the spatial variation of the beam without the sample, and f_{OB} is a scaling factor. The calculations are performed pixel-wise and x and y are the two-dimensional coordinates. The normalization value for f_{OB} accounts for temporal fluctuations of the beam intensity. It is chosen so that the mean transmission in the regions of the image without the sample is 1. For that purpose, a region is defined in the image and compared to the same region in the open beam image. Analogously to Eq. [7], the negative natural logarithm of the corrected image, $I'(x,y)$ yields the attenuation coefficient $\Sigma(x,y)$ of the material summed over the total sample thickness d_{total}.

In addition, to correct for neutron scattering and beam hardening, algorithms such as the quantitative neutron imaging (QNI) are useful (Hassanein, 2006a). The scattering correction is based on an iterative reconstruction of the measured image by overlapping point scattered functions calculated by means of Monte-Carlo simulation of the neutron transport process, depending on sample geometry and material, distance to detector, etc. It eliminates or reduces the additional contribution of scattered neutrons to the intensity measured at the detector (second term on the right-hand side of Eq. [2]). Additionally, the contribution of neutrons scattered from outside the sample (third term in Eq. [2]) can be subtracted, but here the procedure does adopt a simpler approach and subtracts an experimentally derived black body image representing this contribution. Depending on the nature of the sample and the desired interpretation (quantitative vs. qualitative), these corrections can be of high importance. Further effects that are considered inside the QNI correction tool (Hassanein, 2006a) are the energy dependence of the detector efficiency and filtering of zingers.

Once the images are corrected, they can be further used for a quantitative interpretation. Often it is desirable to evaluate a water content map of the sample (cf. Eq. [5]). For a soil sample, this directly applies to the pixels without roots. For root pixels, root water content can also be derived. Also, root structure can be inferred from NT, but it is best done for relatively dry soil conditions to enhance contrast between the roots and the surrounding soil.

The task to extract information (segmentation) from the final image depends on its quality, including contrast, signal to noise ratio, and distribution of attenuation coefficients (i.e., gray values). For example, it is much easier to segment images with good contrast. If the images are poor in contrast, i.e., the gray-level of roots in one part of the image is the same as the background in another part, a sophisticated image analysis algorithm such as retinal blood vessels segmentation (Hoover et al., 2000) can be used. Both plant–root systems and blood vessels are highly branched, allowing this algorithm to be specifically well-suited to determine voxel connectivity (Kaestner et al., 2006; Menon et al., 2007; Moradi et al., 2009a).

Applications

Neutron imaging has been employed to study infiltration, root growth and development, root responses, water uptake (root zone and rhizosphere), and water movement within the plant. Microscopic movement of water within soybean plants was studied by Nakanishi et al. (2003). Nakanishi et al. (2005) qualitatively showed how soil water content changes along a vertical transect from the root surface using thermal neutron beam analysis. Menon et al. (2007) visualized the root growth of lupin (*Lupinus albus* L.) plants over a period of 3 wk and demonstrated the effect of soil heterogeneity on root growth. Substantial details on water movement between soil aggregates (Carminati et al., 2007) and water infiltration in soil and root water uptake (Oswald et al., 2008; Tumlinson et al., 2008; Badorreck et al., 2010; Esser et al., 2010) were shown using neutron imaging. Moradi et al. (2009b) quantitatively studied and tested the capability of neutron radiography for root detection and soil water content measurements over a range of soil materials and soil moisture (Table 6–2). Neutron imaging can also successfully be combined directly with optical imaging techniques, for example, to map oxygen content in the rhizosphere (Rudolph et al., 2012). To

Table 6–2. Properties of various plant growth media with regard to neutron imaging (after Moradi et al., 2009b).

	Bulk density	Water content at −1 bar	Σ of dry material	Σ at −1 bar	Notes
	t m⁻³	m m⁻¹	cm⁻¹	cm⁻¹	
Perlite	0.125	1.08	0.1	4.5	High Σ due to high water holding capacity
Porous glass beads	0.49	0.16	1.6	2.5	High Σ due to boron in glass
Ferrous mine tailings	1.4	0.08	1	1.6	High Σ due to Iron content
Loamy soil	1.2	0.17	0.5	1.5	Ideal root growth condition
Peat	0.58	0.09	0.75	1.4	High Σ due to organic materials
Loamy sand	1.3	0.09	0.3	0.75	Normal root development
Sandy soil	1.35	0.03	0.28	0.36	
Fine quartz sand	1.5	0.01	0.25	0.35	Root development perturbed
Coarse quartz sand	1.45	0.01	0.25	0.3	

date, the ultimate resolution in mapping the rhizosphere in three dimensions was achieved by Moradi et al. (2011), also quantitatively proofing the existence of higher hydrogen (and thus probably water) content in the rhizosphere of single samples of three different plant species. Here we present selected example applications from each category.

Water Flow in Soil

Neutron imaging has been used in studying water movement in soil at the scales of decimeters down to the scale of single aggregates. Carminati et al. (2007) monitored the water redistribution through a series of soil aggregates with a high spatial resolution. Figure 6–7 shows the water redistribution through three soil aggregates after a pulse-injection of approximately 0.015 mL of water. The studies done by Schaap et al. (2008), Oswald et al. (2008), and Badorreck et al. (2010) are examples of the use of neutron radiography and tomography for investigating dynamic infiltration and redistribution of water in heterogeneous soil and sands.

Also, water dynamics in highly heterogeneous or layered soils can be visualized and quantified using neutron imaging. The effect of textural discontinuity on the drying front, hydraulic continuity, and soil evaporation was studied by Shokri et al. (2010). These investigators monitored the liquid phase distribution and air invasion along the fine and coarse sand interface (Fig. 6–8). Before the onset of evaporation, one could establish a defined interface between fine and coarse sand layers by differential saturation with normal (H_2O) and heavy water (D_2O). As the drying front developed and receded into the underlying coarse sandy layer, the interface between the normal and heavy water was displaced upward marking net movement of heavy water from the coarse to the overlying

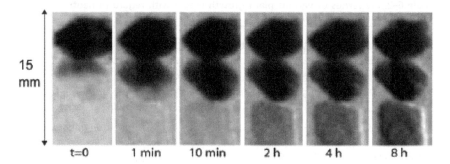

Fig. 6–7. Time-series neutron radiography of water redistribution through three aggregates stacked vertically. The gray values are proportional to the water content. Note the much quicker water exchange between the first two aggregates compared to the slow water exchange between the second and third aggregate. The rate of water exchange was controlled by the aggregate-aggregate contacts, which were suddenly drained as the water potential decreased (after Carminati et al., 2007).

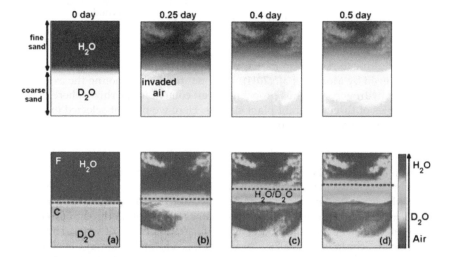

Fig. 6–8. Air invasion into the coarse sand layer saturated with heavy water underlying a layer of fine sand saturated with normal water observed by neutron radiography (field of view is 97-mm height and 75-mm width). (a) Saturated sand column at the beginning of the measurement. (b) Invasion of the underlying coarse layer. (c) and (d) Preferential displacement of the drying front into the underlying coarse sand layer and upward displacement of the interface between normal and heavy water (dashed line). The top row shows the transmission images and the bottom row shows the corresponding images with different colors to highlight the liquid phase distribution during preferential movement of the drying front (after Shokri et al., 2010).

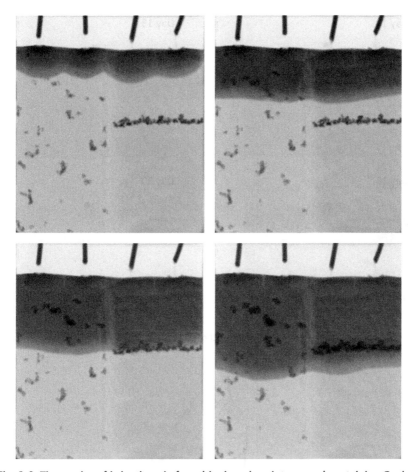

Fig. 6–9. Time series of irrigation via four dripping pipes into a sand containing Geohumus grains (soil enhancing product, storing water by swelling; left side of soil container mixed in homogenously, right side same amount of Geohumus built in as a single layer). Times are about 0.38h, 0.75 h, 1.13 h, and 1.5 h after start of drip irrigation. Uncorrected transmission images, water shows up as darker pixels (lower neutron attenuation).

fine sand layer. While the overlying fine sand layer remained nearly saturated at the interface, air invaded the underlying coarse sand.

A soil enhancing material was investigated by neutron radiography with living roots present. The specific added material was intended to store water when water is abundant and to release it for increasingly dry soil conditions, thereby providing for additional water to delay water stress. An infiltration experiment with the swelling soil enhancer after passage of the wetting front is presented in Fig. 6–9 (S. E. Oswald, unpublished data).

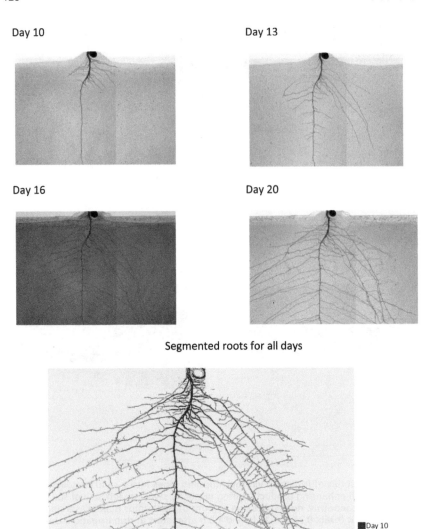

Day 10   Day 13

Day 16   Day 20

Segmented roots for all days

Day 10
Day 13
Day 16
Day 19

Fig. 6–10. Neutron radiograph time-series of root growth of a chick pea plant growing in a sandy soil in a 25 by 15 by 1 cm aluminum box. Roots of single radiographs have been segmented and shown in different colors.

Root Growth and Development

An example of root growth monitoring and root segmentation of a chick pea plant over a period of 12 d is presented in Fig. 6–10. The time series of root growth presented is every third day, but daily or hourly growth can be accomplished as well. Root parameters such as root age, total root length, and average root diameter were obtained from the segmented images. The combination of neutron radiography and image analysis tools has opened new possibilities in studying root–soil inter-actions. It provides a nondestructive and elegant way of quantitatively studying dynamics of root growth and response to various soil parameters. Root prolifera-tion toward resources in soil and or response to contaminated patches in the soil has been quantified (Menon et al., 2007; Moradi et al., 2009a).

Water Uptake and Distribution in Root Zone and Rhizosphere

Root distribution in soil and the dynamics of water in the root zone are the two parameters that contain most of the uncertainty in root water uptake studies. Neutron imaging can simultaneously quantify both parameters. The dynamics of water distribution around roots and in the rhizosphere is of interest to soil and plant scientists who study plant root water uptake mechanisms. Time-series of water content profiles in the root zone with relatively high spatial resolutions can be obtained from neutron tomography and radiography. Oswald et al. (2008) reported the application of neutron imaging to study dynamic infiltration and distribution, root growth, and root water uptake of germinating seeds and well-developed root systems of lupin and maize (*Zea mays* L.). Dynamic processes are studied with the help of real-time imaging (recording images in a time series of the sample) capabilities. Neutron images can be used to study water dynamics at the rhizosphere. Figure 6–11a shows a three-dimensional image of segmented roots of a 12-d-old chick pea grown in a sandy soil, whereas Fig. 6–11b and 6–11c summa-rize the soil water content profiles around the roots for horizontal cross-sections. Although counter-intuitive, the results show increasing soil water content toward the root surface within the rhizosphere. Since roots are the sink for soil water, one would expect decreasing soil water content toward the roots. However, the rhizo-sphere may be characterized by changing soil hydraulic properties as compared with the bulk soil (Carminati et al., 2010) and effectively holding more water than the bulk soil at equal soil water matric potential values. They studied water dynamics in the rhizosphere of lupin plant over a period of drying and rewetting (Fig. 6–12). Their quantified neutron images showed that the rhizosphere holds more water than the bulk soil during the drying period while the opposite was true after rewetting. This demonstrates that water dynamics in the rhizosphere is more complex than we formulate in our modeling simulations and needs further investigation. Also, Tumlinson et al. (2008) used NT to demonstrate the presence of spatially variable soil water content gradients in the rhizosphere and bulk soil for a single root water uptake experiment with a corn seedling.

Water Flow Inside Plants

Using heavy water as a tracer in neutron imaging makes it possible to monitor water in plant tissues. Matsushima et al. (2009) used the differing neutron attenu-ation coefficients of normal water and heavy water to yield contrast in neutron images of plant stems. They calculated heavy water movement inside roses by sequential differential images of the stem over time (Fig. 6–13). Time-evolving flow velocities of water in plant tissues could be estimated. They extended their work to tomato (*Solanum lycopersicum* L.) seedlings and combined water transport in the stem of a hibiscus plant (*Hibiscus rosa-sinensis* L.) with measurements of photosynthetic activity of leafs (Matsushima et al., 2008).

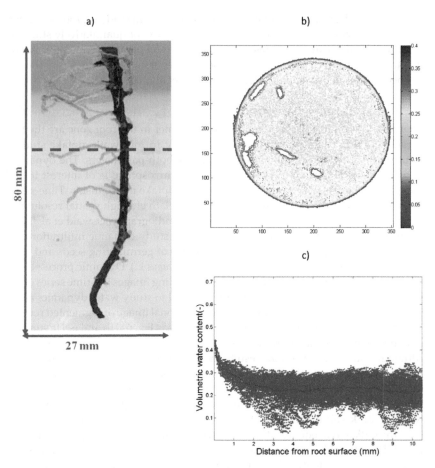

Fig. 6–11. Neutron tomography of a chick pea plant: (a) three-dimensional segmented roots, (b) soil water content around the roots in a horizontal cross-section through the three-dimensional data, and (c) pixel-wise water soil content profile as a function of distance from the root surface. The quantified data shows a higher water content in distance of 2–3 mm from the roots compared to the bulk soil.

Strengths and Weaknesses

Certainly, neutron imaging is a very robust and promising technique to observe root–soil–water processes in situ using two- or three-dimensional imaging. Using real-time imaging, it is possible to study dynamic processes as well. The method is highly complementary to X-ray imaging to study soil processes. In the past decades, resolution of digital images, detector capabilities, and image processing tools has improved substantially and is expected to continue to improve in future years.

Minimum detectable root thickness is a function of plant species, soil type, soil water content, container thickness, and spatial resolution of the image. Moradi et al. (2009b) presented a detailed analysis of the quantification of roots

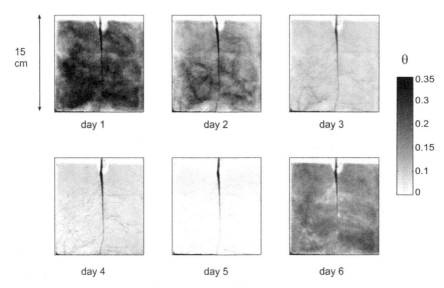

Fig. 6–12. Neutron radiography of a lupin sample grown in sandy soil in aluminum boxes (dimension 15 by 15 by 1.5 cm) during a drying period (Day 1–5) and immediately after rewetting (Day 6). The images are a close-up of the original field of view (27.9 by 27.9 cm) and show the water content at 14:00 for Day 1 to 6 (after Carminati et al., 2010).

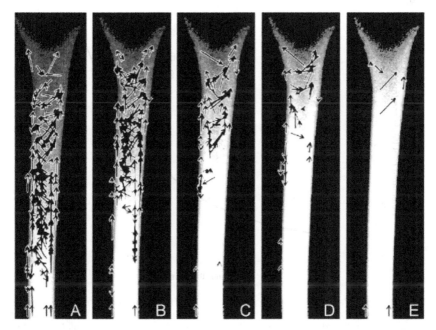

Fig. 6–13. Vector images showing movement of heavy water in the stem of a rose plant over time. The vector images were estimated from differential neutron radiographs of the sample over time (after Matsushima et al., 2009).

a)

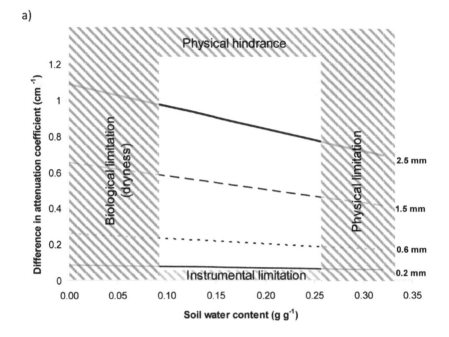

b)

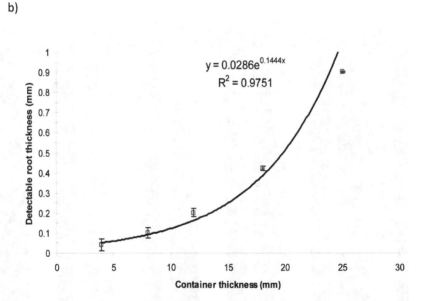

Fig. 6–14. Calculated contrast in attenuation coefficient between roots of various thicknesses and surrounding soil as a function of water contents (a). The associated limitations are shown and minimum detectable root thicknesses are calculated for various container thicknesses (b). The error bars show the variation in detectable root thickness over the range of optimal water contents (q = 0.12–0.18) for root imaging (after Moradi et al., 2009b).

with variable thicknesses from neutron radiographs for a range of soil water content and container thickness values (Fig. 6–14). The minimum root thickness that could be quantified was reported to be at least 0.2 mm under optimum conditions. Using NT, roots of smaller diameter can be detected and segmented into their three-dimensional geometrical structure. However, the size of the sample is heavily restricted because of imaging time and computational constraints.

The optimum condition for root quantification in neutron images does not always correspond with best conditions for plant growth. For example, a loamy soil with relatively high water holding capacity is more suitable for root growth, but a sandy soil with limited water holding capacity yields a better contrast between roots and soil and therefore provides for better root quantification by neutron imaging. However, there is a relatively wide window of opportunity in which soil water content is suitable for root growth and the roots can be imaged with relatively high contrast (Fig. 6–14). Neutron exposure to plants is always a concern by scientists and the rule is the lesser, the better. Most of the studies conducted by authors did not find any serious threats to the life or survival of plants, but additional studies are needed to define threshold levels of radiation exposure based on the experimental conditions and the neutron beam characteristics.

An obvious drawback of neutron imaging is the limited accessibility to a specialized facility of which there are only a few available worldwide. Here, we list selected neutron imaging facilities:

1. NEUTRA, ICON, at SINQ, PSI, Switzerland

2. ANTARES, NECTAR at FRM-2, TU Munich, Germany

3. CONRAD at BER-2, HZB Berlin, Germany

4. ISIS-TS2, Rutherford lab, UK (this is a project in progress)

5. KFKI Budapest, Hungary

6. MNRC/UC Davis, McClellen, USA

7. NIST Gaithersburg, MD, USA

8. HFIR, Oak Ridge, TN, USA

9. SANRAD at SAFARI, South Africa

10. JRR3M, Tokai, Japan

Outlook

It has been a couple of decades since the potential of neutron imaging techniques was realized, allowing us to improve our understanding of how living roots develop in soil and interact with soil and soil water. At that time, imaging with neutrons was in its early stages but just about to start a development of "exponentially" growing applications in soil–root–water systems. Now after a decade of digital neutron imaging, methods development, and diversifying applications, it is time to judge the potential of neutron imaging in its own right and how it fits into the suite of other imaging and tomographic methods that allow for root structure and root water uptake visualization.

It is now clear that neutron imaging complements these other imaging techniques. For most applications, neutron imaging will not be able to provide more information than the combination of X-ray and MRI and light transmission imaging. However, neutron imaging is very much preferred as it has large advantages if both roots and soil water are to be investigated. Neutron imaging has (i) a much higher sensitivity to water than X-ray, (ii) a better signal-to-noise ratio than MRI, while being much more effective at imaging the interfaces of air, water, and soil, and (iii) high transparency to soil minerals (more than an order of magnitude smaller neutron attenuation coefficients than water), which makes quantification of water and roots less biased by soil interferences. Nevertheless, there are limitations to neutron imaging as set by sample size, exposure times, and limited access to neutron beam lines. Access to neutron facilities will likely continue to be limited and expensive. Competition for beam time is fierce, since neutron imaging finds applications in many fields of science. However, the usefulness of neutron imaging indicates that it may follow the same pattern of development as X-ray or MRI facilities. Initially, financial and technical constraints limited their application, but their usefulness ensured subsequent technological development and capital injection, and thus they became more commonplace and routine equipment for specific applications.

At present, two main achievements result from the last decade of methodological and technical improvements of neutron imaging and the exploration of applications. First, it has become clear that neutron imaging can be outstanding in detecting water content of the rhizosphere. Neutron tomography is able to distinguish the maximum volumetric soil water content of around 50% in the rhizosphere to around 90% water content of adjacent plant root cells. Also, high spatial resolution can be achieved to map the rhizosphere with several pixels. The selectivity to hydrogen is high, allowing for targeting water, though the presence of organic molecules in substantial quantities may hamper it occasionally. Limitations are the sample size and the relatively long times needed for a neutron tomography, with two-dimensional applications suffering less in both aspects. Overall, neutron imaging is the most competitive technique, when it comes to application in the area of soil–root processes. A highlight of this development is the work of Moradi et al. (2011), providing evidence for the existence of high water content in the rhizosphere, implying a modification of water-holding capacity of the rhizosphere compared with bulk soil, as presented above. With a neutron imaging approach as used there, the time has come to investigate additional plant species and their genetically modified variants, to select for water-stress tolerance. Neutron imaging will certainly be beneficial for investigating and modifying these mechanisms.

Second, the last few years have seen a diversification of neutron imaging approaches. On one hand, various neutron imaging stations with narrow energy spectrum and specific neutron properties are developing. On the other hand, this is the investigation of modified procedures. For example, heavy water (D_2O) yields a good contrast to H_2O and was demonstrated to be useful to detect the movement of (heavy) water in soil via the rhizosphere into roots, where it is further transported to plant shoots; gadolinium as an element with extremely high neutron attenuation can be applied as tracer, if in complex form, to facilitate mobility in the water while showing less toxicity. Samples with less straightforward geometries are manageable, for example, tomographies of rectangular containers or larger size containers with stitching of images covering only parts of it, done in image processing. These new approaches should now allow targeting additional processes, such as water and solute fluxes in soil and in roots and stem.

Acknowledgments

We are grateful for the technical help we received over these past years from various employees of the Neutron Imaging Group at Paul Scherrer Institute, Villigen, Switzerland; the Neutron Imaging Group at Helmholtz Centre Berlin for Materials and Energy, Berlin, Germany, and the Soil Protection Laboratory at the Institute of Terrestrial Ecosystems, ETH Zurich, Zurich, Switzerland. This study was partly funded by Swiss National Science Foundation, EU Marie Curie Funding, and United States-Israel Binational Agricultural Research and Development Fund.

References

Badorreck, A., H.H. Gerke, and P. Vontobel. 2010. Noninvasive observations of flow patterns in locally heterogeneous mine soils using neutron radiation. Vadose Zone J. 9:362–372. doi:10.2136/vzj2009.0100

Brenner, S. 2002. Life sentences: Detective rummage investigates. Genome Biol. 3:1013.1–1013.2

Carlson, W.D. 2006. Three-dimensional imaging of earth and planetary materials. Earth Planet. Sci. Lett. 249:133–147. doi:10.1016/j.epsl.2006.06.020

Carminati, A., A. Kaestner, O. Ippisch, A. Koliji, P. Lehmann, R. Hassanein, P. Vontobel, E. Lehmann, L. Laloui, L. Vulliet, and H. Fluhler. 2007. Water flow between soil aggregates. Transp. Porous Media 68:219–236. doi:10.1007/s11242-006-9041-z

Carminati, A., A.B. Moradi, D. Vetterlein, P. Vontobel, E. Lehmann, U. Weller, H.J. Vogel, and S.E. Oswald. 2010. Dynamics of soil water content in the rhizosphere. Plant Soil 332:163–176. doi:10.1007/s11104-010-0283-8

Deinert, M.R., J.Y. Parlange, T. Steenhuis, J. Throop, K. Unlu, and K.B. Cady. 2004. Measurement of fluid contents and wetting front profiles by real-time neutron radiography. J. Hydrol. 290:192–201. doi:10.1016/j.jhydrol.2003.11.018

Esser, H.G., A. Carminati, P. Vontobel, E. Lehmann, and S.E. Oswald. 2010. Neutron radiography and tomography of water distribution in the root zone. J. Plant Nutr. Soil Sci. 173:757–764. doi:10.1002/jpln.200900188

Furukawa, J., T.M. Nakanishi, and H. Matsubayashi. 1999. Neutron radiography of a root growing in soil with vanadium. Nucl. Instrum. Methods Phys. Res. Sect. A. 424: 116–121.

Hall, L.D., M.H.G. Amin, E. Dougherty, M. Sanda, J. Votrubova, K.S. Richards, R.J. Chorley, and M. Cislerova. 1997. MR properties of water in saturated soils and resulting loss of MRI signal in water content detection at 2 tesla. Geoderma 80:431–448. doi:10.1016/S0016-7061(97)00065-7

Hassanein, R.K. 2006a. Correction methods for the quantitative evaluation of thermal neutron tomography. ETH Zurich, Switzerland.

Hassanein, R.K. 2006b. Correction methods for the quantitative evaluation of thermal neutron tomography. Ph.D. diss., ETH, Zurich, Switzerland.

Hoover, A., V. Kouznetsova, and M. Goldbaum. 2000. Locating blood vessels in retinal images by piecewise threshold probing of a matched filter response. IEEE Trans. Med. Imaging 19:203–210. doi:10.1109/42.845178

Kaestner, A., M. Schneebeli, and F. Graf. 2006. Visualizing three-dimensional root networks using computed tomography. Geoderma 136:459–469. doi:10.1016/j.geoderma.2006.04.009

Korosi, F., M. Balasko, and E. Svab. 1999. A distribution pattern of cadmium, gadolinium and samarium in *Phaseolus vulgaris* (L) plants as assessed by dynamic neutron radiography. Nucl. Instrum. Methods Phys. Res. Sect. A. 424:129–135.

Lehmann, E., H. Pleinert, and L. Wiezel. 1996. Design of a neutron radiography facility at the spallation source SINQ. Nucl. Instrum. Methods Phys. Res. Sect. A. 377:11–15.

Mason, B., and C. Moore. 1982. Principles of geochemistry. John Wiley & Sons, Hong Kong.

Matsushima, U., W.B. Herppich, N. Kardjilov, W. Graf, A. Hilger, and I. Manke. 2009. Estimation of water flow velocity in small plants using cold neutron imaging with D(2)O tracer. Nucl. Instrum. Methods Phys. Res. Sect. A. 605: 146–149.

Matsushima, U., N. Kardjilov, A. Hilger, W. Graf, and W.B. Herppich. 2008. Application potential of cold neutron radiography in plant science research. J. Appl. Bot. Food Qual.82:90–98.

Menon, M., B. Robinson, S.E. Oswald, A. Kaestner, K.C. Abbaspour, E. Lehmann, and R. Schulin. 2007. Visualization of root growth in heterogeneously contaminated soil using neutron radiography. Eur. J. Soil Sci. 58:802–810. doi:10.1111/j.1365-2389.2006.00870.x

Moradi, A.B., A. Carminati, D. Vetterlein, P. Vontobel, E. Lehmann, U. Weller, J.W. Hopmans, H.-J. Vogel, and S.E. Oswald. 2011. Three-dimensional visualization and quantification of water content in rhizosphere. New Phytol. 192:653–663. doi:10.1111/j.1469-8137.2011.03826.x

Moradi, A.B., H.M. Conesa, B.H. Robinson, E. Lehmann, A. Kaestner, and R. Schulin. 2009a. Root responses to soil Ni heterogeneity in a hyperaccumulator and a non-accumulator species. Environ. Pollut. 157:2189–2196. doi:10.1016/j.envpol.2009.04.015

Moradi, A.B., H.M. Conesa, B.H. Robinson, E. Lehmann, G. Kühne, A. Kaestner, and R. Schulin. 2009b. Neutron radiography as a tool for revealing root development in soil: Capabilities and limitations. Plant Soil 318:243–255. doi:10.1007/s11104-008-9834-7

Nakanishi, T.M., Y. Okuni, J. Furukawa, K. Tanoi, H. Yokota, N. Ikeue, M. Matsubayashi, H. Uchida, and A. Tsiji. 2003. Water movement in a plant sample by neutron beam analysis as well as positron emission tracer imaging system. J. Radioanal. Nucl. Chem. 255:149–153. doi:10.1023/A:1022252419649

Nakanishi, T.M., Y. Okuni, Y. Hayashi, and H. Nishiyama. 2005. Water gradient profiles at bean plant roots determined by neutron beam analysis. J. Radioanal. Nucl. Chem. 264:313–317. doi:10.1007/s10967-005-0713-x

Oswald, S.E., M. Menon, B. Robinson, A. Carminati, P. Vontobel, E. Lehmann, and R. Schulin. 2008. Quantitative imaging of infiltration, root growth, and root water uptake via neutron radiography. Vadose Zone J. 7:1035–1047. doi:10.2136/vzj2007.0156

Pansu, M., and J. Gautheyrou. 2006. Handbook of soil analysis: Mineralogical, organic and inorganic methods. Springer, The Netherlands.

Rudolph, N., H.G. Esser, A. Carminati, A.B. Moradi, N. Kardjilov, S. Nagl, and O. Se. 2012. Dynamic oxygen mapping in the root zone by fluorescence dye imaging combined with neutron radiography. J. Soils Sediments 12:63–74. doi: 10.1007/s11368-011-0407-7.

Salisbury, F.B., and C.W. Ross. 1992. Plant physiology. Wadsworth, Belmont, CA.

Schaap, J.D., P. Lehmann, A. Kaestner, P. Vontobel, R. Hassanein, G. Frei, G.H. de Rooij, E. Lehmann, and H. Fluhler. 2008. Measuring the effect of structural connectivity on the water dynamics in heterogeneous porous media using speedy neutron tomography. Adv. Water Resour. 31:1233–1241. doi:10.1016/j.advwatres.2008.04.014

Shokri, N., P. Lehmann, and D. Or. 2010. Evaporation from layered porous media. J. Geophys. Res. Solid Earth 115: B06204. doi:10.1029/2009JB006743.

Solymar, M., E. Lehmann, P. Vontabel, and A. Nordlund. 2003. Relating variations in water saturation of a sandstone sample to pore geometry by neutron tomography and image analysis of thin sections. Bull. Eng. Geol. Environ. 62:85–88.

Strobl, M., I. Manke, N. Kardjilov, A. Hilger, M. Dawson, and J. Banhart. 2009. Advances in neutron radiography and tomography. J. Phys. D Appl. Phys.42: 243001. doi:10.1088/0022-3727/42/24/243001

Tumlinson, L.G., H.Y. Liu, W.K. Silk, and J.W. Hopmans. 2008. Thermal neutron computed tomography of soil water and plant roots. Soil Sci. Soc. Am. J. 72:1234–1242. doi:10.2136/sssaj2007.0302

Willatt, S.T., R.G. Struss, and H.M. Taylor. 1978. In situ root studies using neutron radiography. Agron. J. 70:581–586. doi:10.2134/agronj1978.00021962007000040016x

7

Magnetic Resonance Imaging Techniques for Visualization of Root Growth and Root Water Uptake Processes

A. Pohlmeier,* S. Haber-Pohlmeier, M. Javaux, and H. Vereecken

Abstract

Root growth and water uptake processes in the subsurface are hard to observe due to the opaque nature of soil. Classical methods are either invasive or restricted to model setups like two-dimensional rhizoboxes and transparent media. Therefore, during the past two decades, noninvasive methods for monitoring root–soil processes have become popular. Among these, magnetic resonance imaging (MRI) is the most versatile one. It allows one to visualize root features like anatomy and root system architecture, water content distribution in the surrounding soil, and tracer movements in soil and roots. In this chapter, the principles of MRI are introduced first, followed by a short description of necessary hardware components. The third section reviews investigations of root systems and water uptake patterns in soils using MRI and discusses the effect of experimental parameters. In the fourth section, the usage of contrast agents for the investigation of root water processes is demonstrated. This also includes our recent results on the visualization of transport processes in root soil systems using GdDTPA as MRI contrast agent. With the example of maize (*Zea Mays* L.) and lupin (*Lupinus albus* L.) in natural sand we can show that this tracer is enriched in the cortex region under high transpiration rate conditions, but is also transported upward. The fifth section is a look at special MRI methods for a direct determination of flow velocities and MRI sequences with ultra-short detection time which are convenient for water content determination in porous media with fast relaxation times like soils with high clay contents.

Abbreviations: FFT, fast Fourier transformation; FID, free induction decay; FOV, field of view; MRI, magnetic resonance imaging; NMR, nuclear magnetic resonance; PGSE, pulsed gradient spin echo; PGSTE, pulsed gradient stimulated echo; SEMS, single echo multislice sequence; SPI, single point imaging; SPRITE, single point imaging sequence.

A. Pohlmeier, M. Javaux, and H. Vereecken, IBG-3, Agrosphere Institute, Research Center Jülich, 52425 Jülich, Germany. S. Haber-Pohlmeier, ITMC, RWTH Aachen University, Worringer Weg 1, 52074 Aachen, Germany. M. Javaux, ELI, Université Catholique de Louvain-la-Neuve, Croix du Sud 2, 1348 Louvain-la-Neuve, Belgium. *Corresponding author (a.pohlmeier@fz-juelich.de).

doi:10.2136/sssaspecpub61.c7

Soil–Water–Root Processes: Advances in Tomography and Imaging. SSSA Special Publication 61. S.H. Anderson and J.W. Hopmans, editors. © 2013. SSSA, 5585 Guilford Rd., Madison, WI 53711, USA.

Magnetic resonance imaging (MRI) is best known

from medical diagnostics, where it has been routinely used for more than three decades. Since then, it is also more and more frequently applied for imaging structures and processes in geo- and material sciences. Its huge advantages are that it is noninvasive, it is sensitive to water, and it offers a great variability of signal information by variation of experimental parameters. All three issues make MRI a valuable tool for imaging of root–soil processes. The first section of this chapter, "MRI Basics," introduces the basic physics of nuclear magnetic resonance (NMR) and MRI. More details are found in several textbooks (e.g., Callaghan, 1991; Coates and Xiao, 1999; Haake et al., 1999; Blümich, 2000; Bakhmutov, 2004; Hornak, 2004; Stapf and Han, 2006; Levitt, 2008; Keeler, 2010). The "MRI of Root–Soil Interactions" and "Contrast Agents" sections summarize the literature about application of MRI on root–soil interaction and present some recent results using contrast agent application in root soil systems. The "Outlook: Special MRI Methods" section describes new MRI methods, which will become necessary for imaging of soils.

MRI Basics

MRI is based on the physical effect of NMR when applied with spatial encoding. The basics of NMR and MRI are extensively reviewed in textbooks and on-line texts, so only the most relevant definitions and equations are given here. Many atomic nuclei possess a quantum mechanical property called "spin," which is coupled with a magnetic moment. In geosciences, mostly the behavior of ^{1}H, which possesses a spin quantum number of 1/2, in water and hydrocarbons in natural porous media is investigated, besides some rare applications of ^{19}F and ^{23}Na. A rigorous treatment of spin dynamics is given by quantum mechanics (Hanson, 2008; Levitt, 2008). However, most aspects of NMR and MRI can be understood by classical physics regarding ensembles instead of individual spins. However, two results from quantum mechanics should be considered: the first result is the Zemann splitting of energy eigenstates in an external magnetic field $\mathbf{B}_0$ leading to an energy difference of $\Delta E = \gamma\, h |\mathbf{B}_0|$ which is proportional to $|\mathbf{B}_0|$ (see Fig. 1a)[1]. The second result is that for an ensemble of spins in thermal equilibrium the population of these states obeys Boltzmann's law, Eq. [1]:

$$\frac{n_\beta}{n_\alpha} = \exp\left(-\frac{\Delta E}{k_B T}\right) = \exp\left(-\frac{\gamma h |\mathbf{B}_0|}{2\pi k_B T}\right) = \exp\left(-\frac{h\nu_0}{k_B T}\right). \qquad [1]$$

In Eq. [1], n_α and n_β are the populations of low and high energy eigenstates, ΔE is the energy difference, k_B the Boltzmann constant, T the temperature, γ the gyromagnetic ratio which is a constant for a given nucleus (for ^{1}H $\gamma = 2.68 \times 10^8$ rad T^{-1} s^{-1}), h the Planck constant, and ν_0 the resonance frequency in units of Hertz, also termed *Larmor frequency*. The surplus of n_α with respect to n_β and normalized on all spins is called the macroscopic magnetization of the sample $\mathbf{M}_0$ (Fig. 1b), and it is typically in the range of some ppm.

---

[1] Although physically not correct, in colloquial language the term *spin* is sometimes also used for the microscopic magnetic moment which is coupled to the quantum mechanical property "spin."

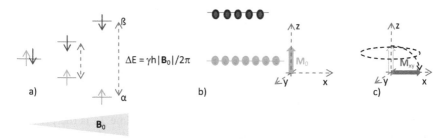

Fig. 7–1. (a) Zeemann splitting of energy levels for a spin-1/2 system in an external magnetic field. $\mathbf{B_0}$. (b) Boltzmann distribution of an ensemble of spins and resulting equilibrium macroscopic magnetization $\mathbf{M_0}$. (c) Macroscopic magnetization in thermal equilibrium (vector model). (d) Flip of the macroscopic magnetization into the xy-plane after excitation by a 90° radiofrequency pulse (rf-pulse) and precession around the z-axis. The spiral indicates the motion of $\mathbf{M}$ during the application of the rf-pulse.

Excitation

Like in optical spectroscopy, spins from the lower energy level can be excited to the upper level by absorption of electromagnetic radiation of the resonance frequency ν_0. This frequency[2] depends linearly on $|\mathbf{B_0}|$:

$$\nu_0 = \gamma |\mathbf{B_0}| / 2\pi. \tag{2}$$

As an example, for a conventional MRI scanner (superconducting magnet) with $|\mathbf{B_0}|$ = 4.7T, one obtains a frequency of ν_0 = 200.5 MHz; for a NMR relaxometer using an array of permanent magnets with $|\mathbf{B_0}|$ = 0.5T, the corresponding frequency is 21.3 MHz. Typically, the excitation is applied by irradiation with a short pulse emitted by a so-called rf-coil placed around the sample. Since the resonance frequency is in the radiofrequency range, the pulse is termed *rf-pulse*. The consequence for the macroscopic magnetization $\mathbf{M_0}$ is that it is flipped away from the alignment with the z-axis by a *flip-angle* α which is proportional to the length and power of the rf-pulse. Typical pulse flip angles are 90 and 180° which flip the magnetization into the xy plane and into the z direction, respectively. Smaller flip angles are used especially for gradient echo pulse sequences and ultra short detection pulse sequences; see below. According to classical physics, the flip of $\mathbf{M_0}$ into the xy plane results in a precession of $\mathbf{M_{xy}}$ around the z-axis due to angular momentum conservation (Fig. 1c). It can be demonstrated that the frequency of this precession is exactly the Larmor frequency ν_0. Now we have the situation of a rotating magnetization in the rf-coil that creates an alternating electrical current in the coil according to the principle of induction. This *signal* contains information about the average spin-density, which is the total amount of proton spins and is proportional to the water content. In imaging mode, the signal is proportional to the amount of protons in the voxel.

---

[2] In most textbooks the Larmor equation is written in terms of angular frequency: $\omega_0 = -\gamma |\mathbf{B_0}|$. The minus sign stems from the convention that the precession around z-axis is clockwise looking from the top on the z-axis. Therefore the Larmor frequency in Hz units should be defined as $\nu_0 = |\omega_0|/2\pi$ to avoid negative frequencies.

Since the signal decay is also controlled by local relaxation times, it further reflects dynamic properties of the water which are influenced by the physical and chemical surrounding in the porous system (see below).

Relaxation in Porous Media

After excitation, the ensemble of spins responds by relaxation to equilibrium. Basically, there are two different relaxation mechanisms: *spin-lattice relaxation* or *longitudinal relaxation*, characterized by the relaxation time T_1, is the process where the original magnetization in the z direction $\mathbf{M}_0$ is restored. In the simplest case, this proceeds according to:

$$\mathbf{M}(t) = \mathbf{M}_0\left[1 - \exp(-t/T_1)\right], \text{ with } 1/T_1 = 1/T_{1,bulk} + 1/T_{1,surface} \cdot \text{ [3]}$$

$T_{1,bulk}$ is the relaxation time of bulk water, which is typically in the range of some seconds, and $T_{1,surface}$ is the surface influenced acceleration of the relaxation time in the porous medium due to more or less frequent collisions of water with the pore walls and paramagnetic impurities (Brownstein and Tarr 1977; Kleinberg et al., 1994; Van As and van Dusschoten 1997; Coates and Xiao 1999; Barrie 2000). Summarized, $T_{1,surface}$ gives information about the pore size and its distribution; see also Eq. [5] below.

To understand the second relaxation mechanism, also termed *spin-spin relaxation* or *transversal relaxation*, one must recall that $\mathbf{M}_{xy}$ is the vector sum over many single spins. With respect to the molecular dynamics, they may interact mutually or with local dipoles and are affected by local inhomogeneities of $\mathbf{B}_0$. These interactions change local Larmor frequencies, and the individual spin packets lose their coherence; i.e., their relative phase angles change. This *dephasing* smears out the magnetization $\mathbf{M}_{xy}$ and the detected signal in the rf-coil decreases. This dephasing, characterized by the relaxation time T_2^* and the signal, is termed *free induction decay* (FID). Regarding the relaxation rates the different contributions are additive and result in:

$$1/T_2^* = 1/T_{2,bulk} + 1/T_{2,inhomo} + 1/T_{2,surface} + 1/T_{2,\,diffusion} \cdot \text{ [4]}$$

$T_{2,bulk}$ is the relaxation time in free solution and $T_{2,inhomo}$ describes the loss of coherence (dephasing) due to interaction of M_{xy} with static inhomogeneities of $\mathbf{B}_0$. Important is that the latter effect can be compensated for by a prominent NMR method, called *spin-echo* or *Hahn-echo*. Motion of water molecules as diffusion or flow in $\mathbf{B}_0$ inhomogeneities caused by the magnet design or susceptibility differences between solid, liquid, and gas phases in the porous media contributes to $1/T_{2,diffusion}$. Its influence can be minimized by using short echo times (Van As and van Dusschoten 1997). $1/T_{2,surface}$ comprises all relaxation effects occurring when water molecules approach or hit interfaces, where proton spins interact with paramagnetic ions located there. However, T_2 (and also T_1, see Eq. [3]) relaxation times are influenced by the pore size, local water content, and the surface relaxivity $\rho_{1,2}$ according to Eq. [5]:

$$1/T_{1,2,surface} = \rho_{1,2}S_{pore}/V_{pore} \qquad \qquad [5]$$

where S_{pore}/V_{pore} is the average surface/volume ratio of the pores in a voxel (Brownstein and Tarr, 1977; Coates and Xiao, 1999; Barrie, 2000; Cameron and Buchan, 2006; Mohnke and Yaramanci, 2008; Schoenfelder et al., 2008; Jaeger et al., 2009; Stingaciu et al., 2010). Often it is convenient to integrate the contributions into the effective relaxation time $T_{2,eff}$:

$$1/T_{2,eff} = 1/T_{2,bulk} + 1/T_{2,surface} + 1/T_{2,\,diffusion} \qquad [6]$$

so that the magnetization $\mathbf{M}_{xy}$ after a spin-echo decays according to Eq. [7]:

$$\mathbf{M}_{xy}(t) = \mathbf{M}_{xy}\exp\left(-t_E / T_{2,eff}\right), \qquad [7]$$

where t_E is the echo-time. Such an echo at time t_E is created by application of an 180° rf-pulse at a time $t_E/2$ after the initial 90° rf-pulse, which is part of many MRI pulse sequences. It should be noted that in soil material $T_{2,eff}$ is dominated by surface and diffusion effects and may become very short. Values ranging between sub-millisecond up to some hundred milliseconds are reported in the literature (Hall et al., 1997; Jaeger et al., 2009; Haber-Pohlmeier et al., 2010a; Stingaciu et al., 2010). Especially material with $T_{2,eff} < 3$ ms can cause problems for imaging, since in conventional pulse sequences echo times smaller than 1.5 ms are hard to achieve.

Spatial Encoding

Generally MRI means that the Larmor frequency (Eq. [2]) now is dependent on the position r by additional switching of magnetic field gradients G according to Eq. [8]:

$$v_0(\mathbf{r}) = \gamma\left|\mathbf{B}_0 + \mathbf{r}\,\mathbf{G}\right| / 2\pi. \qquad [8]$$

This has two consequences for the NMR signal: frequency and phase are space variables, and if the orthogonal gradients are switched in appropriate sequence, the whole space can be probed. The interplay of rf-pulses and gradient switching is often represented in a so-called pulse sequence diagram. Figure 2 shows an example diagram for the frequently used single-echo multi-slice pulse sequence. In MRI, many special pulse sequences are used; however, their presentation would exceed the frame of this article by far. A good overview of acronyms used is given by Nitz (1999). Only one classification should be kept in mind: Two-dimensional sequences excite a single slice or multiple slices in sequential mode; three-dimensional sequences excite the whole field of view at once. Within an excited slice, the signal is encoded in two further dimensions by additional orthogonal magnetic field gradients which change local Larmor frequency and phase. Note that the signal is recorded and stored in a reciprocal space, the so-called k-space, whose coordinates are termed k_x and k_y (and k_z for a three-dimensional sequence). The image in real space is then obtained routinely by inverse Fourier transformation in two or three dimensions, often combined with convenient filters. Important image parameters are the matrix size, i.e., the number of pixels in each direction,

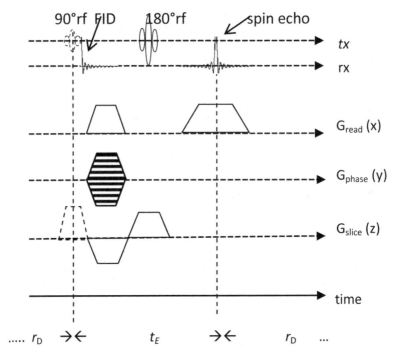

Fig. 7–2. Schematic spin echo imaging pulse sequence. t_E and r_D are echo time and repetition delay, after which a new cycle starts with different G_{phase}. The total repetiton time t_R is here approximately $t_E + r_D$. After excitation by a rf-pulse (tx) with a flip angle of 90° with simultaneous switching of a slice selective gradient G_{slice} in the z direction the phase encoding gradient G_{phase} is applied in the y direction. It follows the rf-pulse (tx) with a flip angle of 180° which refocuses the signal decay due to magnetic field inhomogeneities and creates the echo after $t = t_E$. This echo is recorded under additional switching of the read gradient G_{read} in the x direction, which makes the frequency of the signal position dependent according to Eq. [8]. "FID" is the free induction decay that is created by any 90° rf-pulse, but not monitored here. "Spin echo" is the spin echo that is recorded by this sequence.

and the field of view (FOV) of the three-dimensional image or two-dimensional slice. As stated above, many MRI pulse sequences rely on the creation of an echo. This is the recovery of a part of the already T_2^*-relaxed signal by application of an 180° rf pulse, which refocuses the fractional relaxation part $(1/T_{2,inhomo})$. All other contributions to the $T_{2,eff}$ part of T_2^* (Eq. [4] and [5]) cannot be refocused. This makes the echo signal sensitive to the local dynamics of water in the porous medium and opens doors for creation of different contrasts.

Signal Manipulation

The dependence of the signal on different relaxation processes makes MRI versatile. Equation [9] describes simplified the dependence of the signal intensity on the most important parameters: the repetition time between two successive excitations t_R, and the echo time t_E, (assuming $t_R \gg t_E$ and flip angle = 90°).

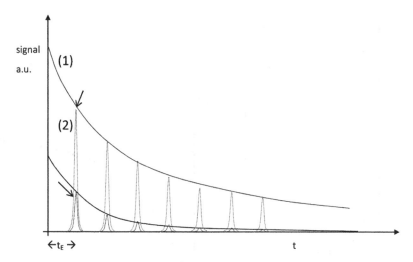

Fig. 7–3. Schematic relaxation in a two compartment system containing fast (2) and slow (1) relaxing components such as soil material and root tissue. Also included are series of spin echos (arrows) whose decays are caused by differently fast T_2 decay processes, e.g., fast decay in soil pores (red) and slower decay in root tissue (blue).

$$\frac{S}{S_0} = \left[1 - \exp\left(-\frac{t_R}{T_1}\right)\right]\exp\left(-\frac{t_E}{T_2}\right) \qquad [9]$$

In Fig. 3 schematic relaxation curves for a system with a slow relaxing component like root tissue and a fast relaxing component like soil material are plotted. By the choice of a short echo time $t_E \ll T_2$ and a long repetition time $t_R > 5T_1$, the signal is rather insensitive on T_2 and one obtains a proton density weighted image. Vice versa, if one uses a comparative long t_E, the signal is T_2 weighted and voxels containing fast relaxing components show comparative less signal. In this way, root tissue can be separated from the surrounding soil. Examples are given later. Another opportunity is weighting T_1 by keeping the repetition delay t_R short. After the detection of the signal, most magnetization is still in the xy plane. Before new excitation, the original z-magnetization must be recovered to avoid saturation effects. This proceeds with the relaxation time T_1 which is generally longer than T_2. So, if one keeps t_R short, only fast T_1 relaxing components are already completely relaxed whereas slower components have not yet recovered to the original z-magnetization (Fig. 4). T_1 contrast can also be created by using variation of the flip angle at constant, short repetition time. This is mostly done in gradient echo sequences like FLASH. For both methods of T_1 weighting voxels containing a T_1–shortening agent like GdDTPA^{2-} appear bright compared to the T_1–agent free surroundings (Greiner et al., 1997; Van As and van Dusschoten 1997; Haake et al., 1999; Nestle et al., 2008; Haber-Pohlmeier et al., 2010b).

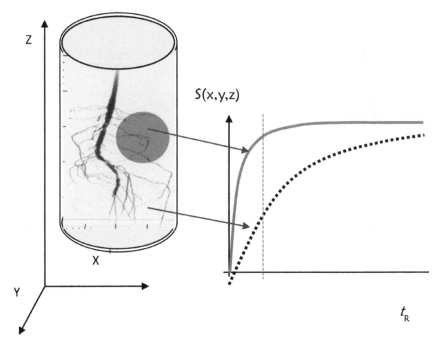

Fig. 7–4. Signal intensity as a function of the repetition time t_R between two successive excitations in a spin echo multi slice imaging pulse sequence. T_1 weighting is achieved by the choice of a relatively short t_R, indicated by the vertical dotted line yielding optimal contrast (modified, from Haber-Pohlmeier et al., 2010a).

Hardware

MRI requires different hardware components. First, a magnet is needed to create a sufficiently homogeneous magnetic field $\mathbf{B}_0$ over the volume of the sample. This is, in most cases, a superconducting magnet. For many applications, conventional medical scanners having a horizontal bore of about 60 cm diameter are sufficient. However, in soil science, the majority of fluxes are affected by gravity. As such, an MRI scanner with vertical bore is generally preferable since it allows the vertical introduction of test columns. However, vertical systems with larger dimensions are still rare to date.

Second, for imaging purposes, coils in the three Cartesian directions are needed for the creation of magnetic field gradients $\mathbf{G}_{x,y,z}$. Good resolution is achieved with gradient systems of some hundreds mT/m.

Third, convenient rf-coils are part of the hardware for both the excitation and detection of the NMR signal. The direction of rf-coils must be perpendicular or at least tilted with respect to $\mathbf{B}_0$. Therefore, volume coils are mostly cylindrical of the birdcage type. Besides these, several special coils like surface coils or separate receive coils exist.

The whole system is operated by a so-called console, consisting of a spectrometer, which controls the pulse sequence by a pulse generator or the creation of shaped rf-pulses, a high power rf amplifier, the magnetic field gradients, and high power gradient amplifiers. These components must fit to the required rf frequency range and dimensions of the rf-coil and gradient systems.

In the future, mobile low field scanners with B_0 in the range of some tens of MHz can become an alternative to stationary MRI scanners. Advantages are lower price, portability, and that permanent cooling is not necessary. Nowadays a Halbach scanner with $B_0 \approx 20$ MHz is already available for such purposes, which has still the limitation with respect to the diameter of the probe up to 4 cm (Blümich et al., 2009; Danieli et al., 2009).

MRI of Root–Soil Interactions

In an early NMR study, Bacic and Ratcovic, (1987) investigated water dynamics in maize roots by T_1 and T_2 relaxometry and concluded by evaluation of H_2O/D_2O exchange kinetics on the cellular pathway as dominant transport path. The MRI of the root system started when (Omasa et al., 1985) imaged roots in a soil material. This was continued later by another group who elucidated the use of MRI for imaging the root system of *Vicia faba* in different soil materials (Bottomley et al., 1986; Rogers and Bottomley, 1987). The relatively long echo time of 12 ms in a spin echo pulse sequence (see Fig. 2) faded out the signal from soil completely and only the root system was imaged. Moreover, the authors reported partially strong distortion of the signals due to ferromagnetic impurities in the soil material. In a second set of experiments, the motion of a $CuSO_4$ solution as a contrast agent was imaged by the introduction of T_1 weighting using short $t_R = 0.2$ s. In a third experiment, water was injected in the bottom of the soil material "peatlite" and uptake was monitored indirectly during a sequence of illumination and induced transpiration and subsequent restoration in dark. Monitoring of root water uptake by *Zea mays* grown in an artificial foam medium was also attempted by Brown et al. (1990), who imaged single selected slices through the root systems using a spin echo sequence ($t_E = 20$ ms, FOV 20 mm, matrix size 256 by 256 pixels, slice thickness = 2 mm). They reported that the substrate showed nearly no signal in the MRI images, whereas the internal structures of the main root, lateral roots, and the rhizosphere were clearly visible. When the water content decreased, the signal intensities in the roots also decreased. After 4 d, only the vascular bundle was still visible.

In parallel, MacFall (MacFall et al., 1990, 1991) continued with MRI on root water uptake processes of loblolly pine (*Pinus taeda* L.) seedlings in fine sand. Due to the small cuvette size, they obtained an in-slice resolution of 0.097 mm using a single slice single echo imaging sequence with a matrix size of 256 by 256 pixels. Due to $t_E = 10$ ms, some signal from the sand matrix was visible. Over a period from 1 to 16 h, they could observe the evolution of zones around the roots, which appeared dark in the image, and disappeared after rewatering the system (see Fig. 5). MacFall et al. interpreted these zones as water depletion due to root uptake. Later, similar dark zones were reported by Segal et al. (2008) for barley roots in fine sand. They were also interpreted as water depletion around the root. The comparison of a wild-type with root hairs to a bald genotype without root hairs showed the influence of root hairs in a more pronounced depletion zone extending over about 0.5 mm. These dark zones are in contradiction to the observations made by high resolution neutron tomography of lupin, maize, and chickpea (*Cicer*

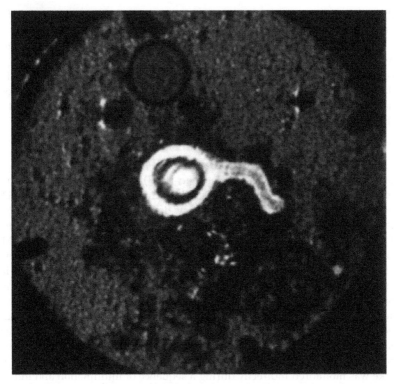

Fig. 7–5. Water depletion zone around the taproot and a lateral root of loblolly pine seedling in fine sand monitored by MRI using a two-dimensional spin-echo single slice sequence. Axial cross-section, matrix size: 256 by 256 (24.8 by 24.8 mm), slice thickness 1 mm. From MacFall et al. (1991), with permission.

*arietinum_*L.) grown in fine sand (Carminati et al., 2010; Moradi et al., 2011), who found zones of increasing water content around the roots.

So far, we reported only two-dimensional single slices from the root–soil system. However, MRI is also able to visualize transport processes that take place in three dimensions. A first step was the three-dimensional visualization of the entire root system by MacFall and Van As (1996) on a loblolly pine seedling in sand. Over 6 wk significant root growth was observed so that finally a dense network was formed. Later analogous investigations were refined by Pohlmeier et al. (2008). The aim was the simultaneous three-dimensional determination of water content change and root system architecture over a period of 2 wk with increasing drought after initial saturation. A single point imaging sequence (SPRITE) was used that abandons the time-consuming creation of an echo and probes the FID signal at a point in time "t_p" after the excitation by a small flip angle α. Images at different values of t_p were recorded and the absolute water content was determined by fitting an exponential decay function and normalizing the amplitude on a standard with known water content. With proceeding transpiration, the largest changes of water content with respect to the initial saturation took place in the lower central part of the plant container, a region where the root density was also the highest. However,

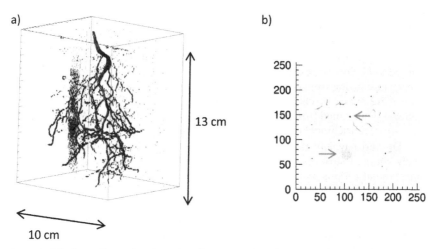

Fig. 7–6. (a) Three-dimensional rendered root system of approximately 3-wk-old *Lupinus albus* in medium sand. MRI sequence: Single echo multi slice recorded at B_0 = 4.7T using t_E = 20 ms, t_R = 20 s, matrix size = 256 by 256, resolution = 0.39 mm, slice thickness = 1 mm. (b) selected central axial slice through the root system showing the marker ($\rightarrow$), the taproot ($\leftarrow$), as well as fine lateral roots.

this was not due to enhanced root water uptake since here the initial water content was already higher. A continuation of this study used a multi-slice, multi-echo sequence, which probes the $T_{2,\text{eff}}$ relaxation instead of T_2^* and which is much faster than SPRITE with comparable resolution. Furthermore, here the largest changes in water content were observed in the neighborhood of the main roots (Pohlmeier et al., 2010). Note that in both studies, distinct depletion zones as observed by MacFall for loblolly pine are below the resolution used in this work. In the future, the roots and their surroundings must be imaged with higher resolution to observe such small-scale processes. This depends mainly on the size of the FOV into which the plant–soil vessel must fit in. For example, for a vessel with 8-cm diameter and an FOV of 10 cm, one obtains a resolution of 0.39 mm, which is sufficient to show the greatest part of the roots; see Fig. 6 (Pohlmeier et al., 2012).

Contrast Agents

The papers presented so far investigated the root water uptake processes on a long-term scale. On a short-term scale even uptake of larger amounts of water may not change significantly the water content in the neighborhood if local uptake can be rapidly equilibrated by supply from the bulk soil. For such situations, the local water fluxes and uptake paths can be visualized using contrast agents. Under natural conditions, certain plants adapted to grow on contaminated sites can take up dissolved heavy metal ions like Ni^{2+}. Fortunately, Ni^{2+} is also a strong paramagnetic ion that reduces the local T_1 relaxation time in its neighborhood. This fact was utilized by (Moradi et al., 2009) who analyzed the spatial distribution of Ni^{2+} around *Berkheya coddii* in a two-compartment rhizobox. In one compartment, only roots grew hydroponically, whereas the adjacent compartment was filled with a porous glass bead packing. After equilibration with water, a Ni^{2+} containing solution was injected and the developing Ni concentration profiles due to root water uptake were monitored by the calculation of two-dimensional T_1 maps. Instead of the time-consuming use of a spin echo sequence, T_1 was determined by variation of the flip angle instead of the repetition time or inversion time during the preparation period before the excitation pulse. This is equivalent to variation of t_R, but the overall measuring time is shorter. They obtained a large matrix size in this single slice of 512 by 512 pixels corresponding to a resolution of 0.21 mm. Ni^{2+} concentration was determined from local T_1 relaxation times by calibration. With proceeding time strong enrichment of Ni^{2+} in the vicinity of the roots was measured.

Di- and trivalent transition metal cations are prone for adsorption at negatively charged surfaces of soil mineral grains. Therefore, instead of using transition metal cations as contrast agents, the use of negatively charged metal ion chelate complexes is preferred for monitoring root water uptake processes. The prerequisite for such contrast agents is their chemical inertness and conservative transport properties in natural porous media. This makes $GdDTPA^{2-}$, well known from medical diagnostics as "Magnevist" formulation, a convenient compound. (Haber-Pohlmeier et al., 2010b) proved the expected behavior by comparing transport experiments and numerical model. This makes $GdDTPA^{2-}$ a convenient tracer for monitoring transport processes in root-soil systems. In a second application the motion of this tracer was employed for monitoring root water uptake processes which is described in the following section (Haber-Pohlmeier et al., 2010c).

GdDTPA Uptake

Maize

Zea mays was germinated in wet paper tissue and replanted in a Perspex MRI column (8-cm inner diameter, 10-cm length) filled with medium sand (Frechen, Germany). The plant grew for 16 d at an average water content of $\theta = 0.16$ cm^3 cm^{-3} under a 12 h 12 h^{-1} day night light cycle. Tracer uptake was started by imbibition through a hole from the bottom by raising the outer water table to 1.5 cm with 2.5 mM GdDTPA solution. Motion of the tracer plume was monitored under illumination in a 4.7T ultra wide bore scanner operated by a Varian console using a strong T_1 weighted single echo multi-slice sequence (SEMS) with following parameters: matrix size of 128 by 128 pixels, FOV 120 by 150 mm, 5 slices per 1 mm; $t_E = 6$ ms, $t_R = 0.47$ s.

Figure 7 shows the imbibition of the tracer plume into the maize root system in sandy soil. Starting under quite dry conditions ($\theta = 0.12$ cm^3 cm^{-3}) the outer water table was raised instantaneously to + 1.5 cm at the lower boundary and tracer solution could penetrate into the system. Initially, the signal intensity in the system is generally low, and the root is hardly visible. The reason is the strong T_1 weighting in combination with long T_1 relaxation times in absence of the contrast agent. During the first 5 min, the solution infiltrated quickly into the soil. What is remarkable is the dark zone around the root in the left half of the images at 3 and 5 min. After 8 min, the situation reverts: now the zone around the root as well as the root itself appeared much brighter than the surrounding soil. The tracer was taken up and appeared also inside the roots. The dark zone around the roots can be caused by two reasons. Either local water content is lower and it is then restored during the imbibition with delay, or locally smaller pore sizes cause accelerated fast T_2 relaxation reducing the signal intensity. Similar dark zones have been observed in other experiments on white lupin in sand with and with-

a)   b)   c)   d)   e)

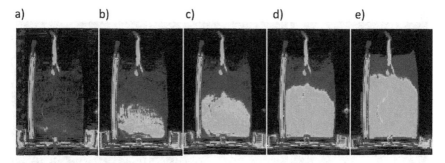

Fig. 7–7. Series of images of GdDTPA^{2-} solution imbibition into maize root system grown in natural medium sand. Initial water conten was $\theta = 0.12$ cm^3 cm^{-3}. Left is a marker tube filled with 30% H$_2$O/D$_2$O and 2.5 mM GdDTPA. Time points: 1, 3, 5, 8, and 20 min. Single slice single echo imaging sequence, T_1 weighted: $t_R = 0.47$ s, matrix size: 128 by 128, resolution 0.78 mm, slice thickness 1 mm, $t_E = 6$ ms.

out tracer motion and T_1 sensitivity as well as under spin-density weighted conditions. They appear especially under re-wetting conditions. The fact that tracer was taken up proves that this zone is hydraulically conductive and does not obstruct the water flow to the root. However, to investigate if the tracer is pushed into the root by the rapid change of water saturation or if it is taken up by root water uptake, further tracer uptake experiments were performed with a lupin plant in natural sand under nearly steady state water flow conditions.

Lupin

White lupin (*Lupinus albus*) was germinated in wet paper tissue and replanted in a Perspex MRI column filled with medium sand FH31 [Quartzwerke Frechen, Germany (3-cm inner diameter, 12.2-cm height)]. After 1 wk of growth, tracer uptake was started by injection of 0.7 mL of 5 mM GdDTPA solution into the lower part of the container at a water content of $\theta = 0.16$ cm^3 cm^{-3}. Motion of the tracer plume was monitored in a 7T wide bore scanner operated by a Varian console using a strong T_1 weighted single echo multislice sequence (SEMS) with following parameters: matrix size of 256 by 256; (FOV 40 by 40) 0.16 by 0.16 mm, 4 slices per 1 mm; $t_E = 3.4$ ms, $t_R = 0.2$ s. During an equilibration period of 16 h, practically no uptake was observed, just diffusive spreading of the plume in the porous system (Fig. 8a and 8b). Slight enrichment is observed in the outermost zone of the root and in the vascular bundle area. Please note that this took place only in the lower part of taproot and the lateral roots, not in the upper taproot. Five hours after starting transpiration to about 10 cm^3 d^{-1} by illumination, a significant amount of tracer was taken up and appears mainly in cortex of the roots (Fig. 8c). Now the vascular bundle region is much darker than the cortex. This behavior is qualititatively explained as follows: during dark, when only slow transpiration took place, the tracer is enriched in the root hair zone around the younger taproot and the lateral roots. It is transported into the vascular bundle by diffusion between soil and root xylem, and it is slowly transported upward due to slow transpiration flux and diffusion. In the full transpiration period, the tracer is transported quickly into the cortex region and delayed at the Caspari strip. Eventually Gd was transported

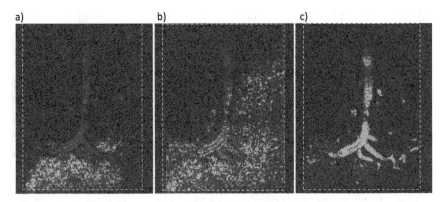

Fig. 7–8. Tracer transport in a lupin–medium sand system. T_1 weighted single echo multislice imaging sequence: $t_R = 0.2$ ms, matrix size: 256 by 256, resolution 0.16 mm, slice thickness 1 mm, $t_E = 3.4$ ms. (a) immediately after injection of 0.7 mL 5 mM GdDTPA solution into the lower part of the column. (b) after 16 h in the dark, where about 2 cm^3 d^{-1} water was taken up. (c) 5 h after starting transpiration to 10 cm^3 d^{-1} by illumination. Final water content was 0.13 cm^3 cm^{-3}.

upward as could be proven by chemical analytics, where decreasing concentrations in the taproot, shoot and also in the leaves were determined.

Outlook: Special MRI Methods
Water Motion Detection

MRI signals are made of two components: real and imaginary part, or, in other words, magnitude and phase difference to a given reference signal. When magnetic field gradients are present, motion of spins (water molecules) during the different periods of the pulse sequence affects the phase of the signal. If the motion is random, the different phases of different spin sub-ensembles cancel each other and no net phase shift is detected; however, a general decay of the magnitude is observed. Diffusion as random motion reduces the signal magnitude or intensity. As such, diffusion encoding can be introduced into the sequence by switching gradients into the preparation and evolution periods. In case of directed flow, the individual phase shifts do not cancel, but sum up to a macroscopic phase shift which is proportional to the average velocity (Kimmich, 1997; Baumann et al., 2000). This method was applied for example in the study of flow in tree stems, where the authors could differentiate between xylem and phloem flux processes (Windt et al., 2006; Homan et al., 2007; Van As, 2007); see Fig. 9. In porous media, these methods face the problem of short relaxation times. Since the evolution period is comparably long at the maximum gradients strengths presently available (see hardware section), the T_1 relaxation time must be sufficiently longer than the evolution time. Therefore the method has been applied up to now only for flux in saturated medium sand (Li et al., 2009; Spindler et al., 2011) but not yet for root uptake processes, which will be the next step. The lowest detectable average flow velocities in a voxel are in the order of some tens of microseconds per second (some m d^{-1}).

Average linear velocity per pixel
(mm s⁻¹)

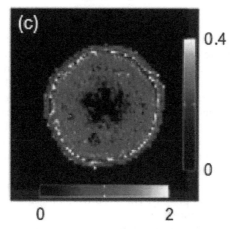

Fig. 7–9. MRI flow velocity determination in a poplar stem using pulsed gradient spin echo (PGSE) and pulsed gradient stimulated echo (PGSTE) sequences, from Windt et al. (2006), with permission.

Ultra-Short Detection

As mentioned in the introduction, the shortening of T_2 by para- and ferromagnetic material in soils is the greatest challenge for MRI in such systems. One way out is the ultra-fast detection of the NMR signal after or simultaneous to the excitation (Fagan et al., 2003; Robson et al., 2003). In addition, the already mentioned single point imaging method (SPI) or derivatives like SPRITE and spiral SPRITE is one way out. In these methods, which were originally developed for imaging in rocks, no echo is created; instead the FID is probed directly (see also Fig. 2). Single point means that, in k-space, one point is addressed directly during the excitation by the combination of gradient pulses either directly (SPI) or stepwise ramped (SPRITE) (Balcom et al., 1996; Szomolanyi et al., 2001; Marica et al., 2006; Li et al., 2009). SPRITE was already used for imaging root soil interaction as described in the preceding chapter. The disadvantage of the SPI methods is the comparative long measurement time. Since each data point is addressed separately in k-space, the measurement time scales to the third with the number of voxels instead of to the second when using echo methods. Practically, only a matrix size of 64 by 64 by 64 voxels can be achieved, slice selection is also not possible. A discussion about sensitivity of SPI and fast spin echo methods is found in Van Duynhoven et al. (2009).

A fundamental different approach is the quasi-simultaneous detection of the signal during excitation by so-called adiabatic pulses (Idiyatullin et al., 2006; Idiyatullin et al., 2008). Here the exciting pulse is gapped, and during the gap, a data point is recorded after a few ms of the excitation and very short dead times. By this, a pseudo echo is recorded for short flip angles which must be deconvoluted from the excitation function. Since k-space is probed non-Cartesian, simple fast Fourier transformation (FFT) cannot be applied for the reconstruction of the image. The total measurement time scales similarly to echo methods and a 256 by 256 by 256 image can be recorded in few minutes. Up to now, this method has been applied to biological objects but not yet to soils or rocks.

Discussion and Outlook

In summary, the state of MRI research allows the noninvasive three-dimensional imaging of root systems as well as single roots grown in artificial soils and some natural soil materials. The highest resolution depends mainly on the sample size (FOV) divided by matrix size which can be 256 to 512 pixel/direction. Water content changes due to root water uptake are monitorable over periods from some hours to weeks, where most investigations have been performed only in pure sands and sandy soils. Short-term monitoring of water fluxes is also possible by tracking plumes of contrast agents in soils as well as inside roots using T_1 sensitive pulse sequences. This allows the determination of flow rates and thus conclusions about the potential pathways participating in the transport. Direct flow velocity imaging is still restricted to saturated model media and flow velocities above some meters per day. This is sufficient for flow in plant structures but in soils only for macropore flow.

At present, we identify the following major questions, which should be addressed in the research of root soil interactions by MRI in the near future:

(i) Until today, there has been a contradiction between MRI experiments which determined regions of low intensity around the roots and neutron tomography experiments where regions of high intensity are found. The question is still open whether regions of low intensity in MRI are due to low spin density (water content), fast relaxation times, or even a combination of both. Neutron imaging determines proton density maps whereas the voxel intensities in MRI images are influenced by both proton density and relaxation times. This can be separated by the variation of t_E. However, there is a lower limit of t_E, and water fractions that relax with $T_2 < t_E$ might get lost. On the other hand, such fast relaxing water might be bound to the matrix and is not available for uptake. The important question is the assignment of water fractions to relaxation times. If this is answered, MRI might become a tool for the identification of bioavailable water like it is successfully applied in petroleum exploration for the determination of bound and movable liquids by their respective relaxation times (Coates and Xiao, 1999). The next step is a joint experiment with neutron tomography and different MRI methods with comparable resolution using identical soil–plant setups.

(ii) Derivation of an unambiguous three-dimensional root system skeleton which is necessary for coupling MRI (and other tomographic) images with model calculations to validate theoretical concepts of root water uptake with experimental evidence (Hopmans and Bristow, 2002; Javaux et al., 2008; Schröder et al., 2008, 2009). Image processing methods are needed that translate the three-dimensional volume graphics MRI data into an array of nodes and segments describing the hierarchy of the root system network and that is usable as input for modeling. This will help to base model calculations on real root systems (Stingaciu et al., 2012). Additionally, the root systems should be imaged with higher resolution, since fine roots with diameter smaller than 0.5 mm play the major role in water uptake.

(iii) Coupling of MRI images of the root system, water content distributions, and water flux maps with three-dimensional modeling, which integrates theoretical concepts of water transport in the soil–plant atmosphere continuum with experimental findings (Hopmans and Bristow, 2002; Javaux et al., 2008). This will allow for inverse modeling to obtain important transport parameters.

(iv) Magnetic resonance microscopy imaging to obtain better resolutions. This is necessary for a better understanding of processes in the rhizosphere and at the root surface.

(v) Imaging of water content and dynamics in natural soil material and natural soil cores is still a challenge due to the inherently short T_2 relaxation. This therefore requires imaging pulse sequences with ultra-short detection times as well as compensation procedures for correction of dipolar artifacts caused by susceptibility differences at liquid–solid interfaces and due to the presence of ferromagnetic or metallic particles in the soil. Convenient pulse sequences are single point imaging (SPI) and its derivatives like SPRITE and adiabatic excitation sequences with quasi simultaneous excitation and detection like SWIFT. Such sequences will be implemented on convenient MRI scanners with vertical bores and applied on natural soil material and natural soil cores. An alternative and complementation could be the application of portable low field scanners, since the T_2 relaxation gets longer with decreasing $\mathbf{B}_0$. At present, only scanners with a bore of about 5 cm are available, so one improvement would be the development of scanners with larger bores. Furthermore, with these scanners, the application of ultrafast detection sequences will be profitable.

(vi) Imaging of dynamic processes (water flux and diffusion and/or dispersion) inside the roots and the surrounding soil. Inside roots, this is presently possible using existing pulsed gradient sequences. In the soil material, this requires additional technical (higher gradient strengths) and sequence design efforts due to fast relaxation times.

Acknowledgments

The authors thank Deutsche Forschungsgemeinschaft (DFG) and Helmholtz Gemeinschaft (HGF) for financial support (SFB TR32 and Po 746-2/1), Central Division of Analytical Chemistry of the Research Centre Jülich, and D. van Dusschoten for assistance with high field MRI scanners at Forschungszentrum Jülich.

References

Bacic, G., and S. Ratcovic. 1987. NMR studies of radial exchange and distribution of water in maize roots: The relevance of exchange kinetics. J. Exp. Bot. 38:1284–1297. doi:10.1093/jxb/38.8.1284

Bakhmutov, W.I. 2004. Practical NMR relaxation for the chemist. John Wiley & Sons, New York.

Balcom, B.J., R.P. MacGregor, S.D. Beyea, D.P. Green, R.L. Armstrong, and T.W. Bremner. 1996. Single-point ramped imaging with T-1 enhancement (SPRITE). J. Magn. Reson. A 123(1):131–134. doi:10.1006/jmra.1996.0225

Barrie, P.J. 2000. Characterization of porous media using NMR methods. Annu. Rep. NMR Spectrosc. 41:265–316. doi:10.1016/S0066-4103(00)41011-2

Baumann, T., R. Petsch, and R. Niessner. 2000. Direct 3-d measurement of the flow velocity in porous media using magnetic resonance tomography. Environ. Sci. Technol. 34(19):4242–4248. doi:10.1021/es991124i

Blümich, B. 2000. NMR imaging of materials. Clarendon Press, Oxford.

Blümich, B., J. Mauler, A. Haber, J. Perlo, E. Danieli, and F. Casanova. 2009. Mobile NMR for geophysical analysis and materials testing. Petrol. Sci. 6:1–7. doi:10.1007/s12182-009-0001-4

Bottomley, P.A., H.H. Rogers, and T.H. Foster. 1986. NMR imaging shows water distribution and transport in plant root systems in situ. Proc. Natl. Acad. Sci. USA 83:87–89. doi:10.1073/pnas.83.1.87

Brown, J.M., P.J. Kramer, G.P. Cofer, and G.A. Johnson. 1990. Use of nuclear magnetic resonance microscopy for noninvasive observations of root-soil water relations. Theor. Appl. Climatol. 42:229–236. doi:10.1007/BF00865983

Brownstein, K.R., and C.E. Tarr. 1977. Spin-lattice relaxation in a system governed by diffusion. J. Magn. Reson. 26:17–24.

Callaghan, P.T. 1991. Principles of nuclear magnetic resonance microscopy. Oxford Univ. Press, Oxford, UK.

Cameron, K.C., and G.D. Buchan. 2006. Porosity and Pore size distribution. Encyclopedia of soil science. Taylor and Francis, London. p. 1350–1353.

Carminati, A., A. Moradi, D. Vetterlein, P. Vontobel, E. Lehmann, U. Weller, H.J. Vogel, and S. Oswald. 2010. Dynamics of soil water content in the rhizosphere. Plant Soil 332:163–176. doi:10.1007/s11104-010-0283-8

Coates, G.R., and L.P. Xiao. M.G. 1999. NMR logging principles and applications. Halliburton Energy Services, Houston.

Danieli, E., J. Mauler, J. Perlo, B. Blümich, and F. Casanova. 2009. Mobile sensor for high resolution NMR spectroscopy and imaging. J. Magn. Reson. 198:80–87. doi:10.1016/j.jmr.2009.01.022

Fagan, A.J., G.R. Davies, J.M.S. Hutchison, and D.J. Lurie. 2003. Continuous wave MRI of heterogeneous materials. J. Magn. Reson. 163(2):318–324. doi:10.1016/S1090-7807(03)00128-9

Greiner, A., W. Schreiber, G. Brix, and W. Kinzelbach. 1997. Magnetic resonance imaging of paramagnetic tracers in porous media: Quantification of flow and transport parameters. Water Resour. Res. 33(6):1461–1473. doi:10.1029/97WR00657

Haake, E.M., R.W. Brown, M.R. Thompson, and R. Venkatesan. 1999. Magnetic rresonance imaging. Physical principles and sequence design. Wiley-Liss, New York.

Haber-Pohlmeier, S., S. Stapf, D. Van Dusschoten, and A. Pohlmeier. 2010a. Relaxation in a natural soil: Comparison of relaxometric imaging, T1– T2 correlation and fast-field cycling NMR. Open Magn. Reson. J. 3:57–62.

Haber-Pohlmeier, S., M. Bechtold, S. Stapf, and A. Pohlmeier. 2010b. Waterflow monitored by tracer transport in natural porous media using MRI. Vadose Zone J. 9(4):835–845. doi:10.2136/vzj2009.0177

Haber-Pohlmeier, S., M. Javaux, and A. Pohlmeier. 2010c. Water fluxes in root-soil-systems investigated by magnetic resonance imaging. Geophys. Res. Abstr.12:7207–2012.

Hall, L.D., M.H.G. Amin, E. Dougherty, M. Sanda, J. Votrubova, K.S. Richards, R.J. Chorley, and M. Cislerova. 1997. MR properties of water in saturated soils and resulting loss of MRI signal in water content detection at 2 tesla. Geoderma 80(3–4):431–448. doi:10.1016/S0016-7061(97)00065-7

Hanson, L.G. 2008. Is quantum mechanics necessary for understanding magnetic resonance? Concepts Magn. Reson. A 32A(5):329–340. doi:10.1002/cmr.a.20123

Homan, N.M., C.W. Windt, F.J. Vergeldt, E. Gerkema, and H. Van As. 2007. 0.7 and 3 T MRI and sap flow in intact trees: Xylem and phloem in action. Appl. Magn. Reson. 32(1–2):157–170. doi:10.1007/s00723-007-0014-3

Hopmans, J.W., and K.L. Bristow. 2002. Current capabilities and future needs of root water and nutrient uptake modeling. Adv. Agron. 77:103–183. doi:10.1016/S0065-2113(02)77014-4

Hornak, J.P. 2004. The basics of MRI. http://www.cis.rit.edu/htbooks/mri/inside.htm (accessed 9 Jan. 2013).

Idiyatullin, D., C. Corum, S. Moeller, and M. Garwood. 2008. Gapped pulses for frequency-swept MRI. J. Magn. Reson. 193(2):267–273. doi:10.1016/j.jmr.2008.05.009

Idiyatullin, D., C. Corum, J.Y. Park, and M. Garwood. 2006. Fast and quiet MRI using a swept radiofrequency. J. Magn. Reson. 181(2):342–349. doi:10.1016/j.jmr.2006.05.014

Jaeger, F., S. Bowe, H. Van As, and G.E. Schaumann. 2009. Evaluation of 1H NMR relaxometry for the assessment of pore size distribution in soil samples. Eur. J. Soil Sci. 60:1052–1064. doi:10.1111/j.1365-2389.2009.01192.x

Javaux, M., T. Schröder, J. Vanderborght, and H. Vereecken. 2008. Use of a Three-Dimensional Detailed Modeling Approach for Predicting Root Water Uptake. Vadose Zone J. 7(3):1079–1088. doi:10.2136/vzj2007.0115

Keeler, J. 2010. Understanding NMR spectroscopy. John Wiley & Sons, Hoboken, NJ.

Kimmich, R. 1997. NMR Tomography, Diffusometry, Relaxometry. Springer, Berlin.

Kleinberg, R.L., W.E. Kenyon, and P.P. Mitra. 1994. Mechanism of NMR relaxation of fluids in rocks. J. Magn. Reson. A.108:206–214

Levitt, M.H. 2008. Spin dynamics. John Wiley & Sons, Chichester, UK.

Li, L.Q., H. Han, and B.J. Balcom. 2009. Spin echo SPI methods for quantitative analysis of fluids in porous media. J. Magn. Reson. 198(2):252–260. doi:10.1016/j.jmr.2009.03.002

MacFall, J.S., G.A. Johnson, and P.J. Kramer. 1990. Observation of a water depletion region surrounding loblooly pine roots by magnetic resonance imaging. Proc. Natl. Acad. Sci. USA 87:1203–1207. doi:10.1073/pnas.87.3.1203

MacFall, J.S., G.A. Johnson, and P.J. Kramer. 1991. Comparative water-uptake by roots of different ages in seedlings of loblolly-pine (Pinus-Taeda L). New Phytol. 119(4):551–560. doi:10.1111/j.1469-8137.1991.tb01047.x

MacFall, J.S., and H. Van As. 1996. Magnetic resonance imaging of plants. In: Y. Shachar-Hill and P.E. Pfeffer, editors, Nuclear magnetic resonance in plant biology. Plant Physiol. Soc. Am., Rockville, MD.

Marica, F., Q. Chen, A. Hamilton, C. Hall, T. Al, and B.J. Balcom. 2006. Spatially resolved measurement of rock core porosity. J. Magn. Reson. 178(1):136–141. doi:10.1016/j.jmr.2005.09.003

Mohnke, O., and U. Yaramanci. 2008. Pore size distributions and hydraulic conductivities of rocks derived from Magnetic Resonance Sounding relaxation data using multi-exponential decay time inversion. J. Appl. Geophys. 66(3–4):73–81. doi:10.1016/j.jappgeo.2008.05.002

Moradi, A., A. Carminati, D. Vetterlein, P. Vontobel, E. Lehmann, U. Weller, J.W. Hopmans, H.J. Vogel, and S. Oswald. 2011. Three-dimensional visualization and quantificatioin of water content in the rhizosphere. New Phytol. 192(3):653–663. doi:10.1111/j.1469-8137.2011.03826.x

Moradi, A.B., S.E. Oswald, J.A. Nordmeyer-Massner, K.P. Pruessmann, B.H. Robinson, and R. Schulin. 2009. Analysis of nickel concentration profiles around the roots of the hyperaccumulator plant Berkheya coddii using MRI and numerical simulations. Plant Soil 328(1–2):291–302. doi:10.1007/s11104-009-0109-8

Nestle, N., A. Wunderlich, and T. Baumann. 2008. MRI studies of flow and dislocation of model NAPL in saturated and unsaturated sediments. Eur. J. Soil Sci. 59(3):559–571. doi:10.1111/j.1365-2389.2008.01040.x

Nitz, W.R. 1999. MR imaging: Acronyms and clinical applications. Eur. Radiol. 9:979–997. doi:10.1007/s003300050780

Omasa, K., M. Onoe, and H. Yamada. 1985. NMR imaging for measuring root systems and soil water content. Environ. Control Biol. 23:99–102. doi:10.2525/ecb1963.23.99

Pohlmeier, A., S. Haber-Pohlmeier, and H. Vereecken. 2012. Root system growth and water uptake of lupin in natural sand: 3D visualization by MRI. Magn. Reson. Imaging.(submitted)

Pohlmeier, A., A.M. Oros-Peusquens, M. Javaux, M.I. Menzel, H. Vereecken, and N.J. Shah. 2008. Changes in soil water content resulting from ricinus root uptake monitored by magnetic resonance imaging. Vadose Zone J. 7:1010–1017. doi:10.2136/vzj2007.0110

Pohlmeier, A., F. Vergeldt, E. Gerkema, D. Van Dusschoten, and H. Vereecken. 2010. MRI in Soils: Determination of water content changes due to root water uptake by means of Multi-Slice-Multi-Echo sequence (MSME). Open Magn. Reson. J. 3:69–74.

Robson, M.D., P.D. Gatehouse, M. Bydder, and G.M. Bydder. 2003. Magnetic resonance: An introduction to ultrashort TE (UTE) imaging. J. Comput. Assist. Tomogr. 27(6):825–846. doi:10.1097/00004728-200311000-00001

Rogers, H.H., and P.A. Bottomley. 1987. Insitu nuclear-magnetic-resonance imaging of roots-influence of soil type, ferromagnetic particle content, and soil-water. Agron. J. 79(6):957–965. doi:10.2134/agronj1987.00021962007900060003x

Schoenfelder, W., H.R. Glaeser, I. Mitreiter, and F. Stallmach. 2008. Two-dimensional NMR relaxometry study of pore space characteristics of carbonate rocks from a Permian aquifer. J. Appl. Geophys. 65(1):21–29. doi:10.1016/j.jappgeo.2008.03.005

Schröder, T., M. Javaux, J. Vanderborght, B. Koerfgen, and H. Vereecken. 2009. Implementation of a microscopic soil-root hydraulic conductivity drop function in a three-dimensional soil-root architecture water transfer model. Vadose Zone J. 8(3):783–792. doi:10.2136/vzj2008.0116

Schröder, T., M. Javaux, J. Vanderborght, B. Körfgen, and H. Verreecken. 2008. Effect of local soil hydraulic conductivity drop using a three dimensional root water uptake model. Vadose Zone J. 7(3):1089–1098. doi:10.2136/vzj2007.0114

Segal, E., T. Kushnir, Y. Mualem, and U. Shani. 2008. Water uptake and the hydraulics of the root hair rhizosphere. Vadose Zone J. 7:1024–1037.

Spindler, N., P. Galvosas, A. Pohlmeier, and H. Vereecken. 2011. NMR velocimetry with 13-interval stimulated echo multi slice imaging in natural porous media under small flow rates. J. Magn. Reson. 212:216–223. doi:10.1016/j.jmr.2011.07.004

Stapf, S., and S. Han. 2006. NMR imaging in chemical engineering. Wiley-VCH, Weinheim.

Stingaciu, L.R., H. Schulz, A. Pohlmeier, S. Behnke, H. Zilken, M. Javaux, and H. Vereecken. 2013. In situ root system architecture extraction from magnetic resonance imaging for water uptake modeling. Vadose Zone J. (in press).

Stingaciu, L.R., L. Weihermüller, S. Haber-Pohlmeier, S. Stapf, H. Vereecken, and A. Pohlmeier. 2010. Determination of the pore size distribution- a comparison study of laboratory methods. Water Resour. Res. 46:W11510. doi:10.1029/2009WR008686

Szomolanyi, P., D. Goodyear, B. Balcom, and D. Matheson. 2001. SPIRAL-SPRITE: A rapid single point MRI technique for application to porous media. Magn. Reson. Imaging 19(3–4):423–428. doi:10.1016/S0730-725X(01)00260-0

Van As, H. 2007. Intact plant MRI for the study of cell water relations, membrane permeability, cell-to-cell and long-distance water transport. J. Exp. Bot. 58:743–756. doi:10.1093/jxb/erl157

Van As, H., and D. van Dusschoten. 1997. NMR methods for imaging of transport processes in micro-porous systems. Geoderma 80(3–4):389–403. doi:10.1016/S0016-7061(97)00062-1

Van Duynhoven, P.M., G.J.W. Goudappel, W.P. Weglarz, C.W. Windt, P.R. Cabrer, A. Mohoric, and H. Van As. 2009. Noninvasive assessment of moisture migration in food products by MRI. In: S.L. Codd and J.D. Seymour, editors, Spatially resolved NMR techniques and applications. Wiley-VCH, Weinheim. p. 331–351.

Windt, C., F. Vergeldt, P.A. de Jager, and H. Van As. 2006. MRI of long-distance water transport: A comparison of the phloem and xylem flow characteristics and dynamics in poplar, castor bean, tomato and tobacco. Plant Cell Environ. 29:1715–1729. doi:10.1111/j.1365-3040.2006.01544.x

8

Segmentation of X-Ray CT Data of Porous Materials: A Review of Global and Locally Adaptive Algorithms

Markus Tuller,* Ramaprasad Kulkarni, and Wolfgang Fink

Abstract

Recent computational and technological advances in X-ray computed tomography (CT) provide exciting new means for nondestructive, three-dimensional imaging of soil–water–root processes with projects ranging from visualization and characterization of biofilms to quantification of effects of root-induced compaction on rhizosphere hydraulic properties. In contrast to these breath-taking technological improvements, the development and evaluation of appropriate segmentation methods for transformation of three-dimensional grayscale X-ray CT data into a discrete form that allows accurate separation of solid, liquid, and vapor phases for quantitative description of the soil–water–root continuum and provides the basis for modeling of static and dynamic system processes appears to lag behind. This chapter reviews the state-of-the-art and recent developments in image segmentation, covering global and locally adaptive two-phase as well as more complex multiphase algorithms. The need for true three-dimensional segmentation is demonstrated and potential techniques for pre-segmentation enhancement of X-ray CT data are discussed.

In recent years, the application of noninvasive three-dimensional imaging methods such as X-ray computed tomography (CT) gained tremendous popularity in soil science research with many exciting projects ranging from visualization of metal-reducing and biofilm-forming bacteria (Iltis et al., 2011) and observation of micropore development in soil aggregates to assess potential pathways for soluble carbon transport (Smucker et al., 2007; Peth et al., 2008) to studying effects of root-induced compaction on rhizosphere hydraulic properties

Abbreviations: AWGN, additive white Gaussian noise; CT, computed tomography; ICM, iterated conditional modes; MMD, modified metropolis dynamics; MRF, Markov random field; OCR, optical character recognition; PDE, partial differential equations; PDF, probability density function; ROI, region of interest.

M. Tuller, Univ. of Arizona, Dep. of Soil, Water and Environmental Science, Tucson, AZ 85721. R. Kulkarni and W. Fink, The Univ. of Arizona, Dep. of Electrical and Computer Engineering, Tucson, AZ. W. Fink, The Univ. of Arizona, Dep. of Biomedical Engineering, Tucson, AZ. *Corresponding author (mtuller@cals.arizona.edu).

doi:10.2136/sssaspecpub61.c8

Soil–Water–Root Processes: Advances in Tomography and Imaging. SSSA Special Publication 61. S.H. Anderson and J.W. Hopmans, editors. © 2013. SSSA, 5585 Guilford Rd., Madison, WI 53711, USA.

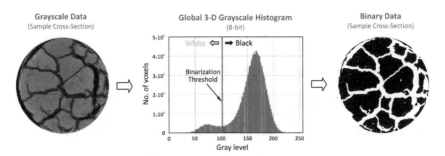

Fig. 8–1. Sketch that illustrates grayscale histogram-based image binarization for a bentonite-sand sample with dehydration cracks. The sample was scanned with an industrial HYTEC Flat Panel Amorphous Silicon High-Resolution Computed Tomography (FlashCT) scanner at a resolution of 180 µm (Gebrenegus et al., 2011).

(Aravena et al., 2011), only to name a few. This surge in CT applications is attributable to steadily increasing computational capabilities, significant advancements of synchrotron and benchtop micro-CT technology (Vaz et al., 2011), and easier access to X-ray micro-CT facilities.

While industry and various research groups invested considerable energy and resources improving micro-CT technology and quantitative micro-scale characterization, significant challenges with image segmentation, which affects all succeeding quantitative analyses efforts, persist. Segmentation of CT data is doubtlessly the most critical step before quantitative analyses. It involves conversion of three-dimensional grayscale CT data into a discrete form to distinguish phases of interest. Binarization is the simplest type of segmentation (Fig. 8–1), with only two phases of interest (e.g., voids and the solid matrix).

Due to the lack of multiphase segmentation capabilities, binarization is also widely applied to porous systems with three or more phases. It is common to either scan the same sample at different liquid saturations (e.g., in dry and partially saturated condition) or at different X-ray energies and successively align and subtract binarized data from individual scans to separate multiple phases for geometrical analysis and to visualize the phase distributions (e.g., Wildenschild et al., 2002; Schnaar and Brusseau, 2006). However, this multilevel approach is prone to operator errors because of the many steps involved.

A comprehensive publication search revealed that there are more than 100 different methods for image segmentation documented in the literature. Most of these methods were specifically developed for optical character recognition (OCR) or medical applications and were not previously applied and evaluated for application to X-ray CT data of soils. Following common classification schemes proposed in the literature (e.g., Haralick and Shapiro, 1985; Pal and Pal, 1993; Pham et al., 2000; Sezgin and Sankur, 2004; Wirjadi, 2007) there are several major approaches to image segmentation (Iassonov et al., 2009). In the following, we provide a brief overview about the most widely applied two- and multiphase segmentation principles.

It needs to be clearly emphasized upfront that not all X-ray CT datasets are suitable for segmentation. In particular, for datasets with insufficient resolution, where voxels suffer from partial volume effects, image segmentation commonly fails.

The most basic two-phase segmentation methods consider a single gray-scale threshold derived from a global linear attenuation coefficient or grayscale histogram such as depicted in Fig. 8–1. If the grayscale intensities of individual image pixels (two-dimensional) or volume voxels (three-dimensional) are below or equal to the threshold's grayscale intensity they are assigned to void space. Vice versa, if the pixel or voxel grayscale intensities are above the threshold, they are assigned to the solid phase (Fig. 8–1). There are several approaches to derive the segmentation threshold from the global grayscale histogram, which include methods based on: (i) similarity of attributes between the segmented and the initial grayscale datasets, (ii) higher-order probability distributions and spatial correlations between image pixels, (iii) analysis of the histogram shape, and (iv) correlation between back- and foreground pixel entropies. For a more detailed discussion of global segmentation methods, interested readers are referred to comprehensive reviews by Pal and Pal (1993) and Sezgin and Sankur (2004). Manual selection of a global segmentation threshold is another widely applied alternative. However, this approach suffers from operator bias (Baveye et al., 2010) and may be inconsistent, with emanating segmentation inaccuracies difficult to predict and correct for.

Another popular group of methods still considers grayscale intensities, but instead of using a single global grayscale threshold, they use a variable thresholding surface calculated from local image characteristics. These locally adaptive methods are essentially an enhancement of global thresholding, utilizing local spatial information such as intensity gradients (Yanowitz and Bruckstein, 1989; Sheppard et al., 2004) or a two-point local covariance function of the image (Mardia and Hainsworth, 1988; Oh and Lindquist, 1999).

An alternative to abovementioned techniques assumes that all pixels or voxels that are part of a specific image object are similar and connected with each other. A general iterative algorithm is used to search the neighborhood of a selected region of interest (ROI) and to apply a specified criterion, usually the similarity between the grayscale level of a considered voxel to the mean grayscale level of the region, to select the voxels from the neighborhood to be added to the ROI (Gonzalez and Wood, 2002). This approach is termed region growing thresholding. Various selection criteria for addition and removal of voxels from the ROI (Wirjadi, 2007) and consideration of local properties such as object boundaries are proposed in the literature (Vlidis and Liow, 1990). An apparent drawback of region growing is the requirement for selection of initial seed regions either manually by an operator or by means of a global thresholding method.

Level set methods (Sethian, 1999) that fall within the larger category of deformable surface methods (e.g., Mumford and Shah, 1989; Cohen et al., 1992; McInerney and Terzopoulos, 1996; Caselles et al., 1997; Montagnat and Delingette, 2001) have been successfully applied to micro-CT data of porous materials. As with region growing thresholding, definition of initial seed regions is required. An implicit iterative approach is applied to achieve optimum segmentation (Mumford and Shah, 1989; Osher and Paragios, 2003; Weeratunga and Kamath, 2004).

The connectivity of image voxels within a cluster (phase) is considered together with their grayscale intensity values in probabilistic fuzzy clustering segmentation methods (Pedrycz, 1996; Rajapakse et al., 1997; Cheng and Chen, 1999; Pham, 2001; Leski, 2003). As soon as the iterative solution converges to a stable configuration of voxel clusters, image segmentation is achieved by thresholding the fuzzy connectedness values (Baraldi and Blonda, 1999; Nyúl et al., 2002). This method yields robust results for noisy and structurally heterogeneous CT data of porous materials (Huang and Wang, 1995; Nefti and Oussalah, 2004; Wirjadi, 2007).

Treating image properties as random variables and deriving a probabilistic model based on Bayesian decision theory (Duda and Hart, 1973) provides the foundation for Bayesian image segmentation. A widely used model for images based on probability of classes or gray levels is referred to as the random field model that was later extended to the Markov random rield (MRF) model (Moussouris, 1974; Kindermann and Snell, 1980; Chellappa and Jain, 1993). The motivation for the application of a stochastic framework is based on the assumption that the variation and interactions between image attributes can be described by probability distributions. Bayesian MRF segmentation is essentially the assignment of labels to image voxels to satisfy an objective function. The labeling problem is commonly solved with deterministic or heuristic combinatorial optimization. The MRF model is inherently powerful for image segmentation as it can generally handle any number of voxel classes [e.g., representing different pore-filling fluids and/or different solid grain materials (Kulkarni et al., 2012)]. However, it must be initialized with reasonable statistics (i.e., mean and variance) for each voxel class (e.g., Berthod et al., 1996, Held et al., 1997; Szirányi et al., 2000; Li, 2001).

Capitalizing on beneficial features of the methods outlined above, several research groups developed hybrid approaches for image segmentation, where, for example, a simple and computationally efficient segmentation method is used to eliminate image voxels related to highly X-ray attenuating materials or to correct for intensity variations within a grayscale CT dataset (Iassonov and Tuller, 2010) before the application of a more complex method for the final segmentation step. Furthermore, comparison and matching results of several methods applied to the same CT dataset (Melgani, 2006) might be a viable alternative. An additional example for a hybrid algorithm is a combination of level set methods with clustering (Li et al., 2008) or Bayesian analysis (Zhou et al., 2008).

The crucial impact of segmentation on subsequent morphometric analysis and fluid dynamics modeling is demonstrated in an interesting study conducted by Baveye et al. (2010), where three test images of porous materials were sent out to thirteen experts, who were asked to apply the segmentation strategies they routinely use for their research (including any pre-segmentation image enhancement methods). In addition, they applied a number of automated segmentation algorithms to portions of the test images to explore the variability in segmentation outcomes. Results of this study revealed vastly different results between all applied manual (observer dependent) and automated methods pointing to the need for new segmentation strategies. The variation of segmentation results is further illustrated in Fig. 8–2, where renderings of a 7.5-cm diameter macroporous loam soil column are shown. The sample was scanned with an industrial HYTEC FlashCT system at 380 keV (San José Martínez et al., 2010). A cone beam reconstruction algorithm was applied to convert the X-ray CT radiographs to 8-bit, grayscale three-dimensional volumes composed of 820 by 820 by 1480 voxels (resolution = 110 μm). Several global and locally adaptive segmentation methods were then applied to convert the grayscale data to binary data, which were rendered to visualize the macropore structure. From visual inspection alone, it is evident that the applied segmentation methods yielded vastly different results significantly impacting all subsequent analyses and modeling steps. This emphasizes the need

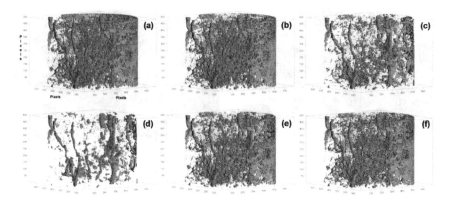

Fig. 8–2. Renderings of a macroporous loam soil column after segmentation with commonly applied methods proposed by Zack et al. (1977) (a), Tsai (1995) (b), Kittler and Illingworth (1986) (c), Otsu (1979) (d), Yen et al. (1995) (e), and Oh and Lindquist (1999) (f).

to further develop and evaluate segmentation algorithms for application in soil and porous media research.

The chapter is organized as follows. The data preprocessing and filtering section discusses the application of filtering and data preprocessing techniques. In the commonly applied segmentation methods in soil science research section, we provide a comparison of commonly applied segmentation algorithms and report about recent advancements. The chapter concludes with a brief summary and outlook on future research.

Data Preprocessing and Filtering

Reconstructed X-ray CT image data commonly contain flaws due to scattered X-rays or poorly centered (aligned) samples, high-frequency noise, beam hardening in case polychromatic X-ray sources are used, ring artifacts due to local scintillator or detector defects, or projection under-sampling (i.e., insufficient exposure time for statistically sound results) (Duliu, 1999; Stock, 1999; Ikeda et al., 2000; Van Geet et al., 2000; Ketcham and Carlson, 2001; Wildenschild et al., 2002). These flaws interfere with segmentation as they may lead to: (i) misinterpretation of attenuation values of a single phase in different locations of the reconstructed CT dataset, or (ii) masking of essential features of the investigated porous material. Furthermore, the radiograph reconstruction method applied to data emanating from X-ray micro-CT scanners with wide cone beam geometry commonly produces distortions that complicate the segmentation stage (Ketcham and Carlson, 2001). Besides the abovementioned problems, there are other potential issues such as image distortions that likely occur when lenses are used for magnification. Note that there are two basic principles used for magnification. Magnification is either achieved optically with magnification lenses or geometrically by moving X-ray source, sample, and detector relative to each other. Noise is commonly modeled as a random process characterized by its probability density function (PDF) (e.g., Gaussian, uniform, or salt and pepper noise).

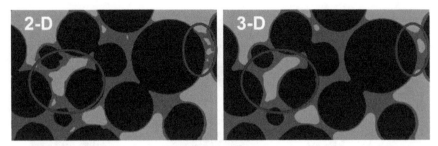

Fig. 8–3. Comparison of partially saturated glass bead sample cross-sections segmented "slice-by-slice" in two-dimensional mode (left) and in true three-dimensional mode (right). The sample was scanned with a high-resolution SkyScan 1172 benchtop scanner (SkyScan, Belgium).

Before proceeding with detailed descriptions of data preprocessing and filtering techniques, it is imperative to mention that all processing steps including segmentation should be performed in three-dimensional mode to achieve realistic representations of phase distributions in porous media. Although X-ray CT provides three-dimensional volume data, two-dimensional "image-by-image" processing is still a widely applied approach in porous media research (Iassonov et al., 2009). Probable motivations for the application of two-dimensional processing are limited computational resources or attempts to circumvent problems arising from vertical image intensity variations. It should be obvious that two-dimensional processing has several drawbacks as important information of voxels neighboring in a three-dimensional space is ignored (Elliot and Heck, 2007). Two-dimensional processing might also lead to a directional bias in subsequent morphometric characterization of observed phase distributions (Iassonov and Tuller, 2010). An instance illustrating problems with two-dimensional processing is shown in Fig. 8–3 where two-dimensional and three-dimensional segmentation results obtained for a glass-bead sample with a recently developed multiphase segmentation method based on a Bayesian Markov random field framework (Kulkarni et al., 2012) are compared. It is clearly visible that if data are processed in two-dimensional mode, phase boundaries appear rugged with unrealistic curvatures. These issues are not apparent in the sample cross-section obtained with true three-dimensional processing.

In the following sections, we describe preprocessing and denoising (filtering) techniques suitable for X-ray CT data considering true three-dimensional processing only.

Intensity Rescaling

Grayscale X-ray micro-CT data emanating from various sources often appear very dark on the computer screen (Fig. 8–4a). This means that the global grayscale distribution is rather narrow occupying only a small portion of an 8-bit (0 to 255 gray levels) or 16-bit (0 to 65,535 gray levels) grayscale range. This can be significantly improved without loss of physical information via intensity rescaling, where the narrow distribution is linearly mapped (rescaled) over the entire 8-bit or 16-bit grayscale range. This yields brighter images as illustrated for a partially saturated glass bead sample in Fig. 8–4b and allows for a better informed manual

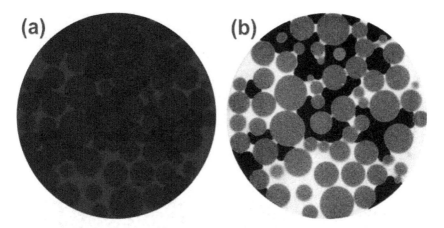

Fig. 8–4. Intensity rescaling applied to an 8-bit grayscale image obtained with synchrotron microtomography (a). The same image after intensity rescaling (b).

operator input if required. Furthermore, intensity rescaling can be useful if integer-based filtering techniques are applied as more gray levels are available for data interpolation. The dataset shown in Fig. 8–4a emanated from experiments conducted with GSECARS bending magnet beam-line at Argonne National Laboratory (Porter et al., 2010). The sample consisted of a packing of glass beads of three different sizes partially saturated (approximately 50%) with an 11% (by weight) potassium iodide solution (three phases). The inverted datasets consisted of 650 by 650 by 427 voxels.

Numerical Beam-Hardening Correction

Beam hardening, a commonly experienced issue with industrial and benchtop X-ray CT systems with polychromatic X-ray sources, can be alleviated already during the scanning and acquisition process by means of metal filter plates placed between the sample and the X-ray source or between the sample and the detector (Vaz et al., 2011). Wedge calibration (Coleman and Sinclair, 1985; Rebuffel and Dinten, 2007) is another method that can be applied for reducing beam-hardening artifacts; however, it is rarely used in standard acquisition protocols (Ketcham and Carlson, 2001).

Beam hardening may be also corrected numerically after data acquisition with a low-pass (smoothing) filter calculating average pixel intensities on a scale larger than the typical pore size. The resulting blurred image is then used as a base map to account for large-scale intensity variations caused by beam hardening. However, this approach is prone to substantial errors because the average pixel intensity in any given part of the image is dependent on the structure of the sample, in particular its local porosity. While it is possible to partially alleviate the problem by using conditional filters, simple techniques such as interpolating maximum local pixel intensity values to obtain a base map are typically not effective due to noise and sample heterogeneities (e.g., Halverson et al., 2005). Iassonov and Tuller (2010) introduced a simple three-dimensional approach to numerically correct for beam-hardening artifacts using sequential segmentation. They

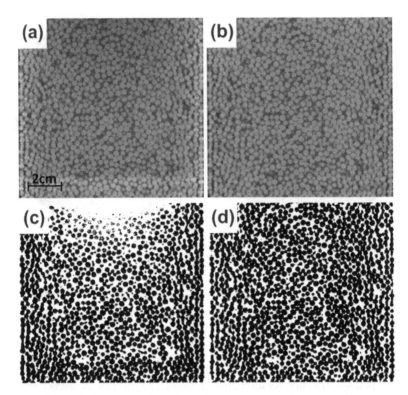

Fig. 8–5. (a) Vertical cross-sections through a micro-computed tomography (CT) volume of packed uniform glass beads with clearly visible density variations due to beam hardening. (b) The same micro-CT volume after sequential segmentation (Iassonov and Tuller, 2010). (c), (d) Associated segmented cross-sections obtained with Otsu's (1979) global three-dimensional segmentation method.

applied image segmentation in conjunction with calculating average intensities for image correction, thereby only selecting voxels that correspond to the phase with large attenuation coefficient and not the void space. As opposed to selecting only a few voxels with locally high intensity, this approach selects a sufficiently large number of voxels to be insensitive to high-frequency noise. One might conclude that there is an apparent flaw associated with this approach, and that the presence of image artifacts such as beam hardening will likely yield poor segmentation results that translate to the subsequent image correction. Fortunately, this problem can be easily solved by iterative application of the thresholding and correction steps, improving image quality with each subsequent iteration. Presented results clearly show that the proposed method leads to significant improvement of grayscale image quality as well as to improvement of the final segmented binary images (Fig. 8–5).

The method can be employed to significantly reduce errors due to beam hardening and similar artifacts typical for industrial CT scanners with polychromatic X-ray sources. Although insensitive to high-frequency noise (typical for CT images), beam-hardening correction does not reduce noise. Therefore, final segmentation may require further image processing for noise reduction, or application of a noise-tolerant segmentation method (e.g., Oh and Lindquist, 1999). These are discussed in the following sections. Besides numerical adjustments after CT-data reconstruction, there are several pre-reconstruction computational methods that are considered superior to most post-reconstruction techniques (e.g., Hsieh et al., 2000). However, it should be mentioned that beam hardening still remains a serious problem with no complete solution.

Neighborhood Statistical Filters

X-ray CT detectors capture (count) the number of incident photons and generate a proportional electrical output signal, with noise caused by capturing an inadequate number of photons within the acquisition timeframe. Because of this, it is a reasonable approach to consider the gray values of neighboring voxels when determining the gray value of a particular voxel to average out the intensity deficiency of the sensor pixel corresponding to that voxel. Such an operation is referred to as a "kernel" operation. A noisy, three-dimensional, X-ray CT dataset can be mathematically defined as (Russ, 2010):

$$I'(x,y,z) = I(x,y,z) + n(x,y,z) \tag{1}$$

where x, y, and z are the coordinates of a voxel, I' is the observed image, n is the noise, and I is the original image. A general kernel operation is to multiply the voxels within a neighborhood with a set of weights, summing the resultant values, and dividing the resultant sum by the sum of weights:

$$\hat{I}(x,y,z) = \frac{\sum_{i=-m,j=-n,k=-p}^{i=+m,j=+n,k=+p} W(i,j,k) \times I'(x+i,y+j,z+k)}{\sum_{i=-m,j=-n,k=-p}^{i=+m,j=+n,k=+p} W(i,j,k)} \tag{2}$$

with $W(i, j, k)$ defined as

$$
\begin{aligned}
-m &\leq i \leq m \\
-n &\leq j \leq n \\
-p &\leq k \leq p
\end{aligned}
\tag{3}
$$

where 2^*m+1, 2^*n+1, and 2^*p+1 are width, depth, and height of the neighborhood window and $\hat{I}$ is the estimate of the original image I from the captured noisy image I'. A mean filter applies a uniform weight to the neighboring voxels (i.e., $W(i, j, k) = 1$), causing a uniform smoothing of the entire dataset. This can cause unwanted blurring of high frequency regions or phase boundaries, where voxel intensities significantly vary over short distances. This, of course, also affects the subsequent segmentation step and might lead to significant shifts of phase boundaries and misclassification of the phase distribution.

Other kernel filters include median, mode, minimum, maximum, and Gaussian filters, where individual voxels are replaced with the median, mode,

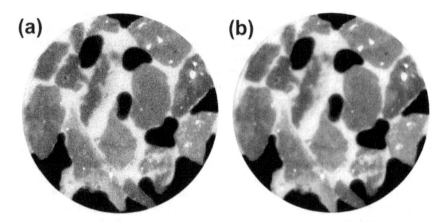

Fig. 8–6. Application of the median filter to an 8-bit, three-dimensional grayscale data-set of crushed volcanic tuff (sample cross-sections are shown). (a) Original image after intensity rescaling. (b) The same image after application of the median filter.

minimum or maximum gray value of the voxel neighborhood or a Gaussian distribution is used to define kernel weights (Russ, 2010). Median filters are nonlinear and widely applied to remove shot or impulse noise. In median filtering, the voxels within a moving neighborhood window are ranked according to their grayscale intensities with the median intensity value assigned to the center voxel (pixel in 2-D). Although computationally more demanding than conventional smoothing filters, median filtering is less sensitive to outlier values, hence they are more efficiently removed than for example with mean filtering. Furthermore, because the intensity value assigned to the center voxel is identical to the value of one of the neighboring voxels, no new "unrealistic" intensity values are created near object boundaries. This leads to better preservation of object edges. Fig. 8–6 depicts the application of the median filter to an 8-bit, three-dimensional grayscale dataset of crushed volcanic tuff.

The mean and Gaussian filters are low pass filters, which retain low frequency regions while reducing the high frequency noise. This means that the entire dataset is smoothed, which especially affects phase boundaries. Below we discuss filtering techniques based on partial differential equations (PDE) that are capable of smoothing image regions while preserving edges (i.e., phase boundaries).

Anisotropic Diffusion Filter

Gaussian filtering can be also expressed in terms of a convolving observed image I' with a Gaussian kernel, $G(x,y;\sigma)$, to obtain an estimate of original image I:

$$\hat{I}(x,y) = G(x,y;\sigma) * I'(x,y) \qquad [4]$$

where $*$ is the convolution operator. Equation [4] is equivalent to the isotropic heat diffusion equation $u = c\Delta u$ where u is the density of diffusing material and c is a constant diffusion coefficient (Perona and Malik, 1990). Considering the general form of the anisotropic diffusion equation:

$$\frac{\partial u(x,t)}{\partial t} = \nabla \cdot [c(u,x)\nabla u(x,t)]$$ [5]

where u is the density of diffusing material at location x and time t, and $c(u,x)$ is the diffusion coefficient for density u at location x. ∇ is a differential operator with respect to space variables ($\nabla = \frac{\partial}{\partial x} + \frac{\partial}{\partial y} + \frac{\partial}{\partial z}$ in three dimensions). When Eq. [5] is applied to a grayscale X-ray CT dataset, the density of diffusing material u is equivalent to the gray scale intensity matrix of the image volume I'. Perona and Malik (1990) define the diffusion coefficient in Eq. [5] as a nonlinear function dependent on the gradient of the image $\nabla I' = \frac{\partial I'}{\partial x} + \frac{\partial I'}{\partial y} + \frac{\partial I'}{\partial z}$, which preserves edges while smoothing homogenous regions:

$$\frac{\partial I'(x,t)}{\partial t} = \nabla \cdot [g(|\nabla I'|)\nabla I'(x,t)], \qquad I'(0) = I'$$ [6]

The function g controls the rate of diffusion and is chosen such that when the gradient $\nabla I'$ is high (i.e., at the edges) the diffusion is low, thus preserving phase boundaries (edges). If $\nabla I'$ is low (i.e., within homogenous regions) the diffusion is high, which leads to smoothing of homogeneous regions. Two commonly used equations for the nonlinear diffusion coefficient are:

$$g(|\nabla I'|) = e^{-\left(\frac{|\nabla I'|}{K}\right)},$$ [7]

and

$$g(|\nabla I'|) = \frac{1}{1 + \left(\frac{|\nabla I'|}{K}\right)^2},$$ [8]

where K is the conduction coefficient that controls conduction as a function of gradient. If K is low, small intensity gradients are able to block conduction and hence diffusion across edges. A large value reduces the influence of intensity gradients on conduction.

The anisotropic diffusion filter can be applied successively for a desired number of iterations. The application of the anisotropic diffusion filter for a single iteration to the grayscale dataset of partially saturated crushed volcanic tuff is illustrated in Fig. 8–7. The dataset originated from experiments conducted with the GSECARS bending magnet beam-line at Argonne National Laboratory. The sample was composed of crushed volcanic tuff that mainly consisted of quartz with minor amounts of feldspar, albite, and volcanic glasses (Wildenschild et al., 2004) and was partially saturated with 11% potassium iodide solution.

Although there is only one parameter, K, in the anisotropic diffusion filter, its value is chosen either experimentally or as a function of image noise. When compared to neighborhood statistical filters, the anisotropic diffusion filter performs better in terms of edge preservation. In fact, with the right choice of nonlinear diffusion coefficient, the filter can be modified to detect edges.

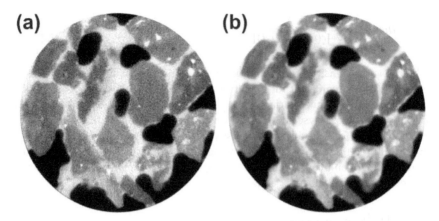

Fig. 8–7. Application of the anisotropic diffusion filter to an 8-bit, three-dimensional grayscale dataset of crushed volcanic tuff (sample cross-sections are shown). Original image after intensity rescaling (a). The same image after application of the diffusion filter (b).

Shock Filter

Similar to the anisotropic diffusion filter the shock filter is based on a PDE, smoothing of homogeneous regions while preserving discontinuities at phase boundaries (edges). The shock filter was first applied to images by Osher and Rudin (1990) with several modified versions proposed since then. The basic filter is defined as:

$$\hat{I} = -sign(\Delta I') | \nabla I' | \qquad [9]$$

where I' is the observed image with noise and $\hat{I}$ is the image after filtering, $sign$ is the signum function, Δ is the Laplacian operator (or second partial derivative, $\Delta = \nabla^2 = \frac{\partial^2}{\partial x^2} + \frac{\partial^2}{\partial y^2} + \frac{\partial^2}{\partial z^2}$), and ∇ denotes the partial derivative. This hyperbolic partial differential equation approximates the deconvolution of Eq. [9] with any general noise in the place of the Gaussian kernel. A drawback of the shock filter in basic form (Eq. [9]) is its sensitivity to additive white Gaussian noise (AWGN). Another major drawback is that in the presence of AWGN, the edge detecting Laplacian term fails due to amplification of noise within anisotropic regions (edges and contours), thus reducing its effectiveness in preserving edges (Ludusan et al., 2010). A potential avenue to address the AWGN issue is to combine an image-driven smoothing method with a noise removal method. Alvarez and Mazorra (1994) proposed a combination of diffusion filter and modified shock filter obtained by convolving the Laplacian by a Gaussian function, basically low pass filtering the Laplacian to smooth the noise:

$$\hat{I} = c\Delta I' - sign(G_\sigma * \Delta I') | \nabla I' | \qquad [10]$$

where G_σ is a Gaussian function with standard deviation σ, and c is a positive constant. The shock filter removes noise from the isotropic regions and the diffusion filter term will overcome the inflection points (the amplification of noise near edges due to the Laplacian term). To address the edge detection limitation and to improve the edge filtering capacity, Gilboa et al. (2002) proposed to consider both magnitude and direction of the image.

A computationally more efficient approach eluding convolution with a Gaussian function was proposed by Gilboa et al. (2002), replacing Eq. [10] with:

$$\hat{I} = -\frac{2}{\pi}\arctan\left(a\,\mathrm{Im}\left(\frac{I'}{\theta}\right)\right)|\nabla I'| + c\Delta I' + \tilde{c}\Delta I' \qquad [11]$$

where a is a parameter that controls the sharpness of the slope near zero, Im() denotes imaginary value, $\theta \in \left(-\frac{\pi}{2}, \frac{\pi}{2}\right]$, c is a complex scalar, and $\tilde{c}$ is a real scalar. Similar to anisotropic diffusion filter, the shock filter can be applied successively.

Inverse Scale Space Filter

Another approach of PDE-based filtering is centered on minimizing the variation within an image by means of a cost function (Rudin et al., 1992). Scherzer and Groetsch (2001) showed that regularization methods (for solving inverse problems) can be used to solve scale space problems by combining diffusion filtering with regularized filtering. They termed this approach inverse scale space filtering. Burger et al. (2005) formulated a simple stopping criterion and extended it for nonlinear scale spaces. The best estimate $\hat{I}$ for the original image I is obtained iteratively from its degraded image I' by a variational method used to minimize the cost function:

$$\underset{\hat{I}}{\arg\min}\left\{ J(\hat{I}) + \lambda H(I', \hat{I}) \right\} \qquad [12]$$

where J is a regularization term, λ a regularization parameter, and H is the fidelity term.

The argument on the right hand side of Eq. [12] is minimum when the following equations are satisfied:

$$\frac{\partial \hat{I}}{\partial t} = -p(\hat{I}) + \lambda q(I' + K, \hat{I}) \qquad [13]$$

$$\frac{\partial K}{\partial t} = \alpha q(I', \hat{I}) \qquad [14]$$

where α is a time scaling constant of the decomposed noise solution K.

Filter behavior depends on the definition of the regularization function $p(\cdot)$ and the fidelity function $q(\cdot, \cdot)$, which are shown by Kaestner et al. (2008) to reduce to:

$$\frac{\partial \hat{I}}{\partial t} = \nabla \cdot \left(\frac{\nabla \hat{I}}{|\nabla \hat{I}|} \right) + \lambda (I' - \hat{I} + K) \tag{15}$$

$$\frac{\partial K}{\partial t} = \alpha (I' - \hat{I}) \tag{16}$$

Unsharp Mask Filter

Previously discussed PDE filtering techniques remove noise by smoothing homogenous regions while preserving phase boundaries (e.g., PDE filters). In case of statistical and Gaussian filters, phase boundaries are also smoothened to a certain extent, leading to a gradual change in voxel gray values in these transitional regions. This creates challenges with the segmentation step, leading to misclassification of phase boundaries, especially in images with multiple (more than two) phases. To overcome these issues and to ensure abrupt changes in voxel values at phase boundaries, an unsharp mask filter can be applied. In contrast to its name, the unsharp mask filter enhances high spatial frequency regions locally by subtracting a smoothed or unsharp version of an image from its original. Another image-sharpening approach is based on convolving an image with a Laplacian filter used to detect edges and high frequency regions. Similar to the kernels of smoothing filters (Neighborhood Statistical Filters), the kernel of a Laplacian filter is symmetric but does not have positive weights in the neighborhood. A simple 3 by 3 Laplacian kernel is given as:

$$W = \begin{bmatrix} -1 & -1 & -1 \\ -1 & +8 & -1 \\ -1 & -1 & -1 \end{bmatrix} \tag{17}$$

Convolving an image with such kernel is essentially subtracting a scaled voxel value from its neighbor voxel values. As a result of convolution, a homogeneous region with the same voxel values results in zero, while regions around line features or edges result in positive or negative values, which are then added with 128 (assuming 8-bit images) to obtain positive values only. A sharpened image is obtained by first convolving the image with a Laplacian kernel and then adding the resultant high frequency version of the image with its original, truncating the minimum and maximum values to 0 and 255. The application of the unsharp filter to a grayscale dataset of partially saturated crushed volcanic tuff is illustrated in Fig. 8–8. The name Laplacian is due to the fact that the convolution operation using the kernel is an approximation to the linear second derivative of voxel values in x, y, and z directions (Russ, 2010):

$$\nabla^2 I' = \frac{\partial^2 I'}{\partial x^2} + \frac{\partial^2 I'}{\partial y^2} + \frac{\partial^2 I'}{\partial z^2} \tag{18}$$

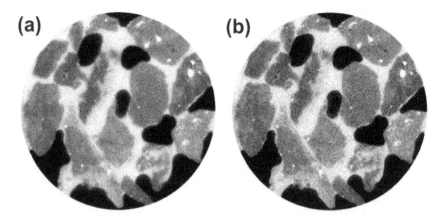

Fig. 8–8. Application of the unsharp filter to an 8-bit, three-dimensional grayscale dataset of crushed volcanic tuff (sample cross-sections are shown). (a) Original image after intensity rescaling. (b) The same image after application of the unsharp filter.

Application of the abovementioned filtering techniques requires special care as they might detrimentally affect the quality of final segmentation results by not only removing the noise contained in the CT dataset, but also true features of the porous material sample under investigation (e.g., blurring of gray scale data). Edge-sharpening filters might also yield undesirable artifacts within the CT grayscale dataset.

Commonly Applied Segmentation Methods in Soil Science Research

Numerous recent studies aimed at quantitative characterization of phase distributions within porous materials based on X-ray CT observations applied basic global segmentation methods, determined the grayscale intensity of the segmentation threshold by matching physically measured material porosities with CT data-derived porosities, or simply determined the segmentation threshold manually. A small minority applied more sophisticated methods, and despite significant enhancements of computational capabilities, it is obvious that two-dimensional "image-by-image" processing is still the prevailing approach to image segmentation (see Table 1 in Iassonov et al., 2009). As discussed in the data preprocessing and filtering section, to obtain realistic segmentation results, it is crucial that all pre- and post-processing operations as well as the segmentation step are performed in "true" three-dimensional mode. This prevents directional bias and assures that all information contained in the three-dimensional voxel neighborhood is considered. Another current limitation is the lack of direct multiphase segmentation capabilities. Most readily available codes are limited to two phases (binarization). This requires porous material samples composed of three or more phases to be scanned at different liquid saturations (e.g., in dry and partially saturated states) or at different X-ray energies to successively align and subtract binarized data from individual scans for phase separation (e.g., Wildenschild et al., 2002). However, this multilevel approach is prone to operator errors because of the many steps involved. Only a very few methods (e.g., Pham, 2001;

Kulkarni et al., 2012) are amenable to direct multiphase segmentation. Below we describe and discuss a selection of established techniques for two- and multiphase segmentation of porous geomaterials.

Global Segmentation Methods

Global segmentation methods derive a single gray level threshold from the grayscale histogram that embodies the distribution of voxels with particular gray level intensities (Fig. 8–1). This sole threshold is applied to the entire CT dataset. Within this group that is directly applicable to two-phase segmentation only, several subcategories can be identified.

The first category contains methods that analyze the shape of the grayscale histogram with simple geometrical techniques such as triangulation (Zack et al., 1977; Rosin, 2001). Analysis of histogram peaks and depressions and curvatures was applied by Tsai (1995) to derive the optimum gray intensity threshold from three-dimensional grayscale histograms.

Clustering techniques are utilized in a second category of global methods to distinguish background and foreground voxels associated with the solid and fluid phases (liquid or vapor) of the investigated porous medium by approximating the grayscale histogram with a combination of two statistical distributions. The approach developed by Otsu (1979), one of the most commonly applied segmentation methods, is based on minimizing the variances within each considered voxel class, which is the same as maximizing the variance between the means of the two voxel clusters (Sund and Eilertsen, 2003). Two Gaussian distributions were fitted to the grayscale histogram and minimum error thresholding was applied by Kittler and Illingworth (1986). An iterative method was applied by Ridler and Calvard (1978) to determine the segmentation threshold as the average (midpoint) between the means of the two voxel clusters. The later method is termed k-means, or iso-data clustering.

Another subcategory of global segmentation assumes image fore- and background voxels as two differing signal sources and determines the segmentation threshold from signal entropies (Shannon and Weaver, 1948; Pal and Pal, 1989; Brink and Pendock, 1996). Kapur et al. (1985) maximized the sum of the foreground and background signal entropies, while Yen et al. (1995) found the optimum grayscale intensity for the global segmentation threshold value from maximizing the entropic correlation between the fore- and background voxel classes.

An additional subcategory contains methods that explore spatial correlations between image voxels and apply higher-order probability distributions to define the optimum segmentation threshold. A two-dimensional histogram that considers the average grayscale intensity of each individual voxel (derived from the intimate voxel neighborhood) together with its local intensity was used by Abutaleb (1989) in conjunction with maximum entropy considerations. This approach was later advanced by Brink (1992), utilizing an optimized solution scheme developed by Chen et al. (1994). Voxel co-occurrence information was used by Pal and Pal (1989) to define the global grayscale segmentation threshold that maximizes the sum of Shannon's entropies of the within-class co-occurrences of fore- and background classes (Chang et al., 2006).

All global segmentation methods discussed above are amenable for two-phase segmentation only and were primarily used in two-dimensional "image-by-image" processing mode. Iassonov et al. (2009) implemented most of the above methods for "true" three-dimensional processing and compared segmentation results for glass bead, macroporous loam soil, and cracked sand-bentonite samples. They found that only the clustering methods proposed by Otsu (1979) and Ridler and Calvard (1978) yielded consistent segmentation outcomes for the majority of tested samples, with micro-CT data-derived medium porosities matching physically measured porosities reasonably well (see Table 2 in Iassonov et al., 2009). All other evaluated global segmentation approaches led to unsatisfactory results for the majority of tested porous material samples, in particular for CT grayscale data of glass beads and macroporous soil with obvious partial volume effects and low contrast between voxel classes. For a detailed discussion of global segmentation results, interested readers are referred to Iassonov et al. (2009).

Locally Adaptive Methods for Image Segmentation

In contrast to applying a single grayscale threshold to the entire CT dataset, locally adaptive methods assign each individual voxel to a particular class (i.e., porous medium phase) based on information contained in the local voxel neighborhood. This generally yields enhanced segmentation outcomes as certain artifacts of the grayscale data such as high-frequency noise are accounted for. Methods falling within this category are computationally more demanding and require significantly larger memory. In the following section, we discuss established and more recently developed methods.

Indicator Kriging

The indicator kriging technique developed by Oh and Lindquist (1999) is one of the most popular techniques applied in soil and porous media research today. It is a two-step method that first delineates two global segmentation thresholds (T_0 and T_1) either by manual selection or by means of a simple global segmentation method (e.g., Ridler and Calvard, 1978). All voxels with a grayscale value lower than the grayscale value of threshold T_0 and larger than the grayscale value of threshold T_1 are automatically allocated either to the background or the foreground voxel classes. After this initial separation step, all the remaining voxels with grayscale intensity values between T_0 and T_1 are allocated either to the foreground or to the background class independently, estimating the maximum likelihood for each voxel from stationary spatial covariance of the CT grayscale image (Oh and Lindquist, 1999). An evaluation of the indicator kriging method by Iassonov et al. (2009) showed that it provides very consistent results based on comparison of micro-CT data-derived and physically measured material porosities. As shown in Fig. 8–9, indicator kriging is relatively tolerant to high-frequency noise. A distinct weakness of the Oh and Lindquist (1999) method is that the final segmentation results seem to be significantly biased by the first stage, where two global thresholds are defined manually or by means of a global segmentation algorithm. While application of the Ridler and Calvard (1978) method for initial global thresholding produced reasonable results for most investigated samples, other tested initial thresholding methods (e.g., Otsu, 1979) did not perform nearly as well. As suggested by Oh and Lindquist (1999) and Prodanović et al. (2006), a careful manual adjustment of the initial global thresholds might potentially result in improvement of segmentation quality. Wang et al. (2011) applied the indicator kriging method to a number of simulated two-dimensional and three-dimensional CT datasets and

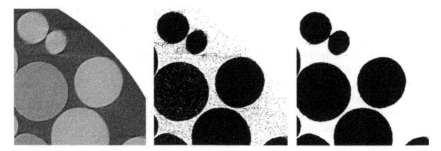

Fig. 8–9. Segmentation of noisy X-ray micro-computed tomography data. Corresponding two-dimensional slices from three-dimensional processing are shown. From left to right: original gray scale image, binary image obtained with manual thresholding, and binary image obtained with indicator kriging (from Iassonov et al., 2009).

concluded that this method yields reasonable results as long as the sample gray-scale histograms were distinctively bimodal.

The Probabilistic Fuzzy C-Means Clustering Method

The probability for each image voxel to belong to a specific voxel cluster is determined based on local spatial information calculated for a small voxel neighborhood by means of indicator kriging and applied in conjunction with its grayscale intensity in a segmentation method proposed by Pham (2001). This so-called probabilistic c-means clustering method is fairly similar to indicator kriging (Oh and Lindquist, 1999), with the exception that the final result is obtained by iteratively updating the fuzzy cluster membership instead of a single-pass evaluation that relies on a priori thresholding information. A drawback of the c-means clustering method is the requirement to provide initial gray intensity values for each cluster center, which might yield operator-biased results. A comparison provided by Iassonov et al. (2009) reveals that probabilistic fuzzy c-means clustering yielded similar results to indicator kriging, but was less sensitive to initialization. This makes this method more convenient for semi-automated segmentation when the optimal threshold is not readily available. However, because of the iterative nature of the algorithm, it is computationally more demanding. Another limitation is associated with the underlying basic c-means algorithm that does not take into account spatial dimensions of each cluster. This might result in mis-classification when cluster centers are not equidistant. For example, while it was possible to segment pore space and the solid phase in macroporous soil images (Iassonov et al., 2009), probabilistic fuzzy c-means clustering was not able to accurately segment dense aggregates of the solid phase at the same time (Fig. 8–10). A considerable advantage of probabilistic fuzzy c-means clustering is that it is amenable for direct segmentation of multiphase systems eliminating the need for wet/dry or dual energy scans, data alignment and subtraction.

The Edge Detection Method

A surface fitting technique applied in conjunction with edge detection was proposed for image segmentation by Yanowitz and Bruckstein (1989). An image edge map is generated by first computing the gradient magnitude of the smoothed data,

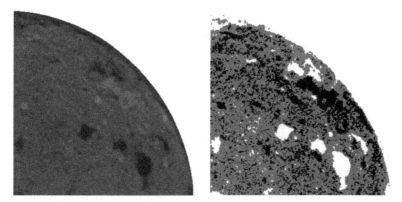

Fig. 8–10. Gray scale X-ray computed tomography image of macroporous loam soil (left). Three-phase segmentation with probabilistic fuzzy c-means clustering (right) (adapted from Iassonov et al., 2009).

which is then truncated and thinned. The grayscale intensities of the edge voxels are assumed as the optimum local grayscale thresholds and are interpolated to the remaining image voxels. The solution of the Laplace equation that matches intensity values of the edge voxels is applied to establish a thresholding surface, which in turn is applied for segmentation of the original smoothed grayscale CT data. Iassonov et al. (2009) implemented and evaluated the edge detection approach using Gaussian smoothing for image filtering, Otsu's (1979) thresholding method as well as the Canny edge detector (Canny, 1986) for thinning. They found that edge detection was the only tested segmentation method that was not affected by global image intensity variations (Fig. 8–11), eliminating the need for pre-segmentation numerical intensity correction (Numerical Beam-Hardening Correction) (Iassonov and Tuller, 2010). However, edge detection is highly sensitive to noise, which requires extensive initial smoothing of considered grayscale CT data with potential loss of important image information. Furthermore, the method is applicable to two-phase systems only. Nevertheless, results presented by Iassonov et al. (2009) reveal that edge detection provided reasonable results for porous media with well pronounced histogram peaks.

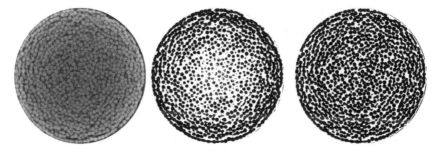

Fig. 8–11. Sample cross-section of grayscale X-ray micro-computed tomography data of 3.5-mm glass beads showing pronounced beam hardening (left). Cross-sections after segmentation with Otsu's (1979) clustering method (middle) and Yanowitz and Bruckstein's (1989) edge detection method (from Iassonov et al., 2009).

The Active Contours Method

Sheppard et al. (2004) proposed a hybrid segmentation approach that combines methods for pre-segmentation CT-data enhancement, converging active contours as well as thresholding. Analogous to several other methods, this hybrid approach attempts to delineate boundaries between various considered voxel classes where intensity gradients are largest. These boundaries are assumed to resemble the interfaces between the phases of the investigated porous material. This approach is similar to region growing with the boundaries progressing from initially defined seed regions with a positive definite speed function (Sethian, 1999). Low- and high-intensity regions are seeded with two contours evolving simultaneously, and the final location of a boundary is automatically positioned with the watershed method where two contours coincide. An evaluation of the active contours method by Iassonov et al. (2009) revealed several challenges. First, this approach is highly sensitive to initial seeding, which frequently leads to a loss or to masking of small features (i.e., small pores or contacts between grains) because of missing seed voxels for these regions. Furthermore, the speed function for the fast marching algorithm that was proposed by Sheppard et al. (2004) consists of three variables that must be determined by trial and error. This potentially leads to biased results and significantly increases the computational demand.

Supervised Bayesian Segmentation Based on a Markov Random Field Framework

Kulkarni et al. (2012) recently implemented and evaluated a three-dimensional multiphase algorithm for supervised Bayesian segmentation of porous media based on MRF following a two-dimensional framework proposed by Berthod et al. (1996). Bayesian MRF segmentation is essentially the assignment of labels to image voxels to satisfy:

$$\hat{L} = \arg\min\left(\sum_{i=1}^{N}\left(\ln\sqrt{2\pi\sigma_L} + \frac{(x_i - \mu_L)^2}{2\sigma_L^2} \right) + \sum_{\{S_i,S_j\}\in C} \beta^* \gamma(l_{S_i}, l_{S_j}) \right) \qquad [19]$$

with

$$\gamma(l_{S_i}, l_{S_j}) = \begin{cases} -1 & if \quad l_{S_i} = l_{S_j} \\ +1 & if \quad l_{S_i} \neq l_{S_j} \end{cases} \qquad [20]$$

where l_{S_i} and l_{S_j} are labels for sites S_i and S_j, respectively, corresponding to voxels x_i and x_j in a three-dimensional space, μ_L is the mean, and σ_L is the standard deviation of labeling L, N is the total number of voxels within the X-ray micro-CT dataset, and β is a constant that represents the homogeneity of individual phases. There are two terms in Eq. [19]. The first term is due to the value at site or gray level for a particular label L, and the second term is due to the interactions with neighboring voxels. Finding a labeling $\hat{L}$ with the constraint of satisfying Eq. [19] is a combinatorial optimization problem. To investigate different optimization approaches, Kulkarni et al. (2012) implemented and evaluated the

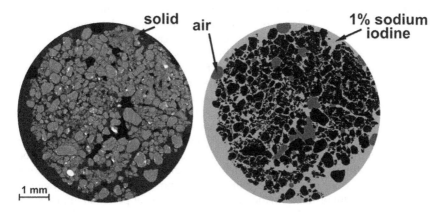

Fig. 8–12. Sample cross-section of grayscale X-ray micro-computed tomography data of a micro-aggregated Oxisol (left). The same cross-section after segmentation with the Kulkarni et al. (2012) Markov random field algorithm (right).

deterministic Iterated Conditional Modes (ICM) algorithm (Besag, 1986) and the heuristic Metropolis (Metropolis et al., 1953; Kirkpatrick et al., 1983) and Modified Metropolis Dynamics (MMD) (Kato et al., 1992) algorithms. All three optimization schemes yielded almost identical segmentation results for all utilized datasets, but ICM was by far the computationally most efficient. Based on comparison of X-ray micro-CT data-derived and physically measured material porosities, Kulkarni et al. (2012) also demonstrated that the homogeneity parameter β (Eq. [19]) can be set to 1, hence eliminated from the MRF equation without impacting segmentation quality. Presented results for glass beads and crushed volcanic tuff showed great potential of the MRF segmentation approach. Besides computational efficiency and stability, the most distinct advantage of the MRF image model is its applicability to multiphase systems as it can consider any number of voxel classes (e.g., representing dissimilar solid grain materials and/or different pore-filling fluids), eliminating the need for wet or dry or dual energy scans, image alignment, and subtraction analysis commonly applied in X-ay micro CT analysis. Segmentation results for a partially saturated micro-aggregated Oxisol are shown in Fig. 8–12. An obvious drawback, as with other locally adaptive methods, is the requirement of initial statistical seeding, which was achieved in the Kulkarni et al. (2012) implementation by manually selecting seed regions from each considered material phase to calculate the initial mean and standard deviation of each phase's grayscale distribution. However, the MRF method has great potential for full automation, which is part of ongoing research. A fully automated version is desired for unsupervised segmentation eliminating any potential bias introduced by a human operator (Baveye et al., 2010).

Summary and Conclusions

This chapter illustrated the crucial importance of image segmentation for subsequent morphometric analysis of porous media and provided a review of global and locally adaptive segmentation techniques applied in soil and porous media research. Various pre-segmentation image enhancement techniques were reviewed and the importance of "true" three-dimensional data processing was demonstrated. Despite recent advances in grayscale image segmentation, primarily in biomedical research, a fully automated (no operator bias), stable, three-dimensional, multiphase algorithm for natural and artificial porous media is still lacking. Although it is doubtful that a "perfect" algorithm applicable to a wide range of porous materials and X-ray CT data sources can be developed, there are some promising advancements with MRF and hybrid algorithms that show potential for unsupervised multiphase segmentation. In addition, there are many image processing and analysis methods such as wavelet-based denoising filters (Figueiredo and Nowak, 2003) or image restoration with Bayesian methods (Awate and Whitaker, 2006) applied in biomedical research that are yet to be evaluated and tested for their applicability to soils and porous materials in general. Baveye et al. (2010) propose further ideas and avenues for potential improvement of the highly complex CT data segmentation issue. It is obvious that substantial efforts need to be devoted in near future to improve current segmentation capabilities. This is especially important in view of the recent surge in X-ray micro-CT applications in soil and porous media research that is attributable to significant advancements of synchrotron and benchtop micro-CT technology, and easier access to X-ray micro-CT scanning facilities. Continually improving X-ray micro-CT detectors that generate ever growing high-resolution datasets (state-of-the-art scanners can output datasets as large as 5×10^{11} voxels) provide another challenge to image segmentation. To handle this growing computational demand, future algorithms should be amenable to parallel computing to, for example, utilize general purpose graphics processing units (GPGPU).

Acknowledgments

The authors gratefully acknowledge support from the U.S. National Science Foundation (NSF) under grant no. EAR-0911242 and from the Arizona Agricultural Experiment Station (AAES). Special thanks go to Dorthe Wildenschild (OSU) and Carlos Vaz (EMBRAPA) for providing some of the datasets used in this chapter.

References

Abutaleb, S.C. 1989. Automatic thresholding of gray-level pictures using two-dimensional entropy. Comput. Vis. Graph. Image Process. 47:22–32. doi:10.1016/0734-189X(89)90051-0

Alvarez, L., and L. Mazorra. 1994. Signal and image restoration using shock filters and anisotropic diffusion. SIAM J. Numer. Anal. 31(2):590–605. doi:10.1137/0731032

Aravena, J.E., M. Berli, T.A. Ghezzehei, and S.W. Tyler. 2011. Effects of root-induced compaction on rhizosphere hydraulic properties- x-ray microtomography imaging and numerical simulations. Environ. Sci. Technol. 45:425–431. doi:10.1021/es102566j

Awate, S.P., and R.T. Whitaker. 2006. Unsupervised, information-theoretic, adaptive image filtering for image restoration. IEEE Trans. Pattern Anal. Mach. Intell. 28(3):364–376. doi:10.1109/TPAMI.2006.64

Baraldi, A., and P. Blonda. 1999. A survey of fuzzy clustering algorithms for pattern recognition–Part II. IEEE Trans. Syst. Man Cybern. 29(6):786–801. doi:10.1109/3477.809033

Baveye, P.C., M. Laba, W. Otten, L. Bouckaert, P. Dello Sterpaio, R.R. Goswami, D. Grinev, A. Houston, Y. Hu, J. Liu, S. Mooney, R. Pajor, S. Sleutel, A. Tarquis, W. Wang, Q. Wei, and M. Sezgin. 2010. Observer-dependent variability of the thresholding step in the quantitative analysis of soil images and X-ray microtomography data. Geoderma 157:51–63. doi:10.1016/j.geoderma.2010.03.015

Berthod, M., Z. Kato, S. Yu, and J. Zerubia. 1996. Bayesian image classification using Markov random fields. Image Vis. Comput. 14:285–295. doi:10.1016/0262-8856(95)01072-6

Besag, J.E. 1986. On the statistical analysis of dirty pictures. J. R. Stat. Soc., B 48(3):259–302.

Brink, A.D. 1992. Thresholding of digital images using two-dimensional entropic thresholding algorithm. Pattern Recognit. 25:803–808. doi:10.1016/0031-3203(92)90034-G

Brink, A.D., and N.E. Pendock. 1996. Minimum cross entropy threshold selection. Pattern Recognit. 29:179–188. doi:10.1016/0031-3203(95)00111-5

Burger, M., S. Osher, J. Xu, and G. Gilboa. 2005. Nonlinear inverse scale space methods for image restoration. Lect. Notes Comput. Sci. 3752:25–36. doi:10.1007/11567646_3

Canny, J. 1986. A computational approach to edge detection. IEEE Trans. Pattern Anal. Mach. Intell. PAMI-8:679–714. doi:10.1109/TPAMI.1986.4767851

Caselles, V., R. Kimmel, G. Sapiro, and C. Sbert. 1997. Minimal surfaces based object segmentation. IEEE Trans. Pattern Anal. Mach. Intell. 19(4):394–398. doi:10.1109/34.588023

Chang, C.-I., Y. Du, J. Wang, S.-M. Guo, and P.D. Thouin. 2006. Survey and comparative analysis of entropy and relative entropy thresholding techniques. IEE Proc. Vis. Image Signal Proess. 153(6):837–850. doi:10.1049/ip-vis:20050032

Chellappa, R., and A. Jain. 1993. Markov random fields–Theory and application. Academic Press, Boston, MA.

Chen, W.-T., C.-H. Wen, and C.-W. Yang. 1994. A fast two-dimensional entropic thresholding algorithm. Pattern Recognit. 27:885–893. doi:10.1016/0031-3203(94)90154-6

Cheng, H.D., and Y.H. Chen. 1999. Fuzzy partition of two-dimensional histogram and its application to thresholding. Pattern Recognit. 32:825–843. doi:10.1016/S0031-3203(98)00080-6

Cohen, I., L.D. Cohen, and N. Ayache. 1992. Using deformable surfaces to segment 3D images and infer differential structures. Computer Vision. Graphics Image Processing: Image Understanding 56(2):242–263.

Coleman, A.J., and M. Sinclair. 1985. A beam-hardening correction using dual-energy computed tomography. Phys. Med. Biol. 30:1251–1256. doi:10.1088/0031-9155/30/11/007

Duda, R.O., and P.E. Hart. 1973. Pattern recognition and scene analysis. Wiley-Interscience, New York.

Duliu, O.G. 1999. Computer axial tomography in geosciences: An overview. Earth Sci. Rev. 48:265–281. doi:10.1016/S0012-8252(99)00056-2

Elliot, T.R., and R.J. Heck. 2007. A comparison of 2D vs. 3D thresholding of X-ray CT imagery. Can. J. Soil Sci. 87:405–412. doi:10.4141/CJSS06017

Figueiredo, M.A.T., and R.D. Nowak. 2003. An EM algorithm for wavelet-based image restoration. IEEE Trans. Image Process. 12(8):906–916. doi:10.1109/TIP.2003.814255

Gebrenegus, T., T.A. Ghezzehei, and M. Tuller. 2011. Physicochemical controls on initiation and evolution of desiccation cracks in sand–bentonite mixtures: X-ray CT imaging and stochastic modeling. J. Contam. Hydrol. 126:100–112.doi:10.1016/j.jconhyd.2011.07.004

Gilboa, G., N. Sochen, and Y.Y. Zeevi. 2002. Regularized shock filters and complex diffusion. Proceedings of the ECCV 7th Europen Conference on Computer Vision Part I. Springer-Verlag, London. p. 399–413.

Gonzalez, R.C., and R.E. Wood. 2002. Digital image processing. Wesley Publishing Company, Boston, MA.

Halverson, C., D.J. White, and J. Gray. 2005. Application of X-ray CT scanning to characterize geomaterials used in transportation construction. Proceedings of the 2005 Mid-Continent Transportation Research Symposium. Iowa State University, Ames, IA.

Haralick, R., and L. Shapiro. 1985. Image segmentation techniques. Comput. Vis. Graph. Image Process. 29(1):100–132. doi:10.1016/S0734-189X(85)90153-7

Held, K., E.R. Kops, B.J. Krause, W.M. Wells, III, R. Kikinis, and H.W. Muller-Gartner. 1997. Markov random field segmentation of brain MR images. IEEE Trans. Med. Imaging 16(6):878–886. doi:10.1109/42.650883

Hsieh, J., R.C. Molthen, C.A. Dawson, and R.H. Johnson. 2000. An iterative approach to the beam hardening correction in cone beam CT. Med. Phys. 27(1):23–29. doi:10.1118/1.598853

Huang, L.K., and M.J.J. Wang. 1995. Image thresholding by minimizing the measures of fuzziness. Pattern Recognit. 28:41–51. doi:10.1016/0031-3203(94)E0043-K

Iassonov, P., T. Gebrenegus, and M. Tuller. 2009. Segmentation of X-Ray CT images of porous materials: A crucial step for characterization and quantitative analysis of pore structures. Water Resour. Res. 45:W09415. doi:10.1029/2009WR008087

Iassonov, P., and M. Tuller. 2010. Application of image segmentation for correction of intensity bias in X-ray CT images. Vadose Zone J. 9:187–191.doi:10.2136/vzj2009.0042

Ikeda, S., T. Nakano, and Y. Nakashima. 2000. Three-dimensional study on the interconnection and shape of crystals in a graphic by X-ray CT and image analysis. Mineral. Mag. 64:945–959. doi:10.1180/002646100549760

Iltis, G.C., R.T. Armstrong, D.P. Jansik, B.D. Wood, and D. Wildenschild. 2011. Imaging biofilm architecture within porous media using synchrotron-based X-ray computed Microtomography. Water Resour. Res. 47:W02601.doi:10.1029/2010WR009410

Kaestner, A., E. Lehmann, and M. Stampanoni. 2008. Imaging and image processing in porous media research. Adv. Water Resour. 31:1174–1187. doi:10.1016/j.advwatres.2008.01.022

Kapur, J.N., P.K. Sahoo, and A.K.C. Wong. 1985. A new method for gray-level picture thresholding using the entropy of the histogram. Graph. Models Image Processing 29:273–285.

Kato, Z., J. Zerubia, and M. Berthod. 1992. Satellite image classification using a modified Metropolis dynamics. IEEE International Conference on Acoustics, Speech, and Signal Processing, San Francisco, CA. p. 573–576.

Ketcham, R.A., and W.D. Carlson. 2001. Acquisition, optimization, and interpretation of X-ray computed tomographic imagery: Applications to the geosciences. Comput. Geosci. 27:380–400.

Kindermann, R., and J.L. Snell. 1980. Markov random fields and their applications. American Mathematical Society, Providence, RI.

Kirkpatrick, S., C.D. Gelatt, and M.P. Vecchi. 1983. Optimization by simulated annealing. Science 220(4598):671–680. doi:10.1126/science.220.4598.671

Kittler, J., and J. Illingworth. 1986. Minimum error thresholding. Pattern Recognit. 19:41–47. doi:10.1016/0031-3203(86)90030-0

Kulkarni, R., M. Tuller, W. Fink, and D. Wildenschild. 2012. Three-dimensional multiphase segmentation of X-Ray CT data of porous materials using a Bayesian Markov random field framework. Vadose Zone J. 11 (available online): doi:10.2136/vzj2011.0082

Leski, J. 2003. Towards a robust fuzzy clustering. Fuzzy Sets Syst. 137(2):215–233. doi:10.1016/S0165-0114(02)00372-X

Li, C., C.-Y. Kao, J. Gore, and Z. Ding. 2008. Minimization of region-scalable fitting energy for image segmentation. IEEE Trans. Image Process. 17(10):1940–1949. doi:10.1109/TIP.2008.2002304

Li, S.Z. 2001. Markov random field modeling in image analysis, 2nd ed. Springer-Verlag, New York.

Ludusan, C., O. Lavialle, R. Terebes, and M. Borda. 2010. Morphological sharpening and denoising using a novel shock filter model. Image Signal Processing 6134:19–27. doi:10.1007/978-3-642-13681-8_3

Mardia, K.V., and T.J. Hainsworth. 1988. A spatial thresholding method for image segmentation. IEEE Trans. Pattern Anal. Mach. Intell. 10:919–927. doi:10.1109/34.9113

McInerney, T., and D. Terzopoulos. 1996. Deformable models in medical image analysis: A survey. Med. Image Anal. 1(2):91–108. doi:10.1016/S1361-8415(96)80007-7

Melgani, F. 2006. Robust image binarization with ensembles of thresholding algorithms. J. Electron. Imaging 15(2):023010. doi:10.1117/1.2194767

Metropolis, N., A.W. Rosenbluth, M.N. Rosenbluth, A.H. Teller, and E. Teller. 1953. Equation of state calculations by fast computing machines. J. Chem. Phys. 21(6):1087–1092. doi:10.1063/1.1699114

Montagnat, J., and H. Delingette. 2001. A review of deformable surfaces: Topology, geometry and deformation. Image Vis. Comput. 19(14):1023–1040. doi:10.1016/S0262-8856(01)00064-6

Moussouris, J. 1974. Gibbs and Markov random systems with constraints. J. Stat. Phys. 10:11–33. doi:10.1007/BF01011714.

Mumford, D., and J. Shah. 1989. Optimal approximations by piece-wise smooth functions and associated variational problems. Commun. Pure Appl. Math. 42:577–685. doi:10.1002/cpa.3160420503

Nefti, S., and M. Oussalah. 2004. Probabilistic-fuzzy clustering algorithm. Proceedings of the IEEE International Conference on Systems, Man and Cybernetics 5:4786–4791.

Nyúl, L.G., A.X. Falcão, and J.K. Udupa. 2002. Fuzzy–connected 3D image segmentation at interactive speeds. Graph. Models 64(5):259–281. doi:10.1016/S1077-3169(02)00005-9

Oh, W., and W.B. Lindquist. 1999. Image thresholding by indicator Kriging. IEEE Trans. Pattern Anal. 21:590–602. doi:10.1109/34.777370

Osher, S., and N. Paragios. 2003. Geometric level set methods in imaging, vision, and graphics. Springer-Verlag, New York.

Osher, S.J., and L.I. Rudin. 1990. Feature-oriented image enhancement using shock filters. SIAM J. Numer. Anal. 27(4):919–940. doi:10.1137/0727053

Otsu, N. 1979. A threshold selection method from gray-level histograms. IEEE Trans. Syst. Man Cybern. 9:62–66. doi:10.1109/TSMC.1979.4310076

Pal, N.R., and S.K. Pal. 1989. Entropic thresholding. Signal Process. 16:97–108. doi:10.1016/0165-1684(89)90090-X

Pal, N.R., and S.K. Pal. 1993. A review on image segmentation techniques. Pattern Recognit. 26(9):1277–1294. doi:10.1016/0031-3203(93)90135-J

Pedrycz, W. 1996. Conditional fuzzy c-means. Pattern Recognit. Lett. 17(6):625–629. doi:10.1016/0167-8655(96)00027-X

Perona, P., and J. Malik. 1990. Scale-space and Edge detection using anisotropic diffusion. IEEE Trans. Pattern Anal. Mach. Intell. 12(7):629–639. doi:10.1109/34.56205

Peth, S., R. Horn, F. Beckmann, T. Donath, J. Fischer, and A.J.M. Smucker. 2008. Three-dimensional quantification of intra-aggregate pore-space features using synchrotron-radiation-based microtomography. Soil Sci. Soc. Am. J. 72:897–907 10.2136/sssaj2007.0130. doi:10.2136/sssaj2007.0130

Pham, D.L., C. Xu, and J.L. Prince. 2000. Current methods in medical image segmentation. Annu. Rev. Biomed. Eng. 2000:315–337. doi:10.1146/annurev.bioeng.2.1.315

Pham, T.D. 2001. Image segmentation using probabilistic fuzzy c-means clustering. Proceedings of the 2001 International Conference on Image Processing 1, Thessaloniki, Greece. p. 722–725.

Porter, M.L., D. Wildenschild, G. Grant, and J.I. Gerhard. 2010. Measurement and prediction of the relationship between capillary pressure, saturation, and interfacial area in a NAPL-water-glass bead system. Water Resour. Res. 46:W08512. doi:10.1029/2009WR007786

Prodanović, M., W.B. Lindquist, and R.S. Seright. 2006. Porous structure and fluid partitioning in polyethylene cores from 3D X-ray microtomographic imaging. J. Colloid Interface Sci. 298:282–297. doi:10.1016/j.jcis.2005.11.053

Rajapakse, J.C., J.N. Giedd, and J.L. Rapoport. 1997. Statistical approach to segmentation of single-channel cerebral MR images. IEEE Trans. Med. Imaging 16(2):176–186. doi:10.1109/42.563663

Rebuffel, V., and J-M. Dinten. 2007. Dual-energy X-ray imaging: Benefits and limits. Insight–Non-Destructive Testing Condition Monit. 49(10):589–594.

Ridler, T.W., and S. Calvard. 1978. Picture thresholding using an iterative selection method. IEEE Trans. Syst., Man, Cybernetics SMC-8:630–632.

Rosin, P.L. 2001. Unimodal thresholding. Pattern Recognit. 34:2083–2096. doi:10.1016/S0031-3203(00)00136-9

Rudin, L., S. Osher, and E. Fatemi. 1992. Nonlinear total variation based noise removal algorithms. Physica D 60:259–268. doi:10.1016/0167-2789(92)90242-F

Russ, J.C. 2010. The image processing handbook. CRC Press, Boca Raton, FL.

San José Martínez, F., M.A. Martín, F.J. Caniego, M. Tuller, A. Guber, Y. Pachepsky, and C. García-Gutiérrez. 2010. Multifractal analysis of discretized X-ray CT images for the characterization of soil macropore structures. Geoderma 156(1–2):32–42. doi:10.1016/j.geoderma.2010.01.004

Scherzer, O., and C. Groetsch. 2001. Inverse scale space theory for inverse problems. Scale-Space and Morphology in Computer Vision. Lect. Notes Comput. Sci. 2106:317–325. doi:10.1007/3-540-47778-0_29

Schnaar, G., and M.L. Brusseau. 2006. Characterizing pore-scale configuration of organic immiscible liquid in multiphase systems with synchrotron X-ray microtomography. Vadose Zone J. 5:641–648. doi:10.2136/vzj2005.0063

Sethian, J.A. 1999. Level set methods and fast marching methods. Cambridge Univ. Press, Cambridge.

Sezgin, M., and B. Sankur. 2004. Survey over image thresholding techniques and quantitative performance evaluation. J. Electron. Imaging 13:146–165. doi:10.1117/1.1631315

Shannon, C.E., and W. Weaver. 1948. The mathematical theory of communication. Bell Syst. Tech. J. 27:379–423.

Sheppard, A.P., R.M. Sok, and H. Averdunk. 2004. Techniques for image enhancement and segmentation of tomographic images of porous materials. Physica A 339:145–151. doi:10.1016/j.physa.2004.03.057

Smucker, A.J.M., E.-J. Park, J. Dorner, and R. Horn. 2007. Soil micropore development and contributions to soluble carbon transport within macroaggregates. Vadose Zone J. 6:282–290. doi:10.2136/vzj2007.0031

Stock, S.R. 1999. X-ray microtomography of materials. Int. Mater. Rev. 44:141–164. doi:10.1179/095066099101528261

Sund, R., and K. Eilertsen. 2003. An algorithm for fast adaptive image binarization with applications in radiotherapy imaging. IEEE Trans. Med. Imaging 22:22–28. doi:10.1109/TMI.2002.806431

Szirányi, T., J. Zerubia, L. Czúni, D. Geldreich, and Z. Kato. 2000. Image segmentation using Markov random field model in fully parallel cellular network architectures. Real-Time Imaging 6:195–211. doi:10.1006/rtim.1998.0159

Tsai, D.M. 1995. A fast thresholding selection procedure for multimodal and unimodal histograms. Pattern Recognit. Lett. 16:653–666. doi:10.1016/0167-8655(95)80011-H

Van Geet, M., R. Swennen, and M. Wevers. 2000. Quantitative analysis of reservoir rocks by microfocus X-ray computerised tomography. Sediment. Geol. 132:25–36. doi:10.1016/S0037-0738(99)00127-X

Vaz, C.M.P., I.C. de Maria, P.O. Lasso, and M. Tuller. 2011. Evaluation of an advanced benchtop micro-computed tomography system for quantifying porosities and pore-size distributions of two brazilian oxisols. Soil Sci. Soc. Am. J. 75(3):832–841. doi:10.2136/sssaj2010.0245

Vlidis, T., and Y.-T. Liow. 1990. Integrating region growing and edge detection. IEEE Trans. Pattern Anal. Mach. Intell. 12(3):208–214.

Wang, W., A.N. Kravchenko, A.J.M. Smucker, and M.L. Rivers. 2011. Comparison of image segmentation methods in simulated 2D and 3D microtomographic images of soil aggregates. Geoderma 162:231–241. doi:10.1016/j.geoderma.2011.01.006

Weeratunga, S., and C. Kamath. 2004. An investigation of implicit active contours for scientific image segmentation. Visual communications and image processing. Proc. SPIE 5308(1):210–221. doi:10.1117/12.527037

Wildenschild, D., K.A. Culligan, and B.S.B. Christensen. 2004. Application of X-ray microtomography to environmental fluid flow problems. In: U. Bonse, editor, Developments in X-ray tomography IV. Proceedings of SPIE 5535. Bellingham, WA. p. 432–441.

Wildenschild, D., J.W. Hopmans, C.M.P. Vaz, M.L. Rivers, D. Rikard, and B.S.B. Christensen. 2002. Using X-ray computed microtomography in hydrology: Systems, resolutions and limitations. J. Hydrol. 267:285–297. doi:10.1016/S0022-1694(02)00157-9

Wirjadi, O. 2007. Survey of 3D image segmentation methods. Tech. Rep. No. 123, Fraunhofer ITWM, Kaiserslautern, Germany.

Yanowitz, S.D., and A.M. Bruckstein. 1989. A new method for image segmentation. Comput. Vis. Graph. Image Process. 46:82–95. doi:10.1016/S0734-189X(89)80017-9

Yen, J.C., F.J. Chang, and S. Chang. 1995. A new criterion for automatic multilevel thresholding. IEEC Trans. Image Processing 4:370–378. doi:10.1109/83.366472

Zack, G.W., W.E. Rogers, and S.A. Latt. 1977. Automatic measurement of a sister chromatid exchange frequency. J. Histochem. Cytochem. 25:741–753. doi:10.1177/25.7.70454

Zhou, H., Y. Yuan, F. Lin, and T. Liu. 2008. Level set image segmentation with Bayesian analysis. Neurocomputing 71(10–12):1994–2000. doi:10.1016/j.neucom.2007.08.035

9

Heterogeneous Porous Media Structure Analysis with Computed Tomography Images and a Multiphase Bin Semivariogram

Glenn O. Brown* and Jason R. Vogel

Abstract

Computed tomography provides detailed images of porous media, but the complexities of the pore geometry and component distributions make simple quantification of most materials problematic. Two related techniques are presented to assist in characterizing measurements: (i) artificial image generation of ideal media geostatistics and (ii) a multiphase bin semivariogram (MBS). Artificial images are used to identify semivariograms associated with several specific characteristic media features, which reduces the ambiguity in interpreting semivariograms generated from real images. The MBS is able to characterize the structural properties of porous media features with discrete bulk density bins. Finally, MBS and artificial image analyses have been combined with histograms and traditional semivariograms to create a realistic stochastic representation of the porous media. MBS analysis has been completed on a Culebra Dolomite core and four concrete cores. These rigid cores have significant component variations that are well defined, and as such provide a robust test of the method. Finally, by combining artificial images, histogram analysis, semivariogram analysis, and MBS analysis, a stochastic representation of porous media has been generated. These representations have the potential to be used in flow and transport models, or may be used as an initial image for an advanced modeling technique such as simulated annealing.

Abbreviations: CT, computed tomography; GPR, ground-penetrating radar; MBS, multiphase bin semivariogram; WIPP, Waste Isolation Pilot Plant.

G.O. Brown and J.R. Vogel, Biosystems and Agricultural Engineering, Oklahoma State University, Stillwater, OK. *Corresponding author (gbrown@okstate.edu).

doi:10.2136/sssaspecpub61.c9

Soil–Water–Root Processes: Advances in Tomography and Imaging. SSSA Special Publication 61. S.H. Anderson and J.W. Hopmans, editors. © 2013. SSSA, 5585 Guilford Rd., Madison, WI 53711, USA.

Soil and groundwater research is increasingly being conducted using computed tomography (CT) images for analysis of porous media properties. Anderson and Hopmans (1994) describe many of these applications. However, while CT provides detailed images of porous media, the complexities of the pore geometry and component distributions make simple quantification of most materials problematic. There is a long-standing and persistent need to develop tools to turn the images provided by CT into useful information. Geostatistical analysis of CT images provides a method for obtaining small-scale structural properties of porous media. Rasiah and Aylmore (1998) and Vogel and Brown (2002) use CT images of γ ray attenuation for geostatistical characterization of soil and rock cores. Tercier et al. (2000) uses ground-penetrating radar (GPR) images to compare the correlation structure of different depositional environments. Two-dimensional geostatistical image analysis and three-dimensional reconstruction is used by Ioannidis et al. (1999) to predict aquifer properties for use in flow and transport models.

Geostatistical analysis can also be used for recognition of the shape of structure within images. Jupp et al. (1989) use semivariogram analysis of remote-sensed images to show that overlapping disks can represent spatial structure. Semivariogram analysis of a moving window on radar images has been used to classify land use (Miranda et al., 1992), and ice crevasse patterns have been characterized using geostatistics by Herzfeld and Zahner (2001). In rock formations, Novakovic et al. (2002) use geostatistics of GPR images to characterize shale outcrops in sandstone for three-dimensional flow modeling, and White et al. (2003) use geostatistics to characterize calcite concretions in sandstone. Swan and Garratt (1995) use semivariance of digitized images of thin rock sections to classify texture. CT images can also be used for density frequency analysis (Hsieh et al., 1998; Clausnitzer and Hopmans, 1999). Indicator semivariograms have also been used for strict linear two-phase geostatistical characterization by Grevers and de Jong (1994) and Tavchandjian et al. (1993) for analysis of soil and rock fractures, and by Ortiz and Deutsch (2004) for spatial characterization of high-grade copper deposits.

Semivariograms can be used as a significant tool to identify and characterize major features in a rock formation. As examples, Özturk and Nasuf (2002) use geostatistics to characterize and predict regionalized zones of mechanical properties in rocks for estimations of net cutting rates required for tunnel excavation, and Pranter et al. (2006) use geostatistics to characterize stratigraphic cyclicity and facies distribution for application in multiphase fluid-flow modeling. Accurately describing the spatial characteristics of porous media at scales of interest is very important for upscaling of the porous media (Tidwell and Wilson, 2000) and fate and transport modeling in the vadose zone (Skøien and Blöschl, 2006). By combining pattern recognition, density frequency analysis for phase differentiation, and semivariograms, the major features in the porous media can be identified to create a stochastic representation of the media. This stochastic representation could then be applied to a flow and transport model or used as an initial image for a simulation model such as simulated annealing (Goovaerts, 2000). Goovaerts (1997) states that the initial image for simulated annealing, which is realistic and meets target constraints, will improve the optimization process. The stochastic representation generated using these methods that used an initial image would be more realistic than a random image that only matches the histogram.

A semivariogram analysis of artificial images is used to identify dominant features of the porous media. Then a Multiphase Bin Semivariogram (MBS) is introduced and used to characterize geostatistical properties of different structures for three different types of heterogeneous porous media. These different "phases" are identified by a strict linear differentiation of the bulk density to define bins which correspond to specific geologic structures. Finally, by combining the artificial image analysis with the results of the MBS, a stochastic representation of the porous media is generated incorporating the geostatistical structure in each of the phase bins.

Theory
Semivariograms

Semivariograms measure the degree of dissimilarity between an unsampled value and a nearby data value. Also called the structure function by Gandin (1963), traditional semivariograms are defined as

$$\gamma(h) = \frac{1}{2N(h)} \sum_{i=1}^{N(h)} (x_i - y_i)^2 \qquad [1]$$

where $\gamma(h)$ is half of the average squared difference, $N(h)$ is the number of pairs, x_i is the value at the start, or tail, y_i is the value at the end, or head, and h is the vector between the two data points.

A sill and range characterize the semivariogram. The sill is a plateau in the semivariogram values that corresponds to the variance of the sampled data. If the semivariogram is standardized, the $\gamma(h)$ value is divided by the variance and the sill value will be equal to one. The range, or correlation length, is the distance at which the semivariogram reaches a plateau. It will be the size of the most common feature in the domain. A non-zero intercept on the semivariogram may exist and is termed the nugget effect. It may result from sampling error or variability at scales less than the smallest sampling interval (Cressie, 1991). Another possible semivariogram characteristic is the hole effect, which is described as a semivariogram that peaks and dips, suggesting that at greater distances, samples are more related. The hole effect may be seen as the rough spacing between adjacent lenses or bedding planes.

Semivariograms are fit by a number of analytical relationships to ensure that the semivariogram model is positive definite. The exponential model is often used by researchers in stochastic hydrology (Woodbury and Sudicky, 1991), and is given by

$$\gamma(h) = 1 - e^{\left(\frac{-3h}{a}\right)} \qquad [2]$$

where a is the range and h is the lag distance. The exponential model reaches its sill asymptotically, with the practical range a defined as that distance at which the semivariogram value is 95% of the sill. Another common semivariogram model is the Gaussian model, which is given by

$$\gamma(h) = 1 - e^{\left(-\frac{3h^2}{a^2}\right)}$$

[3]

and is parabolic in shape. It is often representative of highly regular phenomenon such as elevation of gently undulating hills (Goovaerts, 1997). The final model used is the spherical model,

$$\gamma(h) = \begin{cases} 1.5\dfrac{h}{a} - 0.5\left(\dfrac{h}{a}\right)^3 & \text{if } h < a, \\ 1 & \text{elsewhere.} \end{cases}$$

[4]

The spherical model has a linear behavior near the origin, flattens out at larger distances, and reaches the sill at a (Isaaks and Srivastava, 1989).

Multiple, or nested, sills in the semivariogram may be associated with physical phenomena occurring at different scales. Detection of ranges nested within larger ranges requires sample spacing shorter than the minimum detected range. In general, nested sills are only considered when they can be associated with physical phenomena (Solie et al., 1999). Nested sills can be modeled by using a linear combination of positive definite semivariogram models with positive coefficients. This property results in the family of positive definite models,

$$\gamma(h) = \sum_{i=1}^{n} |w_i| \gamma_i(h)$$

[5]

which are positive definite as long as the n individual models are all positive definite. A weighting function, w_i, is defined for each individual model subject to $\sum_{i=1}^{n} w_i = 1$ (Isaaks and Srivastava, 1989).

Indicator Semivariograms

Data may also be analyzed with an indicator semivariogram to measure how often two values of x_i are on opposite sides of the threshold value z_k (Goovaerts, 1997). For indicator semivariogram analysis, the data is first transformed by coding each datum value $z(x_i)$ as an indicator datum $i(x_i; z_k)$ defined as

$$i(x_i; z_k) = \begin{cases} 1 & \text{if } z(x_i) \leq z_k \\ 0 & \text{otherwise} \end{cases}$$

[6]

The indicator semivariogram is then calculated as

$$\gamma_I(h; z_k) = \frac{1}{2N(h)} \sum_{\alpha=1}^{N(h)} \left[i(u_\alpha; z_k) - i(u_\alpha + h; z_k) \right]^2$$

[7]

Table 9–1. Density mean and standard deviation for each of the major features of the Culebra Dolomite core (Hsieh et al. 1998).

Feature	Mean	SD
	— Mg m^{-3} —	
Solid dolomite	2.52	0.07
Gypsum	2.32	0.07
Mixed voxels (beta distribution)	2.02	0.41
Void	0.0	–

Multiple indicator semivariograms can be calculated for a single data set to identify different structures (Isaaks and Srivastava, 1989). However, this technique works best for structures at the extreme ends of the range of the variable of interest. In the case of bulk density of soil, this would work for identifying voids or dense particles, such as gravel. Some detail from the data is lost in the indicator transformation that may make the semivariogram calculation easier, but which may cause the user to lose information from extreme values and smaller variations within the rock.

Artificial Images

Computed tomography uses artificial or phantom images for algorithm testing and calibration. Phantom images are artificial "scans," which are theoretically generated to see if the reconstruction algorithm can reproduce the original data. A similar approach is used here to show how the characteristics of semivariograms are influenced by various phase structure geometries. Semivariogram analysis of artificial images containing properties common in the Culebra Dolomite formation indicates distinct shapes in the semivariogram, which can be used in interpreting the major characteristics of the rock.

Holt (1997), Hsieh et al. (1998), Brown et al. (2000), and visual inspection of Culebra Dolomite cores indicate that solid dolomite, bedding planes made up of a large number of vugs, vertical gypsum-filled cracks, gypsum infilling, silty dolomite outcrops, and dissolution voids are the major features of this rock formation. The distribution of bulk density of each of these features except mixed voxels is normally distributed and summarized in Table 9–1. Mixed voxels, which are primarily areas with voids smaller than the scanning resolution, are approximated by a β distribution

$$f_m^*\left(\frac{\rho_m}{\rho_{max}}\right) = \left[\left(\frac{\rho_m}{\rho_{max}}\right)^{\alpha-1}\left(1-\frac{\rho_m}{\rho_{max}}\right)^{\beta-1}\right]\frac{\Gamma(\alpha+\beta)}{\Gamma(\alpha)\Gamma(\beta)} \qquad 0 \le \frac{\rho_m}{\rho_{max}} \le 1 \qquad [8]$$

$$f_m^*\left(\frac{\rho_m}{\rho_{max}}\right) = 0 \qquad\qquad\qquad\qquad \text{elsewhere}$$

where $f_m^*(\rho_m/\rho_{max})$ is the true mixed voxel density distribution, Γ is the γ function, α and β are fitting coefficients, and ρ_{max} is the largest density in the distribu-

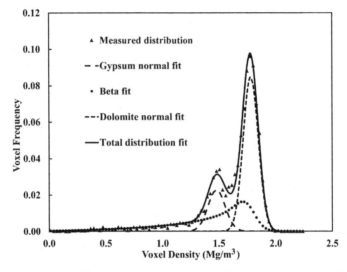

Fig. 9–1. Measured and fitted density frequency distribution for core C2AV, slice 14.

tion. The range of the β distribution (Eq. [8]) for mixed voxels is from zero to 2.52, the bulk density of solid dolomite. The fitting parameter α is estimated to range from 3.30 to 7.80 and fitting parameter β from 0.41 to 1.73 for Culebra Dolomite. Additional information on the use of the β distribution in image segmentation is presented in Hsieh et al. (1998), and Fig. 9–1 shows a typical fitting. Using these properties, two-dimensional, 100 by 100 pixel phantom images were constructed. Random values for the β distribution are generated by

$$y' = \frac{\sum_{i=1}^{2A} R_{N_i}^2}{\sum_{i=1}^{2A} R_{N_i}^2 + \sum_{i=2A+1}^{2A+2B} R_{N_i}^2} \qquad [9]$$

where A and B are integers, and R_{Ni} is a random value from a normal distribution with $A = 0$ and $B = 1$ (Hahn and Shapiro, 1967). Clausnitzer and Hopmans (1999) describe an additional scheme for phase volume differentiation in two-phase systems that fits two normal probability density functions and with an analytical expression to account for mixed voxels.

Knowledge of the semivariograms representative of specific rock characteristics can aid in determining features in specific rock formations. The use of semivariograms, however, is not an absolute, as one semivariogram could be representative of numerous combinations of features. The more complex the porous media, the harder it is to apply these principles. Larger resolution allows for larger scans with the same number of data points and therefore increases the possibility of larger-scale variability represented in the scan, but may not detect smaller resolution variability. Resolution selection should be based on the scale of the problem being investigated.

Multiphase Bin Semivariograms

The ability to differentiate between soil and rock features with different densities would simplify the task of interpreting the semivariogram. The indicator semivariogram allows for differentiation into two phases (Goovaerts, 1997). However, we introduce an MBS to determine the structure of multiple rock features that fall in specified density ranges. An MBS is defined as

$$\gamma_j(h) = \frac{1}{2N_j(h)} \sum_{i=1}^{N(h)} (x_i - y_i)^2 \qquad \text{if } x_i \in X_j \qquad [10]$$

where the value $(x_i - y_i)^2$ is included in $\gamma_j(h)$ if and only if the value of x_i is in the range of bin X_j, and $N_j(h)$ is the number of pairs included in the calculation.

Using this MBS, only the spatial characteristics in the density range represented by that bin are included in the calculation. In effect, the average structure size of that phase can be determined. If intrinsic stationarity were assumed for the traditional semivariogram calculation, the intrinsic assumption would be appropriate within each bin for this calculation as well.

Materials and Methods
Samples and Scanning

To provide a robust test with unambiguous features, consolidated samples were tested instead of soils. A rock core sample (C2AV) was taken from the Culebra Dolomite Member of the Rustler Formation from drill holes near the Waste Isolation Pilot Plant (WIPP) site in southeastern New Mexico. The rock sample was sampled because the Culebra Dolomite is the most transmissive confined unit above the waste repository (Lappin et al., 1989). An additional four concrete cores (A1A, A2A, A3A, and A4A) were analyzed to provide a robust demonstration of the methods. Major features of the concrete cores include small voids, aggregates, and cement. Each sample was scanned on a custom γ scanner (Brown et al., 1993, 1994). Voxel density was computed from linear attenuation using a water calibration. Table 9–1 lists the sample scanning parameters.

Semivariogram Analysis

Deutsch and Journel (1998) provide a package of Fortran 77 geostatistical programs. The GAM code for semivariogram analysis with a regular grid was recompiled using Visual Fortran 6.0 to increase the maximum number of data points which could be analyzed. Semivariogram generation was then completed using the recompiled GAM program. Additional Fortran code was developed and used for MBS analysis.

Results and Discussion
Geostatistics of Artificial Images

Figure 9–1 shows the artificial images of bulk density and the associated semivariograms for several characteristic properties of Culebra Dolomite. The uniform artificial image in Fig. 9–2a is solid dolomite. The bulk density is normally dis-

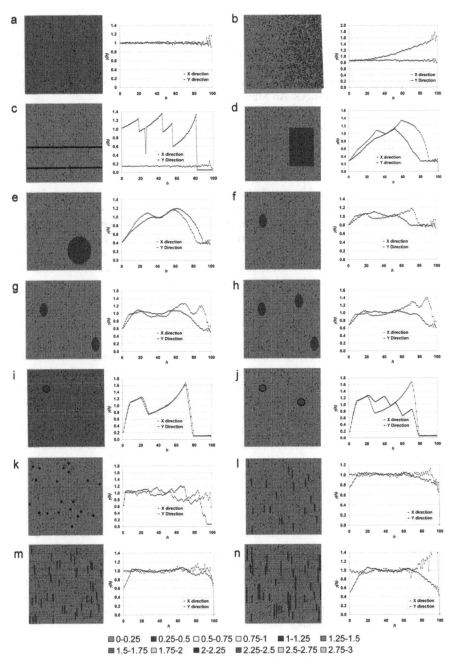

■ 0-0.25   ■ 0.25-0.5   □ 0.5-0.75   □ 0.75-1   ■ 1-1.25   ■ 1.25-1.5
■ 1.5-1.75   □ 1.75-2   ■ 2-2.25   ■ 2.25-2.5   □ 2.5-2.75   □ 2.75-3

Fig. 9–2. Phantom images and semivariograms for (a) solid dolomite, (b) solid dolomite with a mean trend, (c) solid dolomite with vuggy bedding planes, (d) a rectangular gypsum outcrop, (e) an oval gypsum outcrop, (f) one small oval gypsum outcrop, (g) two small gypsum outcrops, (h) three small gypsum outcrops, and (i) one small void. Phantom images and semivariograms for (j) two small voids, (k) many very small voids, (l) twenty 10-pixel gypsum-filled cracks, (m) forty 10-pixel gypsum-filled cracks, and (n) twenty 10-pixel gypsum-filled cracks and twenty 20-pixel gypsum-filled cracks.

tributed with a mean and standard deviation as indicated in Table 9–1. The standardized results indicate a horizontal semivariogram with a sill and nugget at 1.0. The range is smaller than the smallest lag interval of 1 pixel.

Figure 9–2c represents vuggy bedding planes found in solid dolomite. These are voids caused by layers of trapped gases in the sedimentary material. The artificial image has bedding planes in rows 19 and 45. The peaks on the semivariogram are at approximately 19, 45, 55, and 81 pixels. These distances correspond to the distances to each bedding plane from the edges of the scanned area. With regard to the peak locations, the semivariogram is a mirror image at the center (50 pixels). For lags larger than 81 pixels, the semivariogram is linear because this represents comparisons of uniform regions. The semivariogram associated with bedding planes is an example of zonal anisotropy with strata, as the vertical semivariogram reaches a higher sill than the horizontal semivariogram because of larger average differences in the vertical direction.

Gypsum outcrops are also common in this rock formation. Artificial images of rectangular and oval-shaped gypsum outcrops have been generated and analyzed. The rectangular gypsum outcrop in Fig. 9–2d extends from pixel (60, 30) to pixel (90, 80) on the 100. The x direction semivariogram shows peaks at lags of 30 and 60 pixels, which correspond to the width of the uniform regions in that direction. The slope of the x direction semivariogram also changes at a lag of 10 pixels, which represents the width of the smaller uniform region on the right side of the image. The y direction semivariogram shows a peak at a lag of 50 pixels, which corresponds to the height of the outcrop on the artificial image. Changes of the slope of the semivariogram are also indicated at lags of 20 and 30 pixels, corresponding to the heights of the dolomite areas above and below the outcrop. Figure 9–2e represents an oval-shaped gypsum outcrop with a major axis length of 35 pixels in the y direction and a minor axis length of 30 pixels in the x direction and uniform dolomite surrounding. Results similar to the rectangular outcrop with regard to correlation expected ranges were found for the oval gypsum outcrop.

Figures 9–2f through 9–2h represent uniform dolomite with one, two, or three small oval gypsum outcrops, respectively. The outcrops have a major axis of 17 pixels in the y direction and 10 pixels in the x direction. The semivariograms for all three of these representations have similar ranges of 17 and 10 pixels in the y and x direction, respectively. For lags larger than the range in each semivariogram, the three semivariograms vary and are dependent on the physical location of the outcrops in each phantom image. The normalized nugget for the image with one small gypsum outcrop is approximately 0.8, compared to 0.7 for two gypsum outcrops and 0.6 for three outcrops. This indicates that the nugget decreases as the number of outcrops increases.

Another prominent characteristic of the Culebra Dolomite is hollow voids, or vugs, spaced randomly throughout the sample. Figures 9–2 and 9–2j represent solid dolomite with one and two circle voids, respectively. The voids have a 9-pixel diameter and are randomly located in the dolomite. The semivariogram for one vug (Fig. 9–2i) shows an abrupt change of slope in each direction at a 9-pixel lag and peaks at approximately 21 and 70 pixels. The change of slope is associated with the size of the void, and the two peaks are associated with the physical location of the vug in the matrix. The semivariogram in Fig. 9–2j is for two vugs with a diameter of 9 pixels. As in Fig. 9–2i, there is an abrupt change of slope at 9 pixels representing the width of the voids. The y direction semivariogram exhibits pseudocycling with peaks associated with location features of the voids. The x direction variogram would exhibit similar behavior for two holes except for the fact that by coincidence the edges of the two holes are equidistant from the left and right edge of the image. Because of this, the effect of the second void mirrors onto the first and appears to be only one void. Both Fig. 9–2i and 9–2j show a standardized nugget of approximately 0.2.

Figure 9–2k represents 20 very small circular voids with a diameter of 3 pixels in a uniform dolomite formation. The semivariogram clearly indicates the 3-pixel range associated with the voids. At lags greater than 3 pixels, the semivariogram in each direction shows uneven pseudo-cycling associated with location of the voids. The standardized nugget for this phantom image is approximately 0.6.

Figures 9–2l through 9–2n represent vertical gypsum-filled cracks in a uniform dolomite formation. Analysis of the first three figures reveals that differing lengths and densities of cracks provide different ranges and nuggets for the semivariograms. Figure 9–2l represents 20 randomly spaced vertical cracks 10 pixels long. This results in a y direction semivariogram with a range of 10 pixels and a standardized nugget of approximately 0.75. The phantom image in Fig. 9–2m with 20 randomly spaced vertical cracks 20 pixels long results in a y direction semivariogram with a range of 20 pixels and a standardized nugget of approximately 0.6. Figure 9–2n shows a phantom image with 40 randomly spaced vertical cracks 10 pixels long. This results in a y direction semivariogram with a range of 10 pixels and a standardized nugget at approximately 0.6. These results indicate that the range is the length of the cracks in that direction and that as crack density increases the standardized nugget decreases. Figure 9–2n represents a uniform dolomite with 20 randomly spaced vertical gypsum-filled cracks 10 pixels long and 20 randomly spaced vertical gypsum-filled cracks 20 pixels long. The resulting y direction semivariogram has a range of 20 pixels and a standardized nugget of approximately 0.5. There is also a noticeable decrease in the slope of the semivariogram at 10 pixels, corresponding to the length of the shorter cracks. This semivariogram is also more representative of the spherical and exponential models generally used for modeling semivariograms in stochastic hydrology (Isaaks and Srivastava, 1989; Woodbury and Sudicky, 1991).

Sample Statistics

Four concrete cores and one Culebra Dolomite core have been analyzed. Characteristics of these cores and the image scanning performed on each are summarized in Table 9–2. The average bulk densities range from 2.35 g mL^{-1} for sample A1A to 2.42 g mL^{-1} in sample A2A.

Figures 9–3 and 9–4 show the histograms for each of the samples. Among the concrete cores, A1A exhibits the smallest percentage of aggregates and A2A exhibits the largest percentage of aggregates. The percentage of aggregates is similar for samples A3A and A4A. A peak for dolomite and a mixed voxel tail is shown in the C2AV sample histogram (Fig. 9–4).

Traditional semivariograms of these samples are shown in Fig. 9–5 and 9–6. Figure 9–5 shows pseudo-cycling in all directions for the concrete cores. Results of modeling the semivariograms with a combination model are shown in Table 9–3. The specific models used are simply those that produced the least squares best fit. The four concrete core samples were analyzed in two dimensions with

Table 9–2. Sample imaging parameters and measured average bulk density.

Sample	Image voxel dimensions, X ×Y × Z	Voxel size	Bulk density	
			Mean	SD
	voxels	mm	—— Mg m⁻³ ——	
A1A	93 × 44 × 1 (4092)	1.02 × 1.016 × 1.02	2.34	0.20
A2A	94 × 45 × 1 (4230)	1.02 × 1.02 × 1.02	2.42	0.26
A3A	94 × 45 × 1 (4230)	1.02 × 1.02 × 1.02	2.35	0.24
A4A	93 × 44 × 1 (4092)	1.02 × 1.02 × 1.02	2.39	0.23
C2AV	71 × 71 × 46 (231,886)	0.37 × 0.37 × 0.75	2.44	0.30

a single scan. Sample A2A has larger horizontal and vertical correlation lengths than the other three concrete core samples. The C2AV sample was analyzed in three dimensions with multiple scans. Further discussion of the geostatistical characteristics of sample C2AV can be found in Vogel and Brown (2002).

Multiphase Bin Semivariogram Analysis

MBS analysis has been completed on the five samples, and the results are summarized in Table 9–4. Figure 9–7 shows the MBS for a concrete sample (A1A). None of the concrete cores show greater than 2.4% voids. The size of detected voids indicates only a few voids per core. Percent aggregates range from 14.4% in A3A to 29.5% in core A2A. The smallest aggregates were detected in core A4A, and the largest aggregates were shown in core A2A. The Culebra Dolomite core (C2AV) indicated 92.8% solid dolomite and 7.1% mixed voxels. In general, the range for the mixed voxels was larger than the range for the solid dolomite. In addition, voids in the Culebra Dolomite in the x and y directions were best fit with Gaussian semivariogram models, while the semivariograms for all other samples were best fit by spherical, exponential, or combination spherical-exponential models.

Traditional and Multiphase Bin Semivariogram Comparisons

Correlation lengths from traditional semivariograms can be compared with correlation lengths detected by MBS by comparing Tables 9–3 and 9–4. In general, the traditional semivariogram correlation lengths are a weighted average of the MBS ranges that is weighted by a combination of the sill value and the percent of points in the bin.

Stochastic Representation

Finally, a two-dimensional stochastic representation of the concrete core A1A has been generated. A comparison of the generated representation and the actual scan is shown in Fig. 9–8. The representation is generated by placing structures of the size indicated by the MBS of voids and aggregates on a background of cement. The generated density of voids, cement, and concrete are normally distributed

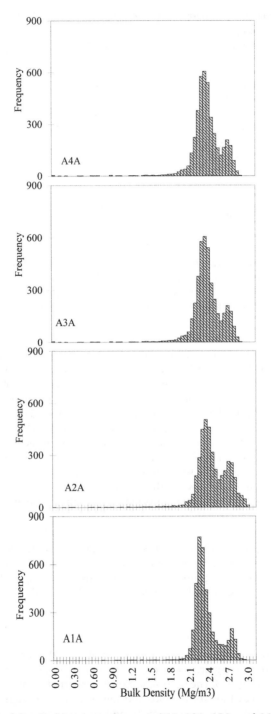

Fig. 9–3. Measured density histograms for cores A1A, A2A, A3A, and A4A.

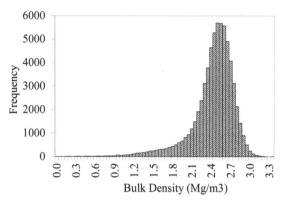

Fig. 9–4. Measured density histogram for core C2AV.

with mean and standard deviation as shown to match the original sample. The histogram showing the relative frequency distribution of the stochastic representation and actual CT data is shown in Fig. 9–9. Relative frequencies are reproduced well using these assumptions for all values except between 2.5 and 2.6 g mL^{-1} , where the actual image indicates larger relative frequencies than the generated image. This may be a result of edge pixels between cement and aggregate structure that are not adequately represented in the generated image. Figure 9–10 shows the semivariograms of the stochastic representation and the actual CT data. The semivariograms show similar structure, although the hole effect in the y direction of the actual scan at approximately 15 mm is not reflected in the semivariogram model and is therefore not found in the stochastic representation.

Stochastic representations are not expected to be a reproduction of the original image, but should show similar characteristics. Minimum acceptance criteria for a stochastic representation [modified from Leuangthong et al. (2004) for geostatistical realizations] would be (i) to reproduce the distribution of the attribute of interest and (ii) to reproduce the spatial continuity characterized by the semivariogram model. In this procedure, the artificial image histogram closely resembles the histogram of the original CT data as it is used to determine outcrop density (i.e., reproduces the distribution of the attribute of interest). The spatial continuity of the original CT data is also reproduced in the stochastic representation. Semivariograms for the real and phantom image were similar, especially at smaller lags. Larger lags may deviate more than smaller lags because these lags are usually more representative of location than characteristics of the rock. It is noted that in instances where the occurrence when the material associated with one phase bin is correlated to the occurrence of a material in another phase bin, such as with some metal ores, additional correlations between phases would need to be incorporated into the stochastic representation or one could apply a traditional conditional indicator simulation to the data as described by Journel and Isaaks (1984).

Successful application of artificial images and semivariograms of data from computed tomography imagery to identify the major features in the porous media could potentially be used to create a stochastic representation of the media which may be more representative of the true porous media structure for contaminant transport modeling than traditional finite difference modeling. This may be

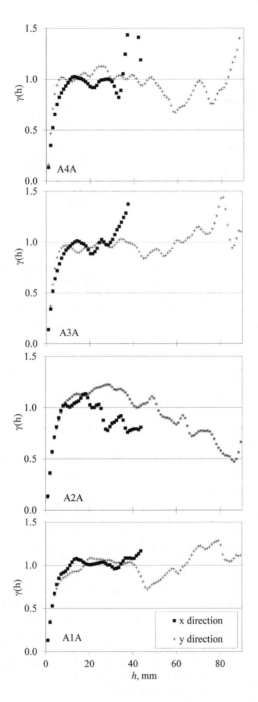

Fig. 9–5. Density semivariograms of cores A1A, A2A, A3A, and A4A.

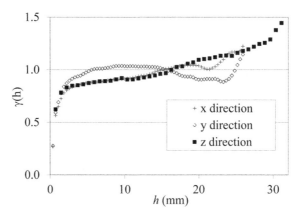

Fig. 9–6. Density semivariograms of core C2AV.

Table 9–3. Properties of standardized semivariograms.†

Sample name and direction		Semivariogram model	Range	Sill, w_i	r^2
			mm		mm §
A1A	x (nested) x	Combination of Two Spherical	6.5 14.3	0.58 0.45	0.973 to 40.6
	y (nested) y	Combination of Two Spherical	6.2 24.1	0.64 0.41	0.987 to 39.6
A2A	x (nested) x	Combination of Two Spherical	7.2 11.6	0.65 0.44	0.977 to 24.4
	y (nested) y	Combination of Spherical and Exponential	8.3 23.4	0.68 0.53	0.991 to 35.6
A3A	x (nested) x	Combination of Two Spherical	5.6 12.3	0.39 0.58	0.970 to 30.5
	y (nested) y	Combination of Spherical and Exponential	6.6 25.0	0.81 0.17	0.964 to 39.6
A4A	x (nested) x	Combination of Two Spherical	6.1 12.0	0.46 0.52	0.982 to 30.5
	y (nested) y	Combination of Two Spherical	6.1 25.0	0.96 0.10	0.968 to 27.4
C2AV	x (nested) x	Combination of Spherical and Exponential	2.3 25.9 ‡	0.83 0.22	0.991 to 13.0
	y (nested) y	Combination of Spherical and Exponential	9.1 2.3	0.17 0.85	0.992 to 15.9
	z (nested) z	Combination of Spherical and Exponential	28.7 ‡ 1.6	0.16 0.83	0.993 to 14.8

† None of the cores were modeled with a nugget.

‡ Range approaching sample width.

§ In this column, the coefficient of determination and the largest lag value used for calculation of the coefficient of determination is reported.

Table 9–4. Properties of conditional semivariograms.†

Sample	Bin range	Direction	Model ‡	a_1	w_1	a_2	w_2
	Mg m^{-3} (% pts in range)			mm		mm	
A1A	0.0–1.9	x	se	4.6	10.0	17.8	4.02
	(0.6%)	y	se	4.5	8.6	25.0	3.23
	1.9–2.6	x	se	6.2	0.35	25.0	0.57
	(84.3%)	y	se	4.9	0.32	25.0	0.48
	>2.6	x	s	10.9	2.02	–§	–
	(15.1%)	y	s	10.0	1.36	–	–
A2A	0.0–1.9	x	se	6.4	9.9	25.0	7.0
	(2.0%)	y	se	8.1	8.4	12.1	3.3
	1.9–2.6	x	ss	4.9	0.34	16.5	0.51
	(68.5%)	y	ss	5.3	0.26	25.0	0.49
	>2.6	x	s	11.1	0.94	–	–
	(29.5%	y	s	14.7	1.49	–	–
A3A	0.0–1.9	x	se	5.3	9.1	25.0	5.84
	(2.1%)	y	se	5.8	9.6	25.0	1.58
	1.9–2.6	x	se	8.5	0.18	12.1	0.42
	(83.5%)	y	s	5.8	0.56	–	–
	>2.6	x	s	13.2	1.46	–	–
	(14.4%)	y	s	11.5	1.82	–	–
A4A	0.0–1.9	x	se	5.9	6.1	17.5	0.18
	(2.4%)	y	s	5.5	8.8	–	–
	1.9–2.6	x	ss	4.8	0.32	15.4	0.47
	(75.7%)	y	ss	5.0	0.56	18.1	0.20
	>2.6	x	s	9.7	1.33	–	–
	(21.9%)	y	s	7.6	1.35	–	–
C2AV	0.0–0.5	x	g	2.8	25.2	–	–
	(0.1%)	y	g	2.8	26.3	–	–
		z	e	5.7	21.5	–	–
	0.5–2.00	x	se	2.2	2.38	6.5	1.19
	(7.1%)	y	se	3.4	3.64	14.7	0.44
		z	se	1.4	2.42	8.9	1.23
	>2.0	x	se	1.7	0.59	16.3	0.13
	(92.8%)	y	se	1.8	0.63	9.9	0.12
		z	se	1.4	0.64	20.0	0.11

† None of the semivariograms were modeled with a nugget.

‡ s = spherical, se = combination spherical and exponential, ss = combination of two spherical, g = gaussian, e = exponential.

§ Large-scale structure not detected or hole effect.

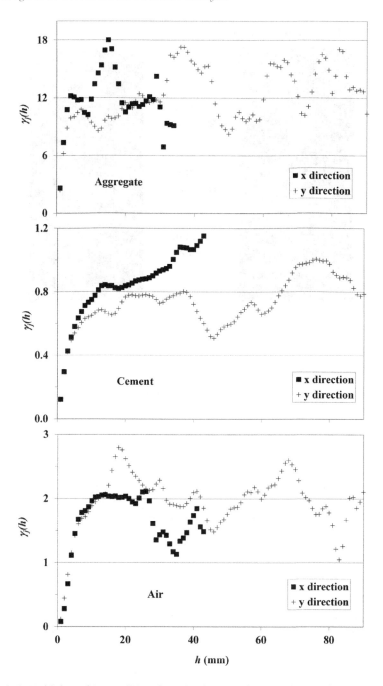

Fig. 9–7. Multiphase bin conditional semivariograms for sample A1A for air, cement, and aggregate.

Stochastic Representation   CT Data

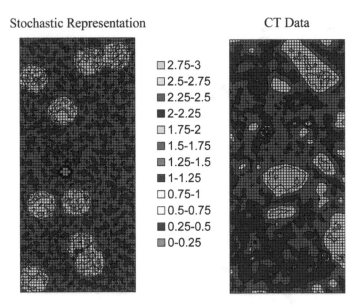

Fig. 9–8. Stochastic representation and computed tomography (CT) density data of core A1A.

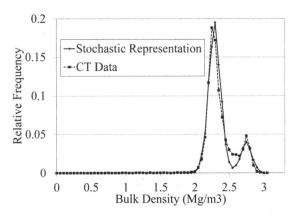

Fig. 9–9. Density histogram of a stochastic representation and the computed tomography (CT) data for core A1A.

especially true for porous media with small hydraulic conductivities where dispersion and diffusion are important in contaminant transport and plume width.

Summary and Conclusions

Artificial images have been utilized to develop signature semivariograms of structural elements common to porous media. Knowledge of semivariograms that are associated with specific structure can be used as a tool to identify that

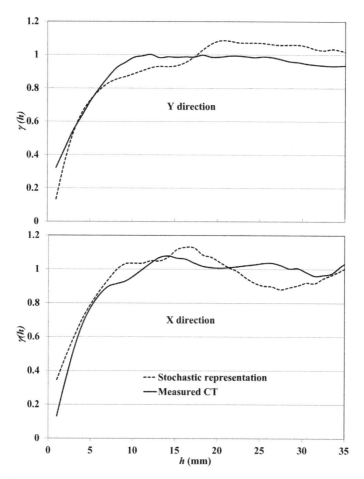

Fig. 9–10. Density semivariograms in two directions of a stochastic representation and the computed tomography (CT) data for core A1A.

structure within the porous media. It should be noted, however, that this is only a tool, and two different structures could possess similar semivariograms.

The multiphase bin semivariogram has been introduced to determine structure size in defined bulk density ranges, or bins. MBS allows geostatistical characterization of different rock types and voids within the porous media from computed tomography imagery. As expected, the traditional semivariogram results are a weighted average of the various MBS.

Finally, by combining artificial images, histogram analysis, semivariogram analysis, and MBS analysis, a stochastic representation of porous media has been generated. These representations have the potential to be used in flow and transport modeling or may be used as an initial image for advanced modeling techniques such as simulated annealing.

References

Anderson, S.H., and J.W. Hopmans, editors. 1994. Tomography of soil-water-root processes. SSSA Spec. Publ. 36. SSSA, Madison, WI.

Brown, G.O., H.T. Hsieh, and D.A. Lucero. 2000. Evaluation of laboratory dolomite core size using representative elementary volume concepts. Water Resour. Res. 36(5):1199–1207. doi:10.1029/2000WR900017

Brown, G.O., M.L. Stone, and J.M. Gazin. 1993. Accuracy of gamma ray computerized tomography in porous media. Water Resour. Res. 29:479–486. doi:10.1029/92WR02151

Brown, G.O., M.L. Stone, J.M. Gazin, and S.R. Clinkscales. 1994. Gamma ray tomography measurements of soil density variation in soil cores. In: S.H. Anderson and J.W. Hopmans, editors, Tomography of soil-water-root-processes. SSSA Spec. Publ. 36. SSSA, Madison, WI. p. 87–97.

Clausnitzer, V., and J.W. Hopmans. 1999. Determination of phase-volume fractions from tomographic measurements in two-phase systems. Adv. Water Resour. 22(6):577–584. doi:10.1016/S0309-1708(98)00040-2

Cressie, N.A. 1991. Statistics for spatial data. John Wiley and Sons, New York.

Deutsch, C.V., and A.G. Journel. 1998. GSLIB: Geostatistical software library and user's guide. 2nd ed. Oxford Univ. Press, New York.

Gandin, L.S. 1963. Objective Analysis of Meteorological Fields. (In Russian, with English translation.) Israel Program for Scientific Translations, Jerusalem. p. 47.

Goovaerts, P. 1997. Geostatistics for natural resources evaluation. Oxford Univ. Press, New York.

Goovaerts, P. 2000. Estimation or simulation of soil properties? An optimization problem with conflicting criteria. Geoderma 97:165–186. doi:10.1016/S0016-7061(00)00037-9

Grevers, M.C.J., and E. de Jong. 1994. Evaluation of soil-pore continuity using geostatistical analysis on macroporosity in serial sections obtained by computed tomography scanning. In: S.H. Anderson and J.W. Hopmans, editors, Tomography of soil-water-root processes. SSSA Spec. Publ. 36. SSSA, Madison, WI. p. 73–86.

Hahn, G.J., and S.S. Shapiro. 1967. Statistical models in engineering. John Wiley & Sons, New York.

Herzfeld, U.C., and O. Zahner. 2001. A connectionist-geostatistical approach to automated image classification, applied to the analysis of crevasse patterns in surging ice. Comput. Geosci. 27:499–512. doi:10.1016/S0098-3004(00)00089-3

Holt, R.M. 1997. Conceptual model of transport processes in the Culebra Dolomite member, Rustler Formation. Contractor Rep. Sandia National Laboratories, Albuquerque, NM.

Hsieh, H.T., G.O. Brown, and M.L. Stone. 1998. Measurement of porous media component content and heterogeneity using gamma ray tomography. Water Resour. Res. 34(3):365–372. doi:10.1029/97WR02228

Ioannidis, M.A., I. Chatzis, and M.J. Kwiecien. 1999. Computer enhanced core analysis for petrophysical properties. J. Can. Petrol. Technol. 38(3):18–24.

Isaaks, E.H., and R.M. Srivastava. 1989. An introduction to applied geostatistics. Oxford Univ. Press, New York.

Journel, A.G., and E.H. Isaaks. 1984. Conditional indicator simulation: Application to a Saskatchewan uranium deposit. Math. Geol. 16(7):685–718. doi:10.1007/BF01033030

Jupp, D.L.B., A.H. Strahler, and C.E. Woodcock. 1989. Autocorrelation and regularization in digital images II. Simple image models. IEEE Trans. Geosci. Rem. Sens. 27(3):247–258. doi:10.1109/36.17666

Lappin, A.R., R.L. Hunter, D.P. Garber, and P.B. Davies. 1989. Systems analysis, long-term radionuclide transport, and dose assessments, Waste Isolation Pilot Plant (WIPP), Southeastern New Mexico, SAND89-0462. Sandia National Laboratories, Albuquerque, NM.

Leuangthong, O., J.A. McLennan, and C.V. Deutsch. 2004. Minimum acceptance criteria for geostatistical realizations. Nat. Resour. Res. 13(3):131–141. doi:10.1023/B:NARR.0000046916.91703.bb

Miranda, F.P., J.A. MacDonald, and J.R. Carr. 1992. Application of the semivariogram textural classifier (STC) for vegetation discrimination using SIR-B data of Borneo. Int. J. Remote Sens. 13(12):2349–2354. doi:10.1080/01431169208904273

Novakovic, D., C.D. White, R.M. Corbeanu, W.S. Hammon, III, J.P. Bhattacharya, and G.A. McMechan. 2002. Hydraulic effects of shales in fluvial-deltaic deposits: Ground-penetrating radar, outcrop observations, geostatistics, and three-dimensional flow modeling for the Ferron sandstone, Utah. Math. Geol. 34(7):857–893. doi:10.1023/A:102098071193

Ortiz, J.M., and C.V. Deutsch. 2004. Indicator simulation accounting for multiple-point statistics. Math. Geol. 36(5):545–565. doi:10.1023/B:MATG.0000037736.00489.b5

Özturk, C.A., and E. Nasuf. 2002. Geostatistical assessment of rock zones for tunneling. Tunneling Underground Space Technol. 17(3):275–285. doi:10.1016/S0886-7798(02)00023-8

Pranter, M.J., Z.A. Reza, and D.A. Budd. 2006. Reservoir-scale characterization and multiphase fluid-flow modeling of lateral petrophysical heterogeneity within dolomite facies of the Madison Formation, Sheep Canyon and Lysite Mountain, Wyoming, USA. Petrol. Geosci. 12:29–40. doi:10.1144/1354-079305-660

Rasiah, V., and L.A.G. Aylmore. 1998. Computed tomography data on soil structural and hydraulic parameters assessed for spatial continuity by semivariogram geostatistics. Aust. J. Soil Res. 36:485–493. doi:10.1071/S97069

Skøien, J.O., and G. Blöschl. 2006. Scale effects in estimating the variogram and implications for soil hydrology. Vadose Zone J. 5(1):153–167. doi:10.2136/vzj2005.0069

Solie, J.B., W.R. Raun, and M.L. Stone. 1999. Submeter spatial variability of selected soil and Bermuda grass production variables. Soil Sci. Soc. Am. J. 63:1724–1733. doi:10.2136/sssaj1999.6361724x

Swan, A.R.H., and J.A. Garratt. 1995. Image analysis of petrographic textures and fabrics using semivariance. Mineral. Mag. 59:189–196. doi:10.1180/minmag.1995.059.395.03

Tavchandjian, O., A. Rouleau, and D. Marcotte. 1993. Indicator approach to characterize fracture spatial distribution in shear zones. In: A. Soares, editor, Geostatistics Troia 1992. Quantitative Geology and Geostatistics Series 1. Kluwer, Boston, MA. p. 965–976.

Tercier, P., R. Knight, and H. Jol. 2000. A comparison of the correlation structure in GPR images of deltaic and barrier-split depositional environments. Geophysics 65(4):1142–1153.

Tidwell, V.C., and J.L. Wilson. 2000. Heterogeneity, permeability patterns, and permeability upscaling: Physical characterization of a block of massillon sandstone exhibiting nested scales of heterogeneity. SPE Reservoir Eval. Eng. 3(4):283–291.

Vogel, J.R., and G.O. Brown. 2002. Geostatistics and the representative elementary volume of gamma ray tomography attenuation in rock cores. In: F. Mees et al., editors, Applications of X-ray computed tomography in geosciences. Geological Society, London, Special Publications 215:81–93.

White, C.D., D. Novakovic, S.P. Dutton, and B.J. Willis. 2003. A geostatistical model for calcite concretions in sandstone. Math. Geol. 35(5):549–575. doi:10.1023/A:1026282602013

Woodbury, A.D., and E.A. Sudicky. 1991. The geostatistical characteristics of the Borden Aquifer. Water Resour. Res. 27(4):533–546. doi:10.1029/90WR02545

10

Characterization of Porous Media Computer Tomography Images Using Semivariogram Families

Jason R. Vogel* and Glenn O. Brown

Abstract

The concept of the semivariogram family has been introduced as an extension of the semivariogram cloud for use with three-dimensional porous media CT images. A semivariogram family is defined as the collection of directional slice semivariograms from slices of a three-dimensional sample. Analysis using semivariogram families of computerized tomography (CT) images of Culebra Dolomite provides improved methods for determining the spatial correlation between structures of the bulk density of porous media. The largest reliable lag in a semivariogram is quantified by analyzing the variance of the semivariogram values within the semivariogram family at each lag. This improves on previous methods for estimating the largest reliable lag that rely on various rules of thumb that are not based on actual characteristics of the data set. In the case of tomography data from Culebra Dolomite cores, the percentage of reliable semivariogram estimates for these samples ranges from 54% to 78%. These are larger than the common rule of thumb of 50%, which allows one to reliably use more of the data set to characterize the porous media. In addition, the distribution of the semivariogram values within a semivariogram family at each lag is analyzed to automatically determine the correlation length (range) values for the porous media. Such a procedure provides range values that are more representative of actual structure caused by differences in porous media bulk density (such as outcrops, vugs, or layering), as opposed to values determined from best-fitting empirical semivariogram models. For these samples, all semivariograms were fit with an R^2 exceeding 0.94. These methods contribute to the goal of utilizing small cores to create small-scale synthetic input data for modeling large-scale porous media systems to predict potential ground water contamination or develop remediation strategies for environmental cleanup.

Abbreviations: CT, computerized tomography; SEM, scanning electron micrographs; WIPP, Waste Isolation Pilot Plant.

J.R. Vogel and G.O. Brown, Biosystems and Agricultural Engineering, Oklahoma State Univ., Stillwater, OK.*Corresponding author (jason.vogel@okstate.edu).

doi:10.2136/sssaspecpub61.c10

Soil–Water–Root Processes: Advances in Tomography and Imaging. SSSA Special Publication 61. S.H. Anderson and J.W. Hopmans, editors. © 2013. SSSA, 5585 Guilford Rd., Madison, WI 53711, USA.

Geostatistics is an important tool for characterizing
spatial properties of porous media. Applications have included investigations of
the spatial variability of subsurface hydraulic conductivity (Schafmeister-Spierling and Pekdeger, 1989; Woodbury and Sudicky, 1991), soil geochemical properties (Miesch, 1975; Solie et al., 1999), rock fractures (Tavchandjian et al., 1993),
soil temperature (Yates et al., 1988), permeability (Goggin et al., 1988), and soil
water content (Greenholtz et al., 1988; Bárdossy and Lehmann, 1998). Geostatistical image analysis also allows for micro-scale nondestructive characterization of
the spatial variability of porous media bulk density, which can in turn be used for
a variety of predictive transport models. Grevers and de Jong (1994) used X-ray
computer tomography (CT) images to analyze soil macroporosity while ground-
penetrating radar images were analyzed by Tercier et al. (2000) to compare the
correlation between structures of different depositional environments. Ioannidis
et al. (1999) used geostatistical analysis of scanning electron micrographs (SEM)
to evaluate various transport and capillary properties of rock samples. Gamma-
ray CT has been used to quantify the spatial variability of soil properties including bulk density, porosity, and unsaturated hydraulic conductivity in soil columns (Rasiah and Aylmore, 1998) and rock cores (Vogel and Brown, 2002). More
recently, Al-Raoush and Willson (2005) used semivariograms to extract pore network properties acquired by synchrotron microtomography images of various
artificial porous media, and Sander et al. (2008) applied geostatistics to bulk density estimated by X-ray CT images of a Chinese soil.

All of these applications depend on reliable determination of the spatial
structure of the soil property examined. Gelhar (1986) pointed out "the key role
that the spatial correlation structure of the input processes plays in the behavior of all the stochastic solutions" in subsurface hydrologic models. In addition,
Gelhar states that "there is a need to find better methods to determine the input
spatial correlation structure of hydraulic parameters. This should include the use
of prior information on geologic and geomorphic conditions as well as improved
methods of statistical analysis to estimate these parameters." This need is reiter-
ated by Gelhar (1993) and Calvete (1997).

Two questions addressed here concerning the reliability of spatial character-
izations are (i) what is the maximum reliable lag on a semivariogram that should
be used? and (ii) which correlation lengths determined from best fit empirical
models are representative of actual structural features? The objective of this study
is to use the large data set of bulk density provided by γ ray CT analysis of Culebra Dolomite cores to introduce semivariogram families and show how they can
be used to identify the maximum lag for reliable semivariogram estimates, and to
automatically detect correlation lengths (ranges) representative of actual structure
for semivariograms derived from computer tomography images of porous media.

Geostatistical Theory
Semivariograms

The semivariogram is one of the most important tools in geostatistics. For an
intrinsically stationary random function $Z(x)$, the semivariogram is the plot of the
semivariogram value, $\gamma(h)$, vs. lag distance, h. The semivariogram value for each
lag distance is defined as

$$\gamma(h) = \frac{1}{2} \text{Var}\big[Z(x) - Z(x+h)\big] \qquad [1]$$

Because $\gamma(h)$ cannot be calculated directly, an unbiased estimator for the semivariogram of a random function is used and is given by

$$\hat{\gamma}(h) = \frac{1}{2N(h)} \sum_{i=1}^{N(h)} \big[Z(x_i + h) - Z(x_i)\big]^2 \qquad [2]$$

where $N(h)$ is the number of pairs of variables at h distance apart (Olea, 1999).

A sill and range characterize the semivariogram. The sill is a plateau in the semivariogram values that corresponds to the variance of the sampled data. If the semivariogram is standardized, the $\gamma(h)$ value is divided by the variance and the sill value will be equal to one. The range, or correlation length, is the distance at which the semivariogram reaches a plateau. One possible semivariogram characteristic is the hole effect, which is described as a semivariogram that peaks and then dips, suggesting that, at greater distances, samples are more related. The hole effect may be seen as the rough spacing between adjacent lenses or bedding planes. Another characteristic similar to the hole effect is pseudocycling. Pseudocycling is the apparent periodic cycling or oscillation of the magnitude of the semivariogram over distance and is common in minerals. However, unless they can be attributed to a physical phenomenon, these changes should be considered random even though they appear periodic (Solie et al., 1999).

Semivariograms are fit by a number of analytical relationships to ensure that the semivariogram's model is positive definite. One of the most common semivariogram fittings used in hydrology is the exponential model (Woodbury and Sudicky, 1991). The model is given by

$$\gamma(h) = 1 - e^{\left(-\frac{3h}{a}\right)} \qquad [3]$$

The exponential model reaches its sill asymptotically, with the practical range a defined as that distance at which the semivariogram value is 95% of the sill (Isaaks and Srivastava, 1989). Agterberg (1970) has shown that a continuous random variable in three-dimensional space has an exponential autocorrelation function if it is subject to a property analogous to the Markov property in time-series analysis.

Multiple, or nested, sills in the semivariogram may be associated with physical phenomena occurring at different scales. Detection of nested sills requires sample spacing shorter than the minimum detected range. As with pseudocycling, nested sills are only considered when they can be associated with physical phenomena (Solie et al., 1999). Nested sills can be modeled by using a linear combination of positive definite semivariogram models with positive coefficients. This property results in the family of models

$$\gamma(h) = \sum_{i=1}^{n} |w_i| \gamma_i(h) \qquad\qquad [4]$$

which are positive definite as long as the n individual models are all positive definite. A weighting function, w_i, is defined for each individual model subject to $\sum_{i=1}^{n} w_i = 1$ (Isaaks and Srivastava, 1989).

Semivariogram Families

Chilès and Delfiner (1999) describe the semivariogram cloud as introduced by Gandin (1963) and used by Chauvet (1982) for meteorological fields. The semivariogram cloud is a plot of all sample pairs (α, β) showing $h_{\alpha\beta} = |x_\beta - x_\alpha|$ on the x axis and the half squared increment, $\frac{1}{2}(z(x_\beta) - z(x_\alpha))^2$, on the y axis. We introduce the semivariogram family as an extension of the semivariogram cloud for use with three-dimensional CT data. A semivariogram family is a plot of the semivariograms for each CT image slice in one direction. It is similar to the semivariogram cloud, except that the semivariogram curve for each slice is plotted. Each point that makes up a semivariogram family "member" therefore represents the mean of all the half squared increments at that lag distance in that slice, as opposed to the point difference represented by the semivariogram cloud.

Semivariogram Estimate Reliability

Besides rules of thumb, there are currently no tests to determine the reliability of semivariogram estimations at large lags (Olea, 1999). Because the accuracy of estimates is proportional to the number of pairs, at greater lags (where there are fewer pairs) the estimates are less reliable. A common practice is to limit semivariogram estimation to lags with a minimum of 30 pairs (Journel and Huijbregts, 1978). Another practical guideline is to limit the lag of the experimental semivariogram to half the extreme distance in the sampling domain for the direction to be analyzed (Journel and Huijbregts, 1978; Clark, 1979).

We determine the largest reliable lag for semivariogram estimation by analyzing the variance of the semivariogram families at each lag distance. A breakpoint exists at which the variance (s^2) starts to increase at a large rate. This could be said to be the point where the variance goes from being a result of structural differences to being a result of too few sample pairs or being at the extreme edges of the sample. For a relatively homogeneous media, this point can be determined by breaking the lag distance, h, versus s^2 plot into two parts and minimizing the sum of the squared differences between the actual s^2 values and a linear approximation of the two parts. However, as demonstrated by the results, this method may not be adequate when analyzing samples with stratification or other irregularities. In these cases the breakpoint is determined visually from the plot.

Automated Range Detection

The correlation length, or range, a, is currently estimated by best-fitting semivariogram models to the estimated semivariogram. However, a method to determine correlation length directly from the data would be more desirable than this empirical curve-fitting method. Estimated semivariogram values from Eq. [2] are distributed by a chi-square distribution (Cressie, 1991). Davis and Borgman (1982) show that the distribution approaches a normal distribution as the number of data pairs goes to infinity for a random, second-order stationary function, or

$$L\left\{\frac{\left[\hat{\gamma}(h)-\gamma(h)\right]}{\sigma\left[\hat{\gamma}(h)\right]}\right\} \to N(0,1) \text{ as } n \to \infty, \qquad [5]$$

where $L(X)$ denotes the law, or distribution of a random variable, $\hat{\gamma}(h)$ is the estimated semivariogram value, σ is the standard deviation, n is the number of data points, and $N(0,1)$ is the standard normal distribution.

Using this property on semivariogram families of CT images, which provide hundreds of thousands of data points, the expected distribution of

$$U(h) = \frac{\left[\hat{\gamma}(h)-\gamma(h)\right]}{\sigma\left[\hat{\gamma}(h)\right]} \qquad [6]$$

at each lag distance h, would be the standard normal. However, if there is significant structure corresponding to correlation length h, the resulting distribution will be multi-modal and not normally distributed. Similarly, the distribution of the $\hat{\gamma}(h)$ values at a particular lag distance can be tested for normality since the $\gamma(h)$ and $\sigma\left[\hat{\gamma}(h)\right]$ values can be considered constant at each lag distance. Therefore, by testing the normality of the $\hat{\gamma}(h)$ values at each lag, we determine if there is structure occurring at that lag. In other words, the range of the semivariogram can be detected based on the distribution of the semivariogram family instead of by empirical curve-fitting of semivariogram models.

To determine normality of the $\hat{\gamma}(h)$ distributions, the Shapiro–Wilk W test (Shapiro and Wilk, 1965), the Anderson–Darling test (Anderson and Darling, 1954), the Martinez–Iglewicz test (Martinez and Iglewicz, 1981), and the D'Agostino skewness test (D'Agostino et al., 1990) have all been employed with a five percent decision criteria. A decision to reject normality from any one of these tests is considered sufficient to determine non-normality.

Materials and Methods

Samples

Three horizontal cores were collected from the Culebra Dolomite Member of the Rustler Formation from the Waste Isolation Pilot Plant (WIPP) site in southeastern New Mexico and are listed in Table 10–1. Core C2611A had a diameter of 147 mm and length of 450 mm. The barrel size for sample collection was selected such that the core was representative with respect to heterogeneity (Lucero et al., 1997). Core sample C2611A had been used in numerous actinide transport experiments and was scanned saturated for operational reasons. Vogel and Brown (2002) indicated that the two halves of this sample have different anisotropy and were therefore analyzed in two sections (slices 30–90 and slices 90–150). The other two samples (C1AV and C2AV) had diameters of 37 mm and lengths of 52 mm and were subsamples of a larger core that visually appeared to have different compositions and differing geostatistical properties. Sample C2AV contains a higher

Table 10–1. Culebra Dolomite sample descriptions.

Sample name	Sample dimensions, X x Y x Z	Voxel size	Attenuation		Bulk density	
			Mean	SD	Mean	SD
	voxels	mm	mm^{-1}		$Mg\ m^{-3}$	
C2611A, slice 30–90	$61 \times 81 \times 61$	$1.5 \times 1.5 \times 3.0$	0.0174	0.0013	2.57	0.19
C2611A, slice 90–150	$61 \times 81 \times 61$	$1.5 \times 1.5 \times 3.0$	0.0171	0.0013	2.53	0.19
C1av	$71 \times 71 \times 46$	$0.37 \times 0.37 \times 0.75$	0.0183	0.0030	2.32	0.38
C2av	$71 \times 71 \times 46$	$0.37 \times 0.37 \times 0.75$	0.0192	0.0024	2.44	0.30

percentage of solid dolomite than C1AV (Vogel and Brown, 2002). A complete geo-statistical characterization and comparison of these cores is found in Vogel and Brown (2002).

Computed Tomography

Each core was scanned in a three-dimensional grid using the custom pencil-beam γ ray CT scanner of Brown et al. (1993). The x, y, and z axes on the horizontal cores are defined as shown in Fig. 10–1. The bedding planes of the Culebra Dolomite are parallel to the x and z axes of the cores. Cores C1AV and C2AV were imaged at 71 uniformly spaced slices, while C2611A was imaged at 150 positions, not all of which were used here. The z-direction voxel spacing was double the spacing in the x and y direction for these cores. The images from C1AV and C2AV have also been used for analyses by Hsieh et al. (1998) and Brown et al. (2000).

CT scanning parameters are listed in Table 10–1. For the analysis here, core C2611A has been separated into two different sections to illustrate differences in slices 30 to 90 and 90 to 150 of the same core. Semivariograms are completed on data in rectangular prisms in the center of the cylinders for ease of computation. The bulk density in Table 10–1 is calculated by

$$\rho_b(x,y,z) = C\mu(x,y,z) \tag{7}$$

where C is the calibration factor and $\mu(x,y,z)$ is the voxel attenuation value (Luo and Wells, 1992). The calibration coefficient, which varies for different collimation, was 127 g mL^{-1} mm^{-1} for cores C1AV and C2AV and 147.8 g mL^{-1} mm^{-1} for core C2611A.

Computerized Tomography Imaging

A three-dimensional grid using the custom pencil-beam γ ray CT scanner of Brown et al. (1993) has been used to scan the four samples. While much slower than X-ray CT, γ CT does not suffer from the beam hardening and photon scattering that distort X-ray images and would make geostatistical analysis problematic. The x and z axes on these samples are defined as horizontal, while the y axis is vertical. Cores C1AV and C2AV were imaged at 71 uniformly spaced slices with a voxel size of 0.37 by 0.37 by 0.75 mm, while core C2611A was imaged at 150 uniformly spaced

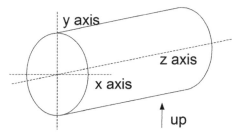

Fig. 10–1. Orientation of the cores.

slices, not all of which were used here, with voxel size of 1.5 by 1.5 by 3.0 mm. The z direction voxel spacing was double the spacing in the x and y direction.

Semivariogram Analysis

The FORTRAN 77 geostatistical programs of Deutsch and Journel (1998) were used for semivariogram analysis of the bulk density data from the soil cores. However, the GAM code was recompiled using Visual Fortran 6.6 to increase the maximum number of data points that could be analyzed. TableCurve 2D was used for best-fitting semivariogram models (TableCurve 2D, 1996) and NCSS (2000) was used for the normality tests.

Results and Discussion
Semivariogram Families

Semivariogram families have been plotted in the x and y direction for the four core samples in Fig. 10–2 and 10–3. The number of members in each family corresponds to the number of voxels in the z direction as shown in Table 10–1. Analysis was not completed in the z direction along the core axis because of the initial slice groupings of the data, but would be possible with further data manipulation to group the slices perpendicular to the x or y direction. These samples do not exhibit a nugget effect in either the x or y direction at these resolutions. Sample C2611A (slice 30–90) (Fig. 10–2b) exhibits pseudo-cycling in the y direction. The pseudo-cycling may indicate layering of rock with different densities at these lag distances in the vertical direction on this core. A hole effect is shown for sample C2611A (slice 90–150) (Fig. 10–2d) in the y direction. This hole effect occurred at a lag distance of approximately 75 mm and is a result of bulk density similarities at this distance in the core; however, the size of the core precludes us from determining whether there is actual layering in the vertical direction at this lag distance. Sample C2611A (slice 90–150) in the x direction (Fig. 10–2c) and C2AV in the y direction (Fig. 10–3d) show slices that may be considered outliers. Further analysis of these slice semivariograms, however, indicates that these semivariograms are for sequential slices. Therefore, these outlier semivariograms are a result of actual micro-scale sample structure, which result in large differences in bulk density, and not a result of operator error during scanning of that slice.

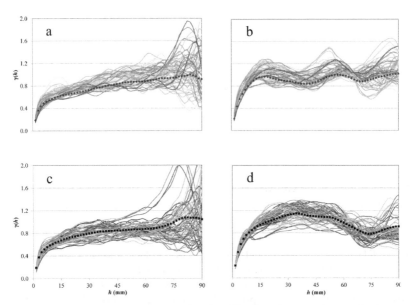

Fig. 10–2. Bulk density semivariogram families (lines) and composite semivariograms (squares) for sample C2611A for (a) slices 30–90 in the *x* direction, (b) slices 30–90 in the *y* direction, (c) slices 90–150 in the *x* direction, and (d) slices 90–150 in the *y* direction.

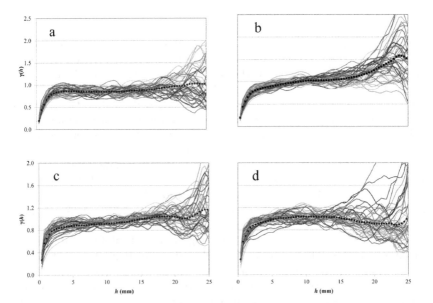

Fig. 10–3. Bulk density semivariogram families (lines) and composite semivariograms (squares) for (a) C1AV in the *x* direction, (b) C1AV in the *y* direction, (c) C2AV in the *x* direction, and (d) C2AV in the *y* direction.

Semivariogram Estimate Reliability

Figures 10–4 and 10–5 show the sample variance (s^2) values of the semivariogram families at each lag and the breakpoint where the variance goes from being a result of sample structure to being a result of too few lags and being at the extreme edges of the sample. The breakpoint is determined by minimizing the sum of the squared differences between the actual s^2 values and a linear approximation of the two parts. However, in directions where the plots are irregular (Fig. 10–4b, 10–4d, and 10–5c), the breakpoint cannot be selected by optimizing the linear models, but rather must be chosen by visually detecting the breakpoint.

The results of the s^2 breakpoint data for determining the maximum lag at which a reliable semivariogram can be estimated are summarized in Table 10–2. The percentage of reliable semivariogram estimates for these samples ranges from 54% to 78% and is generally larger in the x direction than the y direction. These percentages are relatively consistent for each direction, even though two of the cores have different voxel sizes. The differences when comparing the percentage of reliable semivariogram estimates in the x and y directions are likely a result of greater heterogeneity from stratification in the vertical direction. As a check, the sum of the percentage of reliable estimates plus the percentage of pairs needed for a reliable estimate should be slightly greater than 100%. This check is not equal to exactly 100% because the maximum number of pairs will be slightly smaller than the total number of data points compared. These results show more reliable lags than the rule of thumb that considers half of the extreme distance in the sampling domain for that direction as reliable. If the rule of thumb is that 30 pairs were used as an estimate of reliable values, all semivariogram values from

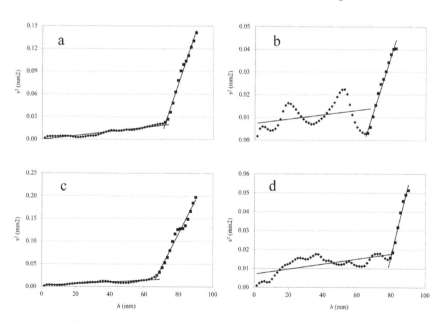

Fig. 10–4. Bulk density variance of the semivariogram family distributions for core C2611A (a) slice 30–90 in the *x* direction, (b) slice 30–90 in the *y* direction, (c) slice 90–150 in the *x* direction, and (d) slice 90–150 in the *y* direction.

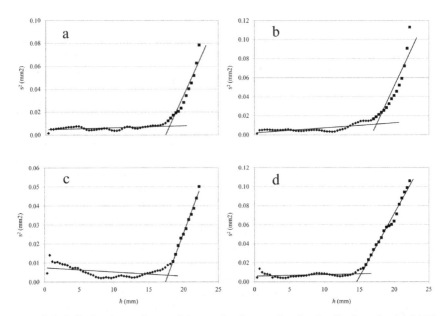

Fig. 10–5. Bulk density variance of the semivariogram family distributions for (a) C1AV
in the x direction, (b) C1AV in the y direction, (c) C2AV in the x direction, and (d) C2AV
in the y direction.

Table 10–2. Summary of s^2 breakpoint data for determining maximum lag at which a
reliable semivariogram can be estimated.

Sample	Direction	Break-point	Reliable estimates	Pairs in $\hat{\gamma}(h)$ at breakpoint	Max. pairs (pairs at $h = 1$)	Pairs for reliable estimate
		mm	%			%
C2611A	x	70.5	78	69,174	296,460	23
(30–90)	y	64.5	54	141,398	297,680	48
C2611A	x	66.0	73	83,997	296,460	28
(90–150)	y	78.0	65	107,909	297,680	36
C1AV	x	17.4	67	78,384	228,620	34
	y	16.7	64	84,916	228,620	37
C2AV	x	18.1	70	71,852	228,620	31
	y	15.2	59	97,980	228,620	43

these CT images would be considered reliable because of the large number of
data points utilized for these comparisons.

Automated Range Detection

By analyzing the distribution of the $\hat{\gamma}(h)$ values at each lag, range values for the
semivariograms of each Culebra Dolomite sample have been determined. These
range values are analogous to the dimension in each direction of various porous

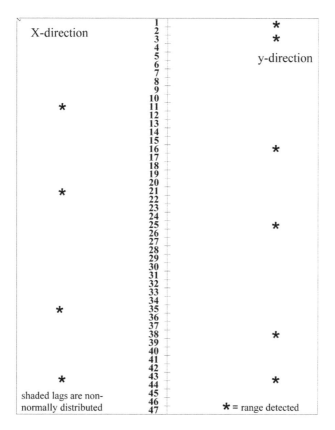

Fig. 10–6. Results of normality tests for each lag (in mm) of the semivariogram families of C2611A (slice 30–90).

media structures that have different bulk densities such as vugs, outcrops, and layers (Brown and Vogel, 2012). Complete results of the normality tests at each lag for sample C2611A (slice 30–90) are shown in Fig. 10–6. Of note are the groupings of non-normal semivariogram family distributions. These groupings could be said to represent elliptical structure with a different bulk density within the porous media such as an outcrop or a large vug. Jupp et al. (1989) and Swan and Garratt (1995) discuss the use of ellipsoids to represent structure in geostatistical image analysis, and Agterberg (1970) describes a three-dimensional autocorrelation function as a hypersurface with an ellipsoid contour. As an example, a directional semivariogram of an ellipsoid with a major axis diameter of 11 lags would result in a grouping of correlation lengths similar to those exhibited in Fig. 10–6, lag = 6 to lag = 11. The directional diameter of this ellipsoid corresponds to the largest value in this range. Single non-normal semivariogram family distributions may be a result of directional cracks or an anomaly of the normality tests at that lag. Examples of the normality plots for a normal and a non-normal lag distance from C2AV in the *y* direction are shown in Fig. 10–7.

Each of the composite estimated semivariograms has been fitted with a combination of exponential semivariograms models using the detected range values.

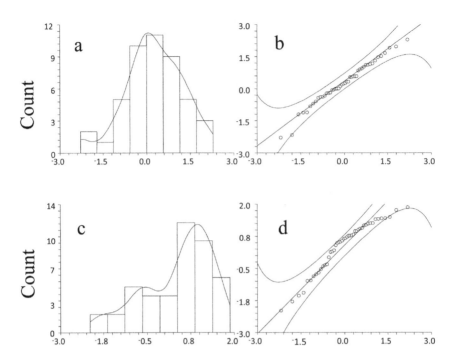

Fig. 10–7. Example bulk density frequency histogram and normality plot of the $\hat{\gamma}(h)$ values for C2AV in the y direction at a lag distance that is normally distributed (a) and (b) and non-normally distributed (c) and (d).

Table 10–3 lists the ranges detected from analysis of semivariogram families, the number of consecutive non-normal semivariogram family distributions contributing the detected range, the sill values (w_i), and the R^2 value between the semivariogram model and the estimated semivariogram. All semivariograms have been fit with an R^2 value of at least 0.94. The number of consecutive non-normal semivariogram family distributions is generally proportional to the w_i values to the extent that a large number of consecutive non-normal distributions corresponds to larger w_i values. The exceptions to this are at small lag distances ($h <$ 4 lags) and in the x direction of C2611A (30–90) at $h =$ 40.5 mm. In this last case, the detected range at $h = 22.5$ mm (15 lags) may represent a slight hole and offset the next effect of additional structure at $h = 40.5$ mm (27 lags) in the best fit semivariogram model. Hole effects were also detected using this method for C2611A (30–90) in the y direction and C2611A (90–150) in the y direction. These slight hole effects may indicate larger-scale layering, be an anomaly in the core, or be an artifact of fewer data points in the comparisons at larger lag distances.

The normality tests do not always detect ranges at small lag distances ($h < 4$ lags). Non-normality is probably not an indicator of a correlation at $h = 1$ lag, as demonstrated by the $w_i = 0.00$ values at that lag for samples C2611A (30–90) in the y direction, C2611A (90–150) in the y direction, and C2AV in the y direction. These results occur due to both the structure of the porous media and to artifacts created by the back-projection reconstruction algorithm and filter function used for

Table 10–3. Range and semivariogram model parameters determined from semivariogram family analysis for Culebra Dolomite samples. Lag distance is 1.5 mm for core C26a and 0.37 mm for cores C1AV and C2AV.

Sample	Direction	Range, *a*	Consecutive non-normal lag distributions	Sill, wi	R^2
		mm			
C2611A	x	4.2[†]	–	0.27	0.97
(30–90)		16.5	6	0.03	
		31.5	1	0.00	
		52.5	11	0.23	
		69.0	7	0.38	
	y	1.5	1	0.00	0.99 [‡]
		4.5	1	0.35	
		24.0	6	0.84	
		37.5	5	hole[‡]	
		57.0	7	pseudo-cycling	
		64.5	2	hole	
C2611A	x	3.0	2	0.31	0.99
(90–150)		22.5	2	0.39	
		40.5	7	0.00	
		66.0	19	0.21	
	y	1.5	1	0.00	0.99 [§]
		6.0	1	0.46	
		18.0	6	0.38	
		43.5	8	0.38	
		64.5	4	hole[§]	
C1AV	x	3.1[†]	–	0.86	0.98
		11.1	3	0.00	
	y	4.1	1	0.73	0.96
		9.3	1	0.00	
		10.0	1	0.30	
C2AV	x	1.4[†]	–	0.58	0.94
		4.1	9	0.11	
		10.4	3	0.00	
		11.1	1	0.07	
		12.6	2	0.22	
	y	0.4	1	0.00	0.98
		1.4[†]	–	0.53	
		5.6	12	0.32	
		10.0	2	0.13	
		11.8	3	0.06	
		13.3	3	0.00	

† These ranges are not detected from semivariogram families because they occur at small lags. They are determined from the best fit of the semivariogram model.

‡ R^2 calculated only for range values up to 24 mm because of pseudo-cycling for this sample in this direction.

§ R^2 calculated only for range values up to 43.5 mm because of the hole effect for this sample in this direction.

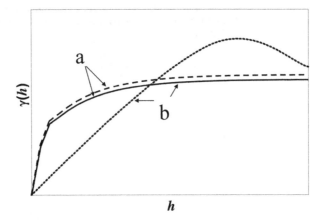

Fig. 10–8. Example bulk density semivariograms for the Wilcoxon matched-pairs signed-ranks test.

CT images (Brown et al., 1993). To offset the lack of automatic detection at these small lags, an additional exponential model is best fit to the estimated semivariogram with a maximum value equal to four lags. This method detected small-scale nested sills in samples C2611A (30–90) in the x direction, C1AV in the x direction, and C2AV in the x and y directions. These nested sills may be indicative of multiple structures when looking at these very short lag distances at the micro scale.

Semivariogram Comparisons

After determining the extent of reliable estimates on a semivariogram, one is tempted to compare semivariograms from two samples with statistical tests to determine if the samples possess similar geostatistical characteristics. Snedecor and Cochran (1967) describe a method for comparing paired samples which may be applicable if the paired samples are normally distributed. However, as shown earlier in this paper, the sample pairs are not normally distributed at lag distances corresponding to the correlation lengths. A nonparametric test such as the Wilcoxon matched-pairs signed-ranks test as described by Sheskin (2000) could therefore be attempted. This test is completed by ranking the absolute values of the difference scores $\left(|D|\right)$ and summing the ranks of the positive scores $\left(\sum R+\right)$ and negative rank scores $\left(\sum R-\right)$ from $h = 0$ to the largest lag distance of a reliable semivariogram estimate. The smaller of these two values is then assigned to the Wilcoxon t test statistic and compared to the critical T value. However, this method fails as well since it discounts obvious trends if one semivariogram contains all values slightly less than the first (Fig. 10–8, pair a), while identifying two semivariograms that cross at the midpoint as similar even though the magnitude of the differences is large (Fig. 10–8, pair b). Comparisons of structure from semivariogram analysis should therefore continue to follow the classical procedure of comparing ranges, sills, and anisotropy.

Conclusions

Semivariogram families have been introduced and shown to provide improved methods for determining the spatial correlation structure of porous media caused by differences in bulk density. By analyzing the variance of the semivariogram values within the semivariogram family at each lag, the maximum reliable lag has been determined. The maximum reliable lag could be said to be the point of the semivariogram where the variance of the semivariogram values at that lag goes from being a result of structural (spatial bulk density) differences to being a result of too few sample pairs. This quantification is an improvement on past methods that rely on rules of thumb to determine the maximum reliable lag on the semivariogram.

By testing the normality of the semivariogram values at each lag within a semivariogram family, range values are automatically detected that represent actual porous media structure. Current methodology is to best-fit empirical semivariogram models that result in range values, which may or may not correspond to actual structure. For our specific samples, the study has indicated that, at these micro scales, similar samples collected near each other may have different geostatistical characteristics.

Finally, the tempting proposition of using the increased characterization of geostatistical properties to apply a statistical test to compare semivariograms from two samples is explored. Because of normality concerns and the inherent nature of nonparametric tests, however, the classical procedure of comparing ranges, sills, and anisotropy is still recommended to compare the spatial correlation structure of two different samples.

These new methods and results contribute toward the ultimate goal of allowing practitioners to analyze small cores and to then create a large representation of the porous media with similar small-scale spatial properties that may be used as synthetic input data for hydrologic modeling. This type of modeling will provide an improvement over some of the arbitrary practices currently used and enhance our ability to predict potential ground water contamination at a site or develop remediation strategies for environmental cleanup.

References

Agterberg, F.P. 1970. Autocorrelation functions in geology. In: D.F. Merriam, editor, Geostatistics. Plenum, New York. p. 113–142.

Al-Raoush, R.I., and C.S. Willson. 2005. Extraction of physically realistic pore network properties from three-dimensional synchrotron X-ray microtomography images of unconsolidated porous media systems. J. Hydrol. 300:44–64. doi:10.1016/j.jhydrol.2004.05.005

Anderson, T.W., and D.A. Darling. 1954. A test of goodness-of-fit. J. Am. Stat. Assoc. 49:765–769. doi:10.1080/01621459.1954.10501232

Bardossy, A., and W. Lehman. 1998. Spatial distribution of soil moisture in small catchment. 1. Geostatistical analysis. J. Hydrol. 206:1-15.

Brown, G.O., H.T. Hsieh, and D.A. Lucero. 2000. Evaluation of laboratory dolomite core size using representative elementary volume concepts. Water Resour. Res. 36(5):1199–1207. doi:10.1029/2000WR900017

Brown, G.O., M.L. Stone, and J.E. Gazin. 1993. Accuracy of gamma ray computerized tomography in porous media. Water Resour. Res. 29(2):479–486. doi:10.1029/92WR02151

Brown, G.O., and J.R. Vogel. 2012. Heterogeneous porous media structure analysis with computerized tomography images and a multi-phase bin semivariogram. In: A. Anderson and J. Hopmans, editors, Tomography and imaging of soil-water-root processes. ASA, CSSA, SSSA, Madison, WI.

Calvete, F.J.S. 1997. Application of geostatistics in subsurface hydrology. In: G. Dagan and S. P. Neuman, editors, Subsurface flow and transport: A stochastic approach. Cambridge Univ. Press, Cambridge, UK. p. 44–61.

Chauvet, P. 1982. The variogram cloud. In: Proceedings of the 17th APCOM International Symposium. Institute of Mining and Metallurgy, Golden, CO. p. 757–764.

Chilès, J.P., and P. Delfiner. 1999. Geostatistics, modeling spatial uncertainty. John Wiley & Sons, New York.

Clark, I. 1979. Practical geostatistics. Applied Science Publishers Ltd., London.

Cressie, N.A.C. 1991. Statistics for spatial data. John Wiley & Sons, New York.

D'Agostino, R.B., A. Belanger, and R.B. D'Agostino, Jr. 1990. A suggestion for using powerful and informative tests of normality. Am. Stat. 44(4):316–321.

Davis, M.D., and L.E. Borgman. 1982. A note on the asymptotic distribution of the sample variogram. J. Int. Assoc. Math. Geol. 14(2):189–193. doi:10.1007/BF01083951

Deutsch, C.V., and A.G. Journel. 1998. GSLIB: Geostatistical software library and user's guide. 2nd ed. Oxford Univ. Press, New York.

Gandin, L.S. 1963. Objective analysis of meteorological fields. (In Russian, with English translation.) Israel Program for Scientific Translations, Jerusalem. p. 47.

Gelhar, L.W. 1993. Stochastic subsurface hydrology. Prentice Hall, Englewood Cliffs, NJ.

Gelhar, L.W. 1986. Stochastic subsurface hydrology from theory to applications. Water Resour. Res. 22(9S):135S–145S. doi:10.1029/WR022i09Sp0135S

Goggin, D.J., M.A. Chandler, G. Kocurek, and L.W. Lake. 1988. Patterns of permeability in eolian deposits. Page Sandstone (Jurassic), Northeastern Arizona. SPE Form. Eval. 3(2):297–306.

Greenholtz, D.E., T.C.J. Yeh, M.S.B. Nash, and P.J. Wierenga. 1988. Geostatistical analysis of soil hydrologic properties in a field plot. J. Contam. Hydrol. 3:227–250. doi:10.1016/0169-7722(88)90033-2

Grevers, M.C.J., and E. de Jong. 1994. Evaluation of soil-pore continuity using geostatistical analysis on macroporosity in serial sections obtained by computed tomography scanning. In: S.H. Anderson and J.W. Hopmans, editors, Tomography of soil-water-root processes. SSSA Spec. Publ. 36. SSSA, Madison, WI. p. 73–86.

Hsieh, H.T., G.O. Brown, M.L. Stone, and D.A. Lucero. 1998. Measurement of porous media component content and heterogeneity using gamma ray tomography. Water Resour. Res. 34(3):365–372. doi:10.1029/97WR02228

Ioannidis, M.A., I. Chatzis, and M.J. Kwiecien. 1999. Computer enhanced core analysis for petrophysical properties. J. Can. Petrol. Technol. 38(3):18–24.

Isaaks, E.H., and R.M. Srivastava. 1989. An introduction to applied geostatistics. Oxford Univ. Press, New York.

Journel, A.G., and C.J. Huijbregts. 1978. Mining geostatistics. Academic Press, London. p. 194.

Jupp, D.L.B., A.H. Strahler, and C.E. Woodcock. 1989. Autocorrelation and regularization in digital images II. Simple image models. IEEE Trans. Geosci. Rem. Sens. 27(3):247–258. doi:10.1109/36.17666

Lucero, D.A., G.O. Brown, and C.E. Heath. 1997. Laboratory column experiments for radionuclide adsorption studies of the Culebra Dolomite member of the Rustler Formation. SAND 97–1763. Sandia National Laboratories, Albuquerque, NM.

Luo, X., and L.G. Wells. 1992. Evaluation of gamma ray attenuation for measuring soil bulk density. I. Laboratory investigation. Trans. ASAE. 35:17-26.

Martinez, J., and B. Iglewicz. 1981. A test for departure from normality based on a biweight estimator of scale. Biometrika 68:331–333. doi:10.1093/biomet/68.1.331

Meisch, A.T. 1975. Variograms and variance components in geochemistry and ore evaluation. Geol. Soc. Am. 142:333–340.

NCSS. 2000. NCSS users guide. NCSS, Kaysville, UT.

Olea, R.A. 1999. Geostatistics for engineers and earth scientists. Kluwer Academic Publishers, Boston. p. 67–75.

Rasiah, V., and L.A.G. Aylmore. 1998. Computed tomography data on soil structural and hydraulic parameters assessed for spatial continuity by semivariogram geostatistics. Aust. J. Soil Res. 36:485–493. doi:10.1071/S97069

Sander, T., H.H. Gerke, and H. Rogasik. 2008. Assessment of Chinese paddy-soil structure using X-ray computed tomography. Geoderma 145:303–314. doi:10.1016/j.geoderma.2008.03.024

Schafmeister-Spierling, M.Th., and A. Pekdeger. 1989. Influence of spatial variability of aquifer properties on groundwater flow and dispersion. In: H.E. Dobus and W. Kinzelbach, editors, Contaminant transport in groundwater. Balkema, Rotterdam, The Netherlands. p. 215–220.

Shapiro, S.S., and M.B. Wilk. 1965. An analysis of variance test for normality. Biometrika 52:591–611.

Sheskin, D.J. 2000. Handbook of parametric and nonparametric statistical procedures. 2nd ed. Chapman and Hall/CRC, Boca Raton, FL. p. 467–477.

Snedecor, G.W., and W.G. Cochran. 1967. Statistical methods. 6th ed. The Iowa State Univ. Press, Ames, IA. p. 91–116.

Solie, J.B., W.R. Raun, and M.L. Stone. 1999. Submeter spatial variability of selected soil and bermuda grass production variables. Soil Sci. Soc. Am. J. 63(6):1724–1733. doi:10.2136/sssaj1999.6361724x

Swan, A.R.H., and J.A. Garratt. 1995. Image analysis of petrographic textures and fabrics using semivariance. Mineral. Mag. 59(2):189–196. doi:10.1180/minmag.1995.059.395.03

Tavchandjian, O., A. Rouleau, and D. Marcotte. 1993. Indicator approach to characterize fracture spatial distribution in shear zones. In: A. Soares, editor, Geostatistics Troia 1992, Quantitative Geology and Geostatistics Series 1. Kluwer, Boston, MA. p. 965–976.

Tercier, P., R. Knight, and H. Jol. 2000. A comparison of the correlation structure in GPR images of deltaic and barrier-split depositional environments. Geophysics 65(4):1142–1153. doi:10.1190/1.1444807

Vogel, J.R., and G.O. Brown. 2002. Geostatistics and the representative elementary volume of gamma ray tomography attenuation in rock cores. In: F. Mees, editor, Applications of X-ray CT on geology and related domains. Geological Society, London.

Woodbury, A.D., and E.A. Sudicky. 1991. The geostatistical characteristics of the Borden Aquifer. Water Resour. Res. 27(4):533–546. doi:10.1029/90WR02545

Yates, S.R., A.W. Warrick, A.D. Matthias, and S. Musil. 1988. Spatial variability of remotely sensed surface temperatures at a field scale. Soil Sci. Soc. Am. J. 52:40–45. doi:10.2136/sssaj1988.03615995005200010007x

11

Electrical Resistivity Tomography of the Root Zone

Alex Furman,* Ali Arnon-Zur, and Shmuel Assouline

Abstract

Understanding root scale processes and primarily those related to uptake in the root zone is very important for proper development of agricultural practices and efficient irrigation management. Such understanding requires an ability to map in high spatial and temporal resolution at the root zone. Electrical resistivity tomography (ERT) is a geophysical method that is very sensitive to the amount of water in the soil, which makes it highly suitable for root zone mapping and monitoring. In this chapter, we review the methodology and some past applications, and through several case studies we demonstrate the limitations of the method for root zone mapping. We identify the cases for which the method is mature or close to mature, that are related mostly to relatively large-scale root zone (trees) and slow (natural) processes, and others (small-scale, high-resolution, and fast processes) for which the technology in our opinion is still limited. Among the major limitations of the technology, we identify several as most important, including the need to develop (water flow) process-based constraints to replace non-process-based ones, the need to develop small scale electrodes for dry conditions, and the need to consider time varying processes in petrophysical relations. We believe that ERT, being a relatively cheap nondestructive noninvasive method for root zone mapping, can already be used at the field scale for various agronomic applications, while as a research tool some improvements are still needed.

Abbreviations: DC, direct current; EM, electromagnetic induction; ERT, electrical resistivity tomography; GPR, ground penetrating radar; IP, induced polarization; TDR, time domain reflectometry; TWW, treated wastewater.

A. Furman, Civil and Environmental Engineering, Technion, Haifa 32000, Israel. A. Arnon-Zur, Geological Survey of Israel, Jerusalem 95501, Israel. S. Assouline, Soil, Water and Environmental Science, Agricultural Research Organization, Bet Dagan 50250, Israel. *Corresponding author (afurman@tx.technion.ac.il).

doi:10.2136/sssaspecpub61.c11

Soil–Water–Root Processes: Advances in Tomography and Imaging. SSSA Special Publication 61. S.H. Anderson and J.W. Hopmans, editors. © 2013. SSSA, 5585 Guilford Rd., Madison, WI 53711, USA.

The ever shrinking availability of arable land, water, nutrients, and economic resources requires efficient agricultural management practices. This is especially true for arid and semi-arid regions. The quest for more efficient irrigation and fertilization techniques led to the development of several important irrigation techniques, such as drip irrigation. More recent important developments include high frequency (continuous) irrigation and fertigation (e.g., Assouline, 2002; Assouline et al., 2002, 2006; Silber et al., 2005; Shock et al., 2005). The common theme in much of the research in irrigation practices and management is the desire to control the environment that exists in the plant's root zone, primarily in terms of water content and its spatiotemporal distribution, but also in terms of other physical and biological parameters, such as pH, redox potentials, nutrient and other solute content, and more.

An efficient control of the root zone environment requires the ability to sense and visualize the root zone environment in relatively high spatial resolution, and in some cases also in high temporal resolution. This is currently true for agricultural and environmental research, but with the increasing awareness of the economic value of precise agriculture, it can be foreseen as the future of at least sections of modern agriculture in general. Typical research practices are somewhat limited in that most sensors (e.g., time domain reflectometry, TDR; microtensiometers) sense relatively small volumes (i.e., provide only point data) and are invasive (and if inserted to an existing root zone, i.e., not pre-installed in site, also destructive). Root excavation and trenching practices are naturally destructive, and in most cases cannot provide any temporal resolution. Rhizotrons and minirhizotrons (e.g., Garré et al., 2011) provide direct access to roots; however, it is not trivial to assume that these roots are a true representation of the root activity.

Clearly there is a need to study root zone states and processes at high spatial and temporal resolution. In the most recent decades, geophysical methods are increasingly used in environmental and agricultural sciences. The improvement of the measuring techniques, both hardware and software, make these capable of studying the root zone at various scales. Some applications of geophysical techniques are presented in this book. The purpose of this chapter is to introduce, through three different applications, electrical resistivity tomography (ERT) as a relatively simple tool for monitoring the root zone at various scales, to discuss its pros and cons, and to point to further research and instrumentation needs. The remainder of this introduction briefly presents root zone processes and modeling concepts as well as the ERT technique in general and specifically for root zone imaging. Later on, three different cases in which ERT was applied to the monitor root zone at three scales—a single plant, a single tree, and a row of trees in an orchard—are presented. Finally, the applicability of the technique for root zone mapping and monitoring is discussed and open challenges are identified.

Root Zone Processes

The root zone hosts a wide variety of physical, chemical, and biological processes that can be examined at various scales. Generally, these include processes related to water flow and to water uptake by the roots, solute transport, including uptake by the roots but also exchange with the soil solid phase, and various bacteria (and larger species) activated processes often grouped under the term biochemical cycling. Each of the processes has its own physical signature on the root zone. For example, Jones (1998) reviews some of the organic acids that are associated with various processes in the root–soil interface. Clearly such acids have their electrical, pH, and redox signature. Most of the sensing and monitoring techniques try to target the unique signature of the process under investigation, but as will be discussed later, in many cases, it is very difficult to differentiate between different signatures, and one senses a mixture of signatures.

This chapter deals with ERT. A wider scope of geophysical methods for root zone tomography is given by Vanderborght et al. (2013). ERT measures the direct current (DC) resistivity of the subsurface, which is sensitive primarily to the amount of water and the amount of solutes in the water (see later for functional relations). Therefore ERT is primarily used to monitor (with respect to the root zone) water content profiles, water content dynamics, and root water uptake. As will be discussed later, less effort is given to solute distribution, solute dynamics, and solute uptake.

All processes can be treated at the pore scale or at a continuum scale. For practical reasons that include both the ability to sense and to model at the pore scale, most research targets the continuum scale. This may change in the not-too-distant future with the introduction of high-resolution sensing technologies such as those presented in other chapters of this book. Root uptake, being perhaps the most important process when talking about the root zone, is also often treated at continuum scale, although the foundations for root-scale treatment were established several decades ago by Gardner (1960), who used a mixed formulation treating the flow in the soil using Richards' equation, but the individual root as a boundary condition. In this approach, the root surface is treated as a boundary condition where flow toward the root is modeled as activated by matric head gradients toward the root (i.e., by setting a low, more negative, matric head as boundary condition at the root surface),

$$\frac{\partial \theta}{\partial t} = \frac{1}{r}\frac{\partial}{\partial r}\left(rK\frac{\partial h}{\partial r}\right) \qquad [1]$$

where θ is the water content, h the matric head, K the hydraulic conductivity that is a function of the water content (or the matric head that is related to it by the water retention function), and r and t are spatial and temporal coordinates, where r is in the radial direction of the root center. Equation [1] is actually a one-dimensional version of the Richards' equation and it describes the flow of water in the soil toward the root vein. The processes that take place at the root–soil interface and in the root "skin" are (to the best knowledge of the authors) much less agreed upon and are out of the scope of this chapter. The reader is referred to Steudle and Peterson (1998) for some concepts for these processes.

Equation [1] and the concept it represents assumes that the spatial distribution of the root system is known in advance. While new techniques may provide that to some spatial resolution (see, e.g., Moradi et al., 2011, and other chapters in this book), it is generally easier to model root uptake in the continuum scale using Richards' equation that includes a sink term to represent root uptake,

$$\frac{\partial \theta}{\partial t} = \frac{\partial}{\partial x}\left(K\frac{\partial h}{\partial x}\right) + \frac{\partial}{\partial y}\left(K\frac{\partial h}{\partial y}\right) + \frac{\partial}{\partial z}\left(K\frac{\partial(h+z)}{\partial z}\right) - S \qquad [2]$$

where S is a sink term representing root uptake, and x, y, and z are spatial coordinates. In the continuum (macroscopic) approach, the individual roots and root veins are not treated explicitly, but in a more statistical way, where root uptake is often related to root length density. For a more comprehensive overview of this modeling approach, including techniques for uptake integration at the whole root scale see, for example, the work by Simunek and Hopmans (2009) and Vrugt et al. (2001).

Root nutrient uptake is much more complex than water uptake. Generally speaking, there are passive and active processes that control nutrient uptake. Passive process means that solutes are carried with water into (and sometimes out of) the root. Active uptake means that the root actively controls the solutes that will pass the plant cell walls directly or with the aid of bacteria (see, e.g., a pioneering review by Bolan, 1991). These processes are out of the scope of this chapter, but it is important to understand that while root water uptake leaves the soil drier than it was, nutrient uptake is more complex and the root not only takes solutes from the soil—it also delivers some back (see, e.g., Hinsinger et al., 2003). Therefore, while using resistivity-based methods to infer root water uptake is natural, using it for nutrient uptake is less trivial. Further discussion is brought later in the text.

Electrical Resistivity Tomography

In recent decades, geophysical methods evolved from a field related primarily to geology to one of much wider scope, including medical, engineering, and environmental applications. Generally speaking, a geophysical investigation involves the recording of an induced or natural physical signature that is related to some other physical property; for example, recording the velocity of the propagation of a mechanical signal and relating that to rock type. Electrical methods, including primarily ERT, induced polarization (IP), and self (streaming) potential, form a subdivision of the geophysical methods that look for the electrical signature of the bulk properties at relatively low frequencies (typically <100 Hz). Electrical signatures are relatively sensitive to water content and solute concentration (compared with that of air and soil solids), which makes geophysical methods of high interest for hydrologists and soil physicists. Other geo-electrical methods [such as ground penetrating radar (GPR), electromagnetic induction (EM), and time domain reflectometry (TDR)] typically use higher frequencies that make it more difficult to decouple the electrical and the magnetic processes. A more detailed description of some of these methods is given by Vanderborght et al. (2013).

Application of ERT (see a more detailed description of resistivity methods in any basic geophysics textbook, e.g., Telford et al., 1990; Butler, 2005) in environmental sciences typically involves three stages:

1.   Application of an electrical current at very low frequency (that can be considered as direct current) through two electrodes while measuring the current and recording the resultant potential drop (voltage) between two

other electrodes, altogether forming a quadripole. This basic measurement is then repeated between a few tens and thousands of times.

2. Using a numerical simulation of current flow (i.e., solution of Laplace equation; note that in the geophysical literature it is common to use Poisson's equation where the injected current is treated as sink and/or source terms) for given boundary conditions and inversion scheme, the electrical resistivity distribution is modeled. This process is mathematically ill-posed, and therefore supporting data is required to reduce the uncertainty in the model.

3. Converting the obtained electrical resistivity values to the soil property of desire (typically water content but sometimes also solute concentration; e.g., Kemna et al., 2002). This is typically done using semi-empiric laws such as Archie's (Archie, 1942).

The measurement stage is often performed using commercial tools (e.g., the Sting family of products, Advanced Geosciences, Inc., USA; the Syscal family of products, Iris Instruments, France; the SAS family, ABEM, Sweden, and more). Most modern commercial systems can handle several tens to hundreds of electrodes and record using more than a single channel, which makes the acquisition semi-automated and relatively fast. The measurement set (the composition of the quadripoles) often follows one or more of classic arrangements (e.g., Wenner, dipole–dipole) but sometimes follows the optimization scheme (see, e.g., Furman et al., 2004; Furman et al., 2007; Stummer et al., 2004).

Inversion (see, e.g., Menke, 1989, for inversion theory) is performed by minimizing the differences between measured surface voltages (or resistances or apparent resistivities) and the model computed values. As the solution of such an inverse problem is inherently ill-posed, most inversion schemes use constraints to eliminate the ill-posedness and assure convergence of the inversion scheme. Generally, the objective function of the geophysical inversion that is to be minimized, ϕ, may be written as

$$\phi = \phi_m + \beta \phi_c \tag{3}$$

where ϕ_m is the data misfit, ϕ_c is the constraint error, and β balances between the two. As in most cases, the constraints are arbitrary and not necessarily physical (e.g., smoothness constraint that requires the solution to avoid sharp changes in resistivity), it inherently introduces bias into the solution, although it does assure a solution in most cases, that otherwise is very far from being assured. In recent years, several attempts were made to constrain the inversion based on the external physical processes of interest [by constraining, e.g., Bamberger, 2011, or by coupled inversion, e.g., Hinnell et al., 2010; Lehikoinen et al. (2009) take this even further and do not consider the process of interest to be stationary so that (at the extreme) individual measurement get their own temporal constraining]; although success was often demonstrated, there is no generic solution to the problem.

The inversion stage is also performed in many cases using commercial software (e.g., RES2DINV, Geotomo, Inc., Malaysia; Earthimager, Advanced Geosciences, Inc., USA; ERTlab, Multi-Phase Technologies LLC., USA and Geostudi Astier srl., Italy) or freeware (e.g., R2, Andrew Binley, Lancaster University, UK; BERT, Thomas Günther, Leibniz Institute for Applied Geosciences, Germany), where typically the commercial systems are more robust but less sophisticated in terms

Fig. 11–1. Sample electrical resistivity tomography inversion cross-section from the second phase of the experiment (left; resistivity in Ω·m) and view of the first phase of the experiment (right).

of the ability to introduce external constraints. Commercial systems are typically also limited in their ability to perform what is called 4-D inversion, that is, invert for multiple times simultaneously (e.g., Karaoulis et al., 2011).

As noted above, the inversion of resistivity data into a profile of the electrical resistivity is inherently an ill-posed problem. We keep these aspects of ERT out of the scope of this paper. The reader is referred to the papers by Day-Lewis et al. (2005) and Nguyen et al. (2009) for further reading. Generally speaking, when an inverse problem is solved, one needs to compute the operator that relates measurements and resistivities (the Jacobian). However, as the last is unknown, it is iteratively approximated. At the bottom line, since this is an approximation and not the true Jacobian, each resistivity value obtained by the inversion has its own reliability. In many cases, this leads to single pixels that stand out of the inverted background and are in fact just an inversion artifact (see for example several very red pixels on the left side of Fig. 11–1; these are probably the combined result of an electrode with bad contact and low sensitivity of a pixel to the measurements).

The separation of the modeled profile of electrical resistivity into its components (i.e., in particular water content and pore water resistivity) is perhaps the most problematic. Generally speaking, the bulk electrical resistivity is a function of the amount of water in the soil and the pore water resistivity, with non-negligible contribution due to solid surface conductivity, significant for clayey soils. Semi-empirically this is expressed as

$$\rho_b = a\rho_w\phi^{-m}S_w^{-n} \qquad [4]$$

where ρ_b is the bulk resistivity, ρ_w is the pore water resistivity, ϕ is the porosity, S_w is the water saturation, a is the tortuosity, and m and n are cementation and saturation exponents, respectively. Equation [4] is also called Archie's law (Archie, 1942) and is the base for most petrophysical relations. Note, however, that this base formulation is limited as it does not account for mineral surface conductivity, not negligible in soils with high silt and clay content, as is the case in one of the studies presented later in this chapter. Practically speaking, while one may get reasonable relations from the literature, it is always better to perform site-specific calibration of petrophysical relations, and this is discussed later around one of the case studies.

The number of applications of ERT in hydrology and soil and agricultural science is rapidly increasing. We focus here on several applications with relevance to the case studies presented later in this chapter. We exclude from the discussion applications of electrical resistivity measurements (as opposed to electrical resistivity tomography). Such applications (e.g., Besson et al., 2010) often do not involve data inversion and typically can be useful in fast acquisition of field-scale data but are relatively weak, in terms of spatial resolution, on the vertical dimension.

Several researchers studied the water content distribution statically and dynamically at the field scale under irrigated and natural conditions using ERT. Michot et al. (2003) used ERT to study the water content dynamics under a sprinkler irrigated corn field. They used a two-dimensional electrode line with electrode separation of 20 cm (corn rows were 80 cm apart) to generate a cross-section of electrical resistivity, later converted to water content using a field calibrated petrophysical equation. Michot et al. (2003) clearly show the dynamics of water content beneath the plant, including relatively dry conditions before irrigation and wetter after. Slower changes are observed between the rows, where roots are less active.

Srayeddin and Doussan (2009) took ERT in field scale one step further and attempted to compute root uptake using successive ERT measurements. In their experiment, they used Wenner α quadripoles with electrode separation of 30 cm (crop rows, corn and sorghum with sprinkler irrigation, were 70 and 50 cm apart, respectively). After calibration of electrical resistivity to water content, they subtracted ERT obtained water content profiles and, assuming vertical flow, estimated the sink term at 5-d intervals. They concluded that ERT can be used to assess root water uptake to limited depth, although better spatial resolution is still required.

While the two examples above deal with root zone tomography at the field scale, Garré et al. (2011) attempted to do that at the laboratory scale. They used ERT to image the three-dimensional water content distribution of a large monolith under drying conditions with weed root (generally horizontally homogeneous) uptake. In parallel, they measured the root length density using several mini-rhizotubes and were able to correlate high water content and high root density to water depletion.

The three examples listed above nicely demonstrate the ability of ERT to measure at the root zone. However, to date, ERT was not often used to map root zone activities in both high spatial and temporal resolution. This may be attributed to many factors, including the need to optimize measurements to increase sensitivity (Srayeddin and Doussan, 2009), the technical complexity of such experiment, cost of equipment, and more. A recent exception is perhaps the work of al Hagrey and Peterson (2011), who applied optimized ERT surveys (surface and borehole) to study the dynamics of the root zone in small scale. The purpose of this chapter is to investigate some of the limiting factors on wider and better application of ERT in root monitoring. This will be done by discussing three case studies, not necessarily successful and not necessarily completed, in which some of the pros and cons of ERT gave rise. In the following, we present the three different applications of ERT in the root zone, including partial results, and discuss issues such as root electrical conductivity, seasonal salinity due to root uptake, ERT in drip irrigated soil, and more.

Case Studies

We present here three different cases where ERT was used to image the root zone of various plants and for various purposes. We start with a three-dimensional imaging of the root zone of a single bell pepper plant that was performed essentially throughout its vegetative growth period, where the goal was to obtain as high as possible spatial and temporal resolution of the water content distribution to study root uptake. We continue through relatively high-resolution two-dimensional imaging of the root zone of an avocado tree, where the goal was to study the effect of prolonged (15–20 yr) irrigation with treated wastewater on soil properties in the root zone. Last, we present a study of a grapefruit orchard where the root zone imaging was performed in relatively coarse resolution as the goal of the study was the deeper vadose zone effects. In all cases, we focus our discussion on the method itself and exclude from the discussion the original objectives of the specific study. We also exclude from the presentation other instrumentation that was used in the study.

The Root Zone of a Single Plant

This study aimed at investigating the spatial and temporal root uptake at relatively high resolution, and to relate the uptake to environmental conditions such as water content and root density distribution. To some degree, this is a combination of the experiment reported by Garré et al. (2011) in terms of spatial resolution ,but with much faster temporal dynamics, with that of Srayeddin and Doussan (2009) in terms of the multi-dimensionality of the root distribution and uptake. ERT here is the main tool for obtaining water content in high temporal and spatial resolution. ERT results will be used at later stages to calibrate Richards' equation based models, from which the root water uptake function will be drawn.

We conducted the experiment twice, with some alternations between the two stages (summer 2008 and spring 2010). Most of the changes were due to mistakes, some foreseeable and some less, made in the first phase. We believe that highlighting these mistakes may help in understanding the difficulties involved in such an experiment. We start by describing the relevant components of the first phase, and then highlight the changes made in the second to correct problems. Below are details of the first phase of the experiment, followed by discussion of changes made for the second phase and further discussion of these changes.

Bell peppers were grown in plastic cylindrical chambers (42 cm in diameter, 55 cm in height; see Fig. 11–1) filled with sand in a greenhouse for about 8 wk after plant stabilization following planting. We considered three different irrigation schemes where all plants received the same daily amount of water through a single dripper located at the center of the sand surface of the chamber, close to the plant stem, but differently distributed throughout the day: (i) all water supplied through a single 2-L-per-h dripper in a single pulse starting 8:00 AM; (ii) water supplied through a 0.25-L-per-h dripper in a single pulse starting 8:00 AM; and (iii) daily amount is equally split between 8 pulses with 30 min separation, starting 8:00 AM, through a 2-L-per-h dripper. The daily amount was set using bucket evaporation and leaf area index. The different schemes were designed to create spatially and temporarily different environmental conditions for comparison. More details are in Stern, 2010. Three chambers (one for each treatment) were connected to a Syscal Switch Pro96 ERT system using 96 electrodes (84 in six rings with the upper three rings at 5 cm vertical separation and the lower three with 10 cm separation, and 12 more electrodes on surface). Similar chambers were used for ERT mapping of plant free soil, and for supporting measurements such as TDR, global and drainage mass and agronomic parameters. ERT measurements were taken once a week. In addition, we destructively mapped root density for 36 other chambers (9 every 2 wk). As suggested above, this phase of ERT measurements was unsuccessful for reasons that will be discussed below, and therefore we do not present any results.

The main problem with the ERT data obtained in the first phase was that it suffered from a very high level of noise, indicated by both high standard deviation of the measurements while "stacking," and many individual extreme measurements (typically of very high resistivity). Analysis of the experimental design suggested that the sand was too coarse, leading to a very narrow wetting plume ("bulb"). That is, as capillary forces were small compared with gravitational forces, over half the electrodes (most surface electrodes and the three upper electrode rings) were too dry. TDR measurements made using a 15-cm probe located 5 cm below surface in similar chambers pointed toward the chamber center starting at its edge (note that the chamber's radius is 21 cm so the TDR probe, with its head, covered practically most of the chamber radius but not the central few centimeters) suggested a water content of 5 to 11%, depending on the exact time and setup. Considering the fact that this is a long-wise average where the inner part of the chamber is probably significantly wetter than the perimeter, this leads to bad contact between the electrodes (5 mm stainless steel rods, inserted about 5 cm into the soil) and the soil. The choice of soil, made primarily to ease on destructive root mapping, was clearly problematic in terms of water flow and of instrumentation sensing. This also created a problem for other instruments such as TDR, as the very sharp wetting front that was created made measurements inaccurate (data not shown).

Contact resistance is a significant issue in application of ERT in dry environments. One of the common practices is to locally wet the electrodes (sometimes with saline water). While in large-scale surface applications, the local change of electrical resistivity is negligible with respect to the problem scale, in small-scale applications this is not trivial (our electrode distance is in many cases 5 cm vertically. It is very difficult at this scale to verify that the water will not create a short-cut between electrodes). This is especially true for wall electrodes where wetting through the plastic container is not easy. One approach that we considered (and not applied eventually) is to use basketball inflating needles as electrodes so that we can occasionally inject few milliliters of water.

In the second phase of the experiment, a major change made was the alternation of the soil type to slightly heavier one. "Hamra," a red sandy soil with about 15% clay, was used instead. This solved the contact problem as most of the chamber was now wet to a reasonable degree. Other changes, less relevant for the key points discussed here, include the reduction of the experiment into two treatments only (as no significant differences were observed, both in terms of plant development as well as in terms of daily mass-balance components) between the low flow (case ii) and the pulse treatment (case iii). For completeness, Fig. 11–1 presents an example of the electrical resistivity distributions in the growth chambers, together with a photo from the first phase of the experiment.

One can easily note the region of high resistivity at the chamber's top. The exact value at each model node needs to be verified against model sensitivity and against external measurements. However, this region roughly coincides with the actual root distribution (see Stern, 2010). While the interpretation of such an image is out of the scope of this paper, it demonstrates the spatial resolution that can be achieved—at the order of the electrode separation.

The Root Zone of an Avocado Tree (Acre)

This study is part of a larger effort to understand long-term effects of irrigation with treated wastewater (TWW) on soil properties. While in the short term (few years) irrigation with TWW was extensively shown to have minimal effect on agricultural productivity, in orchards irrigated with TWW for long terms a reduction in productivity was recorded, accompanied with lessening of other agronomic indices (Assouline and Narkis, 2011). One of the hypotheses raised in this case is that long-term irrigation with high organic loads eventually alters the soil hydraulic properties, leading to too high root-zone water content. The goal of the ERT measurement was to compare the soil water distribution between trees irrigated with freshwater and ones irrigated with treated wastewater. A secondary goal was to study the ability of ERT in very heavy (clayey) organic soils. The research farm is located east of Acco (Acre), Israel, on alluvial clayey soil. The specific plot of the farm is an avocado orchard planted about 20 yr ago with alternating rows of treated wastewater and tap water irrigation. Irrigation is by two drip lines per row, about 30 cm apart on both sides of the tree trunks, with dripper separation of 50 cm. Tree spacing is 4 m (along row) by 6 m (between rows).

Two ERT lines were constructed at the sites along rows 16 (TWW) and 18 (tap water). Stainless steel electrodes, 3 mm in diameter and 15 cm long, were inserted into the soil roughly to half their length along the tree row (in a straight line touching the tree trunks). Altogether, 96 electrodes were used for each line at 10-cm electrode separation. A dipole–dipole survey composed of 7847 measurements (1144 current injections) was conducted for both lines on 13 Apr. 2011, which for the local conditions means that the soil was still fairly wet and that no seasonal associated drip wetting patterns were established.

Inversion was performed using R2 (Andrew Binley, Lancaster University, UK). A rectangular grid of 240 (horizontally) by 80 (vertically) nodes was used with node spacing of 5 cm. Standard smoothening constraints were used and the resistivity of the homogeneous starting model was set to 100 Ωm. Convergence of the inverse solution was achieved in five (TWW) or six (tap water) iterations (desired misfit was set to 1.0; see http://www.es.lancs.ac.uk/people/amb/Freeware/freeware.htm for details).

Figure 11–2 presents the resistivity cross-sections obtained. One can note the high variability at the upper few centimeters, resulting probably from the specific location of drippers (but see next sentence regarding the measurement timing). The images seem quite chaotic; however, this is more than expected as the subsurface has no specific structure and the timing (end of winter) should imply a relatively uniform distribution of the water in the profile. Furthermore, at this timing, the least differences between the treatments are expected as the

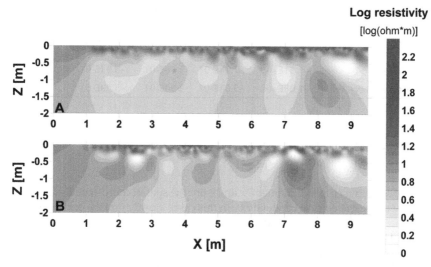

Fig. 11–2. Cross-sections in the Acre orchard: tap water irrigated (A) and treated wastewater (TWW) irrigated (B).

pore water is dominated primarily by the rain water (and not the irrigated water; typically for Israel, TWW are characterized by a 100 mg L^{-1} higher chloride concentration, compared with tap water; specifically for the Acre farm, tap water is characterized by an electrical conductivity of 0.66 to 1.01 dS m^{-1} and a chloride concentration of 44 to 93 mg L^{-1}, compared with 1.7 to 1.88 dS m^{-1} and 170 to 272 mg L^{-1}, respectively, for treated wastewater). The fact that high variability is seen at the soil surface implies that the wetted area beneath the drippers has a more permanent signature. Therefore, comparison of the images on a spatial basis will probably be wrong and we opt for statistical comparison. We use truncated moment analysis (see below) to compare the statistical distribution of the water beneath the plot. As direct conversion to water content is not available at this point, we perform the analysis using the electrical conductivity, the reciprocal of the resistivity, that is directly related (through petrophysical relations) to the water content. It is important to note that as the relations between water content and electrical conductivity are typically nonlinear, one should be cautious when interpreting such an analysis.

Spatial moments and the centers of mass and spatial variances (e.g., Lazarovitch et al., 2007) are computed for the two-dimensional case by

$$M_{ij} = \int_{-\infty}^{\infty} \int_{-\infty}^{\infty} \xi x^i z^j dx dz \qquad [5]$$

$$x_c = \frac{M_{10}}{M_{00}}, \quad z_c = \frac{M_{01}}{M_{00}} \qquad [6]$$

Table 11–1. Spatial moments and vertical centers of mass and variances.

	M_{00}	M_{01}	M_{02}	z_c	σ_z^2
	(m3)	(m4)	(m5)	(m)	(m2)
Tap water	1.31E+00	−7.74E-01	5.38E-01	−5.93E-01	6.10E-02
Treated wastewater	1.08E+00	−5.81E-01	3.90E-01	−5.39E-01	7.16E-02

$$\sigma_x^2 = \frac{M_{20}}{M_{00}} - x_c^2, \quad \sigma_z^2 = \frac{M_{02}}{M_{00}} - z_c^2 \qquad [7]$$

where M_{ij} is the i,j spatial moment, x and z are the spatial coordinate, ξ is the property of interest (electrical conductivity in this case), x_c and z_c are the centers of mass in the x and z directions, and σ_x^2, σ_z^2 are the spatial variances in the x and z directions. The advantage of spatial moments is that they allow averaging behavior over a volume and by that reducing the effect of noise. Compared to simple averaging, spatial moments are superior as they allow full quantification of a very complex behavior. While plotting horizontally averaged profiles would have provided a comparison between the two treatments, moments provide a numerical value for the average depth of the water and the spatial distribution around this point.

Using the spatial moments computed for a truncated domain for each section between $x = 3$ to $x = 9$, and from the ground surface to depth of 1 m, we compute the vertical center of mass and variance, which is listed in Table 11–1. The domain is centered about a tree and spanning half the tree separation in each direction. Note that, as the analysis is performed for the electrical conductivity (and not for example water content), we do not present units. Further note that we treat spatial moments here just as they are defined; that is, the analysis is limited to the spatial distribution of a property in a given domain and we do not attempt to reconstruct process related meaning (e.g., dispersion coefficients) by the moments.

Looking at the values in Table 11–1 yields several interesting observations. First, the tap water irrigated plot is generally more conductive, which implies that it is wetter (see later), and these waters are slightly deeper in the profile. This is in agreement with results from the evaluation of the long-term use of TWW on soil hydraulic properties in general (Levy and Assouline, 2010) and on the properties of this specific site (Assouline and Narkis, 2011). What is more interesting is that the waters in the TWW profile are more variably distributed around the center of mass (seen as higher σ_z^2), indicating perhaps preferential flow paths caused by induced repellency.

The Root Zone of an Orchard (Nir Galim)

The overall objective of this study (Zur, 2009) was to investigate the spatial patterns of moisture and fluxes in the vadose zone and the way they are influenced by near-surface properties and activities. That is, root zone activities, and primarily water uptake, are treated (even if not officially) as "boundary conditions" to the region of interest. The study was conducted between early 2008 and early 2009, with ERT measurements taken more or less once a month.

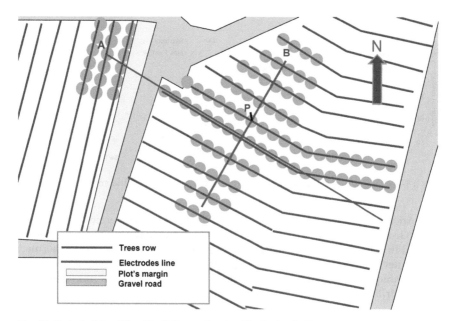

Fig. 11–3. A sketch of the Nir Galim orchard with electrode lines.

The site under investigation is a mature grapefruit orchard near the community of Nir Galim, a few kilometers northeast of the city of Ashdod, Israel. The orchard is located above the southern part of Israel's coastal aquifer, and the groundwater level at the site is about 30 m below the surface, as measured at an observation well located on site (Nativ et al., 2005). The average annual precipitation at this area is 480 mm (Israel Meteorological Service). The vadose zone at the site is composed of two main units: the top 4.5 m from the surface consist of sandy clay loam soil (27% clay, 15% silt, 59% sand) and the rest of the profile down to groundwater level (4.5–30 m deep) consists of a loamy sand soil (11% clay, 6% silt, 83% sand) (Nativ et al., 2005; Rimon et al., 2007). The orchard is irrigated by drip irrigation. The dripper lines are spread parallel to the tree rows at 0.5 m intervals. Tree separation along the row is 4 m and row separation is 6 m. The flux of a single dripper is 4 L per h. The irrigation started at the end of March and ended at the end of September; the irrigation duration was 7 to 10 h at a frequency of two or three times a week. Further and exact details are given by Zur (2009).

Two fixed electrode lines were installed at the site (Fig. 11–3), one generally parallel to the tree rows at an electrode separation of 1 m, and the other generally perpendicular to the tree rows at an electrode separation of 0.5 m. The line parallel to the tree rows extended beyond the orchard edge and covered a part where the rows bend a bit. Both lines were composed of 96 electrodes connected by a wire (Lify 0.25, Helukabel, Germany) to central plugs. Altogether, more than 2300 measurements (quadruples) were conducted in every survey, composed mostly of Schlumberger and inverse Schlumberger quadripoles. Surveys were taken between March 2008 and March 2009 roughly every month, with an additional 4-d continuous survey taken in August 2008 around an irrigation event.

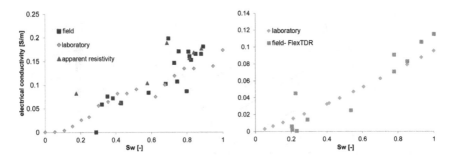

Fig. 11–4. Calibration of petrophysical relations. Clayey soil (left) and sandy soil (right).

After data screening (see details in Zur, 2009) data was inverted using RES2D-INV (Geotomo, Inc.). Each survey was inverted individually without considering any additional external information. The inverted image was then transformed into water content using the following petrophysical relations, obtained in various techniques as described below, partially using soil extracted from boreholes on site (drilled for the study presented in Rimon et al., 2007),

$$\sigma_b = 0.203 S_w^{1.3} + 0.006 \qquad\qquad [8]$$

$$\sigma_b = 0.0937 S_w^{1.2} \qquad\qquad [9]$$

where σ_b is the bulk electrical conductivity (the reciprocal of the resistivity) in S m^{-1}. Equations [8] and [9] are, respectively, for sandy clay loam (27% clay, 15% silt, 59% sand, referred to as clayey layer, down to about 4.5 m) and loamy sand (11% clay, 6% silt, 83% sand, referred to as sandy soil). Petrophysical relations (see Fig. 11–4) in this specific case were performed in three different ways. First, we used subsurface materials in a laboratory apparatus to measure the electrical resistivity as a function of the water content. Subsurface material was collected by Rimon et al. (2007). Measurements were performed for three soil samples, from depths of 4.5, 11, and 20 m. For each sample, about 15 saturation levels were considered where wetting was with tap water.

Second, we compared ERT resistivity data (post inversion) with true field water contents measured independently, using either direct sampling with a nondestructive hand sampler (Eijelcamp Inc., Netherlands) or FlexTDR measurements (Rimon et al., 2007). The FlexTDR system was located beneath the tree row under which our parallel ERT line (A) sat. The last method uses apparent resistivity values measured for dense (single electrode separation) Wenner quadripoles (pre inversion) and comparison with the manual samples.

Calibrated petrophysical relations obtained by the three methods were not identical. Furthermore, while laboratory measurements were relatively smooth, field measurements were very noisy. However, the relations presented above (Eq. [8] and [9]) are the result of a combined fitting of all data. We believe that this method is superior over laboratory exclusive calibration as this way we bring into account

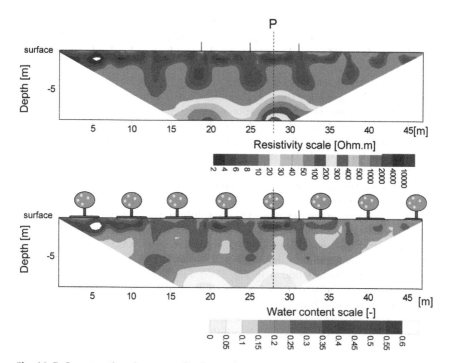

Fig. 11–5. Cross-sections in perpendicular to the tree rows: electrical resistivity (up) and water content (bottom). Images are for survey taken on 1 Apr. 2008. Tree images are used to mark the location of tree rows.

direct measurements performed with true pore water, and using truly undisturbed soil. See further discussion later regarding the importance timing of the calibration.

Figure 11–5 presents a representative cross-section of the orchard in terms of electrical resistivity and the associated water content distribution. One may easily note the repeated structure of very wet region below the tree row (and drip line) but to shallow depth, and less wet regions between the rows, extending deeper into the subsurface. The region under the tree lines at greater depths is typically very dry.

Discussion

The three case studies presented briefly above, together with the examples presented in the introduction chapter of this paper, provide a glimpse to the potential of ERT in root zone monitoring and visualization. In the following, we discuss some of the pros and cons of using ERT in root zone investigations and try to highlight current limitations and desired developments. Naturally, some of the points raised are based on empirical knowledge gathered and not on pure scientific studies.

We start by looking at the accuracy and the spatiotemporal resolution that can be achieved. We presented above two cases where the electrode separation was minimal, between 5 and 10 cm. Electrode separation roughly indicates, as a rule of thumb, the desired spatial resolution of the inverted image, at least close to the electrodes. Naturally, at greater distance from the electrodes the sensitivity of the measurement is reduced (see, e.g., Furman et al., 2003; Friedel, 2003) and although technically one may try to resolve the resistivity at high resolution, in practice, smoothening constraints will dominate the solution and dictate larger features. As can be seen in Fig. 11–1 and 11–2, the size of the inverted features at depth is much larger than those revealed close to the surface. As can be further seen in Fig. 11–1 and 11–2, the variation of resistivity close to the surface is often significant. This is sometimes reality; consider as an extreme example swelling soil that under drying conditions creates cracks and sometimes an artifact that results from, for example, bad electrode contact with the soil.

Most currently available commercial ERT systems are limited to few tens of electrodes. Given this limitation, two-dimensional (and not three-dimensional) surveys are often conducted as a compromise between the desire for a full three-dimensional survey, the desired spatial resolution, and the time needed to conduct the survey. While for static processes one may conduct a rolling survey, this is often impractical for monitoring dynamic processes such as those related to the root zone. To demonstrate this point, in addition to the cross-section presented here (Fig. 11–5, which is line A in Fig. 11–3), Zur (2009) also conducted a cross-section that is parallel to the tree line (line B in Fig. 11–3, not presented here). This cross-section practically averaged the features presented in Fig. 11–5, presenting a close to uniform image that cannot distinguish between different root zones of different trees. For example, such a line will average the water content close to the tree with the water content below the rows adjacent to it. Water content values obtained in such a case are inherently averaged and therefore do not truly represent any point in the orchard. Similar results were obtained in the Acre measurements (Fig. 11–2).

Following the same logic, the sensitivity of an ERT survey is always closer to the electrodes than it is at a given distance. For common applications, especially non-research applications, this means that the sensitivity to the deep parts of the root zone is smaller than the sensitivity close to the soil surface. In that aspect, a significant improvement can be achieved by inserting few deep electrodes. Such an approach was used by Pidlisecky et al. (2006) for much larger scales. We are not aware of any other similar application in general or specifically for the root zone, but in general it can be applied [we have conducted such an experiment relying on a FlexTDR system (Rimon et al., 2007) but have not yet analyzed the apparent superior data].

Conversion of ERT resistivity data into hydrological meaningful data such as water content involved the use of petrophysical relations. The common practice is to perform a laboratory calibration for these empirical relations, but as demonstrated in Fig. 11–4, often a field validation at the scale of the ERT measurements is much more reliable than laboratory-based relations. This is important primarily for two reasons: scale, and the difficulty to truly reconstruct in the laboratory the true chemical composition of the pore water. Figure 11–6 presents representative temporal water content data obtained throughout a year from ERT inverted data such as the one presented in Fig. 11–5 (complete details are given by Zur, 2009). Data is for a representative point, P, marked on Fig. 11–3, while the depths are marked. As one can note, the water content changes significantly throughout the year, with a significant increase of the water content in the summer (irrigation) months in the relatively shallow depths (up to about 1.5 m), and reduction of

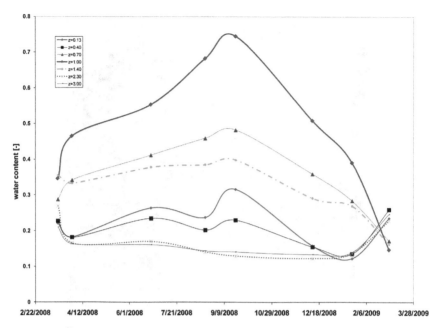

Fig. 11–6. Water content dynamics at point P in the Nir Galim Orchard. Legend indicates depths in meters.

the water content in the same months in deeper layers (and increase in the winter). While one may find a logical explanation for this behavior (related to root uptake), the obtained values are unacceptable (e.g., water contents of about 0.6 in 1-m depth). Furthermore, considering the irrigation dose and its start as early as the beginning of April, one should not expect significant increases in water contents for shallow regions beyond, say, early May. This clearly indicates that the resistivity reading is influenced not only by water.

Dealing with root zone processes, especially in warm and dry environments, means elevated evapotranspiration rates, which implies accumulation of salts, both soluble and in solid state (see Fig. 11–7) close to the surface. As a result, the electrical resistivity of the root zone is significantly reduced during the high evapotranspiration period. In other words, the electrical resistivity of the bulk water is far from being stable during this period, even in commercial agricultural fields where irrigation is performed with a reasonably low leaching factor and with irrigation water with stable electrical conductivity. In other words, when root zone processes are monitored using ERT, one needs to realize that petrophysical relations should be treated as dynamic in time, and that simple static relations may not be sufficient.

Moreover, as can be seen in Fig. 11–6, the apparent water content close to the surface ($z = 0.13$, $z = 0.4$ m) is relatively low and stable due to relatively frequent irrigation (typically every 3 d). It is the apparent water contents at the intermediate depths (roughly 0.5–1.5 m) that depicts elevated values during the summer (salt accumulation is typically highest at the bottom of the root zone, i.e., about 1.5 m). This clearly indicates that the process is a combination of salt accumulation

Fig. 11–7. Salt accumulation near drippers in southern Israel. Note the white color out-side the algae green wet area adjacent to the drippers (right; courtesy Eran Raveh).

due to evapotranspiration and leaching. In other words, in such an environment, one should consider season-based petrophysical relations.

Another issue related to root zone electrical resistivity that, to the best of our knowledge, has not been addressed yet is the role of root connectivity on the electrical properties of the bulk root zone. Volumetrically and by weight, roots account for a very small fraction of the root zone (see, e.g., Rossi et al., 2011). Furthermore, the major root surface area is relatively non-conductive. However, as the root contains a high portion of water and solutes (see, e.g., Nadler and Tyree, 2008, for a study on the relation of the electrical conductivity of the tree trunk in water content and xylem sap mineral properties), it is very conductive. We hypothesize that root conductivity and connectivity may play a non-negligible role in bulk root zone electrical resistivity. Anisotropic inversion may help here, although this needs to be further examined. In addition, it is likely that during growth periods, when root biomass increases (and specifically small veins), the role of root conductivity and connectivity increases.

We have presented earlier in more detail the calibration of petrophysical rela-tions performed using three different methods. We argue that the use of field measurements (and not only laboratory settings) is superior as samples brought to the laboratory (i) are typically small, (ii) are disturbed to some degree, (iii) will be treated with water different in composition from field water, and (iv) do not necessarily truly represent field scale heterogeneity caused by natural or man-induced processes (e.g., wheel tracks, tillage, etc.). Therefore, we believe that petrophysical calibration should be done, if possible, in the field and should include a high number of sampling points that truly represent the field heterogeneity, combining ERT with direct push technology (drilling and or penetrometering; see, e.g., Pidlisecky et al., 2006).

Last, when monitoring root zone processes, one needs to distinguish between close-to-natural settings and intensive irrigation practices. In the former, soil flow processes as well as root uptake are relatively slow and of small pace (e.g., Michot et al., 2003; Srayeddin and Doussan, 2009; Nijland et al., 2010; Garré, 2011). This allows for the consideration of the process of interest to be static during the acquisition time, which is typically on the order of tens of minutes to a few hours. In more intensive practices, such as those presented in the bell pepper experiment above, this assumption is sometimes at the edge of acceptability (an individual survey for that case took just over an hour for a setting that has one daily to eight daily irrigation events). Although not performed here, this clearly requires a different inversion approach that treats the acquisition process as well as the domain as dynamic. These ideas are further elaborated by Seppänen et al. (2001), Furman et al. (2004, 2007), and Lehikoinen et al. (2009).

On the same temporal issue, intensive agricultural practices also imply that farmers may maintain relatively high water content in the root zone throughout long periods. That is, the classic practices often applied in less modern irrigation practices (see, e.g., Zerihun et al., 2005, for short description of surface irrigation), where reapplication of water is set by the time at which the water content reaches some point between field capacity and wilting point, no longer applies. The result is that the temporal changes during short periods are minor and not necessarily observable with relatively coarse tools such as ERT. In the Nir Galim experiment, we conducted a continuous ERT survey for 3 d (starting with an irrigation event ending with the next) and could not visually see any change in the water content distribution. The conclusion is that if high spatiotemporal resolution is desired in addition to high parameter resolution, ERT must be combined with more sensitive techniques, that is, point measurements such as TDR.

Challenges and Concluding Remarks

We have presented and discussed several cases from the literature and from the authors' own experience in which ERT was used to obtain hydrological or agronomic information regarding the root zone. While the literature includes many cases that may create an impression that ERT is already there, that is, ERT is a technology that can be used directly to monitor and map the root zone, the reality is not so optimistic. ERT can be used to obtain a rough mapping of the root zone, but one should not expect it to be too precise and too accurate. It works fine, given that the subsurface is not too heterogeneous. It works well for slow, close-to-natural processes but is challenging for faster processes that characterize modern agriculture. It works well for relatively large scale problems, but small scale is still a challenge, especially when the monitored media is relatively dry and coarse. Several aspects related to ERT application, data processing, and interpretation were not discussed in this chapter but are very important to consider. These not only include primarily mathematical aspects related to ERT inversion and constraining of the inversion, but also technical aspects such as the role of temperature (for that see Hayley et al., 2007, 2010).

Below we list several of the main challenges that need to be addressed to improve the usability of ERT for root zone mapping and monitoring. Several of the challenges are common to other uses of ERT, and primarily the need to introduce process-based constraints instead of arbitrary ones:

1. Develop smaller electrodes with good contact. This is a technical challenge that is extremely important for small-scale experiments, and especially for relatively dry conditions. Dry conditions imply that a longer and thicker electrode is needed, and for small root zones this violates the point size source (and point size measurement) and limits the spatial resolution that can be obtained.

2. Develop a coupling scheme for ERT and root zone processes that will allow enhancement of the spatial resolution and the accuracy of the resistivity inversion, allowing relaxation of nonphysical constraints.

3. Couple ERT with other monitoring tools (geophysical and non-geophysical) to allow separation of water content from solute concentration. This is extremely important for high evaporation regions.

In addition to these technical challenges, we also point out that unlike geological setting or natural hydrological settings; the root zone is an extremely active region in which the conditions are not stable throughout the growth season. This point needs to be considered in experimental design and appropriate measures need to be taken, such as time-variable petrophysical relations.

Despite those important drawbacks of ERT as a tool for root zone investigations, it is clearly the simplest nondestructive noninvasive tool for such tasks, and probably the cheapest. We believe that the tool can be used "as is" for field-scale and coarse investigations in research and even application (e.g., to control irrigation or to identify insufficient of irrigation or nutrition), while improvements at hardware, inversion software, and methodology are needed mostly to enhance its performance as a research tool.

Acknowledgments

This research was supported in part by research grant IL-3823-06 from BARD, the United States-Israel Binational Agricultural Research and Development Fund, by research grant 304-0395-09 from the Chief Scientist of the Israeli Ministry of Agriculture research fund, and by the Stephen and Nancy Grand Water Research Institute.

References

al Hagrey, S.A., and T. Petersen. 2011. Numerical and experimental mapping of small root zones using optimized surface and borehole resistivity tomography. Geophysics 76:G25–G35. doi:10.1190/1.3545067

Archie, G.E. 1942. The electrical resistivity log as an aid in determining some reservoir characteristics. Trans. Am. Inst. Min. Metall. Eng. 146:54–61.

Assouline, S. 2002. The effects of micro-drip and conventional drip irrigation on water distribution and uptake. Soil Sci. Soc. Am. J. 66(5):1630–1636. doi:10.2136/sssaj2002.1630

Assouline, S., S. Cohen, D. Meerbach, T. Harodi, and M. Rozner. 2002. Micro-drip irrigation of field crops: The effect on yield, water uptake and drainage in sweet corn. Soil Sci. Soc. Am. J. 66(1):228–235. doi:10.2136/sssaj2002.0228

Assouline, S., M. Möller, S. Cohen, M. Ben-Hur, A. Grava, K. Narkis, and A. Silber. 2006. Soil-plant response to pulsed drip irrigation and salinity: Bell pepper case study. Soil Sci. Soc. Am. J. 70:1556–1568. doi:10.2136/sssaj2005.0365

Assouline, S., and K. Narkis. 2011. Effects of long-term irrigation with treated wastewater on the hydraulic properties of a clayey soil. Water Resour. Res. 47:W08530. doi:10.1029/2011WR010498

Bamberger, E. 2011. Combined use of ERT and vadose zone flow modeling for improved estimation of soil water content. Ph.D. diss., Technion, Haifa, Israel.

Besson, A., I. Cousin, H. Bourennane, B. Nicoullaud, C. Pasquier, G. Richard, A. Dorigny, and D. King. 2010. The spatial and temporal organization of soil water at the field scale as described by electrical resistivity measurements. Eur. J. Soil Sci. 61:120–132. doi:10.1111/j.1365-2389.2009.01211.x

Bolan, N.S. 1991. A critical-review on the role of mycorrhizal fungi in the uptake of phosphorus by plants. Plant Soil 134(2):189–207. doi:10.1007/BF00012037

Butler, K.D. Editor. 2005. Near-surface geophysics. Society of Exploration Geophysicists, Tulsa, OK.

Day-Lewis, F.D., K. Singha, and A.M. Binley. 2005. Applying petrophysical models to radar traveltime and electrical resistivity tomograms: Resolution-dependent limitations. J. Geophys. Res. 10: B08206. doi:10.1029/2004JB005369

Friedel, S. 2003. Resolution, stability and efficiency of resistivity tomography estimated from a generalized inverse approach. Geophys. J. Int. 153:305–316. doi:10.1046/j.1365-246X.2003.01890.x

Furman, A., T.P.A. Ferré, and G.L. Heath. 2007. Tailoring electrical resistivity surveys. Geophysics 72(2):F65–F73.

Furman, A., T.P.A. Ferré, and A.W. Warrick. 2003. A sensitivity analysis of electrical-resistance-tomography array types using analytical element modeling. Vadose Zone J. 2:416–423.

Furman, A., T.P.A. Ferré, and A.W. Warrick. 2004. Optimization of ERT surveys for monitoring transient hydrological events using perturbation sensitivity and genetic algorithms. Vadose Zone J. 3:1230–1239.

Gardner, W.R. 1960. Dynamic aspects of water availability to plants. Soil Sci. 89(2):63–73. doi:10.1097/00010694-196002000-00001

Garré, S., M. Javaux, J. Vanderborght, L. Pagès, and H. Vereecken. 2011. Three-dimensional electrical resistivity tomography to monitor root zone water dynamics. Vadose Zone J. 10(1):412–424. doi:10.2136/vzj2010.0079

Hayley, K., L.R. Bentley, M. Gharibi, and M. Nightingale. 2007. Low temperature dependence of electrical resistivity: Implications for near-surface geophysical monitoring. Geophys. Res. Lett. 34:L18402. doi:10.1029/2007GL031124

Hayley, K., L.R. Bentley, and A. Pidlisecky. 2010. Compensating for temperature variations in time-lapse electrical resistivity difference imaging. Geophysics 75:WA51–WA59.

Hinnell, A.C., T.P.A. Ferré, J.A. Vrugt, J.A. Huisman, S. Moysey, J. Rings, and M.B. Kowalsky. 2010. Improved extraction of hydrologic information from geophysical data through coupled hydrogeophysical inversion. Water Resour. Res. 46:W00D40. doi:10.1029/2008WR007060

Hinsinger, P., C. Plassard, C.X. Tang, and B. Jaillard. 2003. Origins of root-mediated pH changes in the rhizosphere and their responses to environmental constraints: A review. Plant Soil 248(1–2):43–59. doi:10.1023/A:1022371130939

Jones, D.L. 1998. Organic acids in the rhizosphere- a critical review. Plant Soil 205(1):25–44 10.1023/A:1004356007312. doi:10.1023/A:1004356007312

Karaoulis, M.C., J.-H. Kim, and P.I. Tsourlos. 2011. 4D active time constrained resistivity inversion. J. Appl. Geophys. 73: 25–34. doi:10.1016/j.jappgeo.2010.11.002

Kemna, A., J. Vanderborght, B. Kulessa, and H. Vereecken. 2002. Imaging and characterisation of subsurface solute transport using electrical resistivity tomography (ERT) and equivalent transport models. J. Hydrol. 267:125–146. doi:10.1016/S0022-1694(02)00145-2

Lazarovitch, N., A.W. Warrick, A. Furman, and J. Šimůnek. 2007. Subsurface water distribution from drip irrigation described by moment analyses. Vadose Zone J. 6:116–123. doi:10.2136/vzj2006.0052

Lehikoinen, A., S. Finsterle, A. Voutilainen, M.B. Kowalsky, and J.P. Kaipio. 2009. Dynamical inversion of geophysical ERT data: State estimation in the vadose zone. Inverse Problems Sci. Eng. 17:715–736. doi:10.1080/17415970802475951

Levy, G., and S. Assouline. 2010. Physical aspects. In: G. Levy et al., editors, Use of treated sewage water in agriculture: Impacts on crop and soil environment. Blackwell Publishing, New York.

Menke, W. 1989. Geophysical data analysis: Discrete inverse theory. Academic Press, San Diego, CA.

Michot, D., Y. Benderitter, A. Dorigny, B. Nicoullaud, D. King, and A. Tabbagh. 2003. Spatial and temporal monitoring of soil water content with an irrigated corn crop cover using surface electrical resistivity tomography. Water Resour. Res. 39(5):1138–1158. doi:10.1029/2002WR001581

Moradi, A.B., A. Carminati, D. Vetterlein, P. Vontobel, E. Lehmann, U. Weller, J.H. Hopmans, H.-J. Vogel, and S.E Oswald. 2011. Three-dimensional visualization and quantification of water content in the rhizosphere. New Phytologist 192: 653–656. doi:10.1111/j.1469-8137.2011.03826.x

Nadler, A., and M.T. Tyree. 2008. Substituting stem's water content by electrical conductivity for monitoring water status changes. Soil Sci. Soc. Am. J. 72(4):1006–1013. doi:10.2136/sssaj2007.0244

Nativ, R., O. Dahan, and Y. Rimon. 2005. Quantity and quality of groundwater recharge into the coastal plain aquifer under undeveloped, agricultural and urban setups. Interim Scientific Report for the period Oct 2004–Sep 2005 for the Israel Science Foundation. Soil and Water Science, The Hebrew University of Jerusalem, Israel.

Nguyen, F., A. Kemna, A. Antonsson, P. Engesgaard, O. Kuras, R. Ogilvy, J. Gisbert, S. Jorreto, and A. Pulido–Bosch. 2009. Characterization of seawater intrusion using 2D electrical imaging. Near Surf. Geophys. 7:377–390.

Nijland, W., M. van der Meijde, E.A. Addink, and S.M. de Jong. 2010. Detection of soil moisture and vegetation water abstraction in a Mediterranean natural area using electrical resistivity tomography. Catena 81:209–216. doi:10.1016/j.catena.2010.03.005

Pidlisecky, A., R. Knight, and E. Haber. 2006. Cone-based electrical resistivity tomography. Geophysics 71(4):G157–G167.

Rimon, Y., O. Dahan, R. Nativ, and S. Geyer. 2007. Water percolation through the deep vadose zone and groundwater recharge: Preliminary results based on a new vadose zone monitoring system. Water Resour. Res. 43:W05402. doi:10.1029/2006WR004855

Rossi, R., M. Amato, G. Bitella, R. Bochicchio, J.J. Ferreira Gomes, S. Lovelli, E. Martorella, and P. Favale. 2011. Electrical resistivity tomography as a non-destructive method for mapping root biomass in an orchard. Eur. J. Soil Sci. 62(2):206–215. doi:10.1111/j.1365-2389.2010.01329.x

Seppänen, A., M. Vauhkonen, P.J. Vauhkonen, E. Somersalo, and J.P. Kaipio. 2001. State estimation with fluid dynamical evolution models in process tomography: An application to impedance tomography. Inverse Probl. 17:467–483. doi:10.1088/0266-5611/17/3/307

Shock, C.C., E.B.G. Feibert, and L.D. Saunders. 2005. Onion response to drip irrigation intensity and emitter flow rate. Horttechnology 15(3):652–659.

Silber, A., M. Bruner, E. Kenig, G. Reshef, H. Zohar, I. Posalski, H. Yechezkel, D. Shmuel, S. Cohen, M. Dinar, E. Matan, I. Dinkin, Y. Cohen, L. Karni, B. Aloni, and S. Assouline. 2005. High fertigation frequency and phosphorus level: Effects on summer-grown bell pepper growth and blossom-end rot incidence. Plant Soil 270:135–146. doi:10.1007/s11104-004-1311-3

Simunek, J., and J.W. Hopmans. 2009. Modeling compensated root water and nutrient uptake. Ecol. Modell. 220(4):505–521 . doi:10.1016/j.ecolmodel.2008.11.004

Srayeddin, I., and C. Doussan. 2009. Estimation of the spatial variability of root water uptake of maize and sorghum at the field scale by electrical resistivity tomography. Plant Soil 319(1–2):185–207. doi:10.1007/s11104-008-9860-5

Stern, S. 2010. Characterization of root uptake and its dependence on soil environmental conditions by Geophysical tools. MSc. thesis, Technion, Haifa, Israel (in Hebrew).

Steudle, E., and C.A. Peterson. 1998. How does water get through roots? J. Exp. Bot. 49(322):775–788. doi:10.1093/jexbot/49.322.775

Stummer, P., H. Maurer, and A.G. Green. 2004. Experimental design: Electrical resistivity data sets that provide optimum subsurface information. Geophysics 69:120–139.

Telford, W.M., L.P. Geldart, and R.E. Sheriff. 1990. Applied geophysics. Cambridge Univ. Press, New York.

Vanderborght, J., J.A. Huisman, J. van der Kruk, and H. Vereecken. 2013. Geophysical methods for field-scale imaging of root zone properties and processes. In: S.H. Anderson and J.W. Hopmans, editors, Soil–water–root processes: Advances in tomography and imaging. SSSA Spec. Publ. 61. SSSA, Madison, WI, p. 247–282. doi:10.2136/sssaspecpub61.c12

Vrugt, J.A., J.W. Hopmans, and J. Simunek. 2001. Calibration of a two-dimensional root water uptake model. Soil Sci. Soc. Am. J. 65(4):1027–1037. doi:10.2136/sssaj2001.6541027x

Zerihun, D., A. Furman, A.W. Warrick, and C.A. Sanchez. 2005. A coupled surface-subsurface model for improved basin irrigation management. J. Irrig. Drain. Eng. 131(2):111–128. doi:10.1061/(ASCE)0733-9437(2005)131:2(111)

Zur, A. 2009. Use of electrical resistivity tomography for study of flow and water distribution in the vadose zone. MSc. thesis, Technion, Haifa, Israel.

12

Geophysical Methods for Field-Scale Imaging of Root Zone Properties and Processes

J. Vanderborght,* J.A. Huisman, J. van der Kruk, and H. Vereecken

Abstract

Geophysical methods offer the possibility to image noninvasively three-dimensional subsurface structures of soil properties and associated flow and transport processes at spatial scales ranging from soil columns to whole field plots. Measured geophysical properties can be related to soil state variables, such as soil moisture and salt concentration, soil properties, such as clay content, and root properties, such as root mass and surface area. Because of their noninvasive nature, time lapse geophysical measurements can be used to monitor spatial patterns of processes such as water flow, root water uptake, and solute transport in soils. Geophysical methods have also been used to detect the presence of roots and to characterize the architecture, distribution and density of roots. Exciting research questions dealing with the feedback between vegetation, soil heterogeneity and soil moisture patterns, such as row-inter row variations, effects of three-dimensional root system of trees, redistribution of rainfall by canopies and stem flow, re-infiltration of runoff in vegetated patches, could all be addressed using geophysical methods. In this contribution, we give an overview of geophysical methods that show the greatest promise in plant-root related research: electrical resistivity tomography (ERT), electrical impedance spectroscopy and tomography (EIT), and ground penetrating radar (GPR). For each method, we discuss the geoelectrical (soil) properties they measure, their relation to soil state variables and properties, and their ability to directly detect roots. We will also present examples of applications of these methods to noninvasively study soil–root interactions.

Abbreviations: EIT, electrical impedance spectroscopy and tomography; EM, electromagenetic; ERT, electrical resistivity tomography; GPR, ground penetrating radar; IP, induced polarization; RMD, root mass density; TDR, time domain reflectometry.

Agrosphere Institute, IBG-3, Forschungszentrum Jülich GmbH, D-52425 Jülich, Germany. *Corresponding author (j.vanderborght@fz-juelich.de).

doi:10.2136/sssaspecpub61.c12

Soil–Water–Root Processes: Advances in Tomography and Imaging. SSSA Special Publication 61. S.H. Anderson and J.W. Hopmans, editors. © 2013. SSSA, 5585 Guilford Rd., Madison, WI 53711, USA.

Water uptake by plants is a crucial process in hydrology, meteorology, and climatology. Plant transpiration is a very important component in the hydrological cycle and the associated latent heat flux is an important component of the land surface energy balance. Since crop production is strongly linked to crop transpiration, crop water use and availability of soil moisture for root water uptake is a critical issue in agriculture.

The importance of the parameterization of root density distributions for the prediction of plant transpiration was demonstrated by Feddes et al. (2001). They argued that the identification of robust parameterizations of root density for different vegetation types, soil types, and climates should have high priority to improve hydrological, meteorological, and climate models from local, to regional, to global scales. Until today, the relation between the vertical distributions of water and roots on the one hand, and the effective water uptake by roots on the other hand, is still not well understood. This is reflected in the different model concepts that are currently used in soil-vegetation-atmosphere transfer models to describe root water uptake as a function of root density and the vertical distribution of the water content. In most models, local root water uptake is assumed to be proportional to the root length density, although this is questioned by experimental data (Green and Clothier, 1998; Lai and Katul, 2000). This assumption neglects well-established processes like hydraulic lift (Dawson, 1996) and compensatory root water uptake from deeper in the soil profile (Jarvis, 1989; Simunek and Hopmans, 2009). As a consequence, the role of deep roots on water uptake seems to be underestimated (Feddes et al., 2001). Investigating root water uptake by deep rooting vegetation is hampered by the inaccessibility of the subsurface and by the lateral extent that needs to be covered to obtain a representative average of root water uptake. Noninvasive methods that enable monitoring of soil moisture content along transects of 10^2–10^3 m long to a depth of 10^1 m or deeper are therefore expected to deliver data that provide inside in these processes.

At smaller scale, for instance in orchards (Andreu et al., 1997; Green and Clothier, 1998), in drip irrigated fields (Green and Clothier, 1995; Coelho and Or, 1999; Li et al., 2002), or in row crops (Timlin et al., 2001; Hupet and Vanclooster, 2005), a characterization of the spatial distribution of soil water, root density and root uptake is important for the optimization of water application and for reducing leaching losses of plant nutrients and plant protection products. Noninvasive methods that offer the opportunity to image patterns of water contents and root distributions at this scale (10^{-2}–10^{-1} m spatial resolution and 10^0–10^2 m range) are therefore of importance and will provide insight in the interaction between root water uptake, soil moisture distribution and root growth.

Finally, spatial variation in soil hydraulic properties, which may be influenced by plants, leads to spatially variable water fluxes and soil water content. The interaction between spatial patterns of soil properties, root distributions, vegetation types at different spatial scales, i.e., from the root system of an individual plant to the distribution of vegetation types and vegetation patterns in a landscape, plays a crucial role for the use of soil water resources by plants and vegetation and their potential to survive and grow. To investigate these soil-plant and landscape-vegetation interactions, methods that can image spatial distributions of roots and soil moisture content are required.

In this paper, we will provide an overview of three different geophysical methods: electrical resistivity tomography, electrical impedance spectroscopy, and tomography and ground penetrating radar and their use for soil-plant investigations. We will give a brief introduction to the general principles of the methods and the soil or root parameters that are measured or imaged. For a more detailed discussion on these methods, we refer to specialized text books on geophysics (Reynolds, 1997; Butler, 2005) and hydrogeophysics (Hubbard and Rubin, 2004; Vereecken et al., 2006).

Electrical Resistivity Tomography
General Principles

Using electrical resistivity tomography (ERT), the spatial distribution of the soil bulk electrical conductivity σ_b (S m^{-1}) or its inverse, the bulk soil resistivity, ρ_b (Ω m) $= \sigma_b^{-1}$, is derived. An electric current is injected into the soil using a pair of electrodes and electric potential differences are measured using one or more other electrode pairs. From the ratio of the potential difference to the current flow, the transfer resistance is calculated. This is repeated for a number of electric current injections using different electrode pairs to obtain a set of transfer resistances. Deviations between measured transfer resistances and expected transfer resistances for a homogeneous soil contain information about the heterogeneity of the subsurface so that the spatial distribution of σ_b can be inferred from a dataset of transfer resistances. This step involves an inversion of the Poisson equation that describes the relation between the electric potential and electric current in a heterogeneous electric conductivity field. In the inversion, an objective function which quantifies the deviation between the measured and simulated transfer resistances for a certain distribution of σ_b is minimized. This inversion is, however, ill-posed, which means that several distributions of σ_b may lead to identical measured transfer resistances. Therefore, the set of possible solutions is constrained by adding an extra term in the objective function, typically called the regularization term (Tikhonov and Arsenin, 1977). This term may represent a-priori knowledge about the σ_b distribution, such as the variance and spatial correlation of the σ_b distribution. However, such information is in most cases not available. Therefore, the smoothness of the σ_b distribution is often used as a constraint so that the optimized σ_b distribution corresponds with the smoothest distribution that represents the measured transfer resistances (Constable et al., 1987).

Relation between Imaged σ_b, Soil State Variables, Soil Properties, and Root Densities

The spatial distribution of bulk soil electrical conductivity σ_b or resistivity ρ_b obtained from electrical resistivity tomography (ERT) needs to be related to variables that are relevant for soil-plant processes. The bulk soil electrical conductivity depends on the amount of electrical charge carriers (i.e., ions) that are available in a bulk soil volume, their mobility and the spatial configuration of conducting and non-conducting regions. The mobility of the ions depends on the solute concentration and composition and more importantly on the soil temperature. The effect of soil temperature on bulk electrical conductivity can be described with sufficient accuracy by functions that do not require a specific calibration for a certain soil type (Keller and Frischknecht, 1966). In combination with the fact that soil temperature profiles can be measured or modeled quite easily, this means that corrections for the effect of temperature on the spatial distribution of bulk soil electrical conductivity are relatively straightforward.

The amount of charge carriers in a soil volume obviously depends on the equivalent ion concentration in the soil solution and the volumetric water content. The equivalent ion concentration of the soil solution is directly related with the electrical conductivity of the pore water: σ_w. Keeping σ_w constant, a larger volumetric water content θ leads to a larger amount of charge carriers and therefore a larger bulk soil electrical conductivity. Since with an increase in water content, the electrically conductive regions in the soil volume become better connected, the increase in σ_b with θ is more than linear. A commonly used empirical relation between bulk soil electrical conductivity, water content, and electrical conductivity of the pore water is the Archie equation (Archie, 1942):

$$\sigma_b = a'\phi^m \left(\frac{\theta}{\phi}\right)^n \sigma_w \qquad [1]$$

where ϕ is the porosity and a', m, and n are petrophysical parameters, also called cementation and saturation exponents, that are characteristics of the porous medium. The factor $a'\phi^m$ is often replaced by $1/F$, where F is the so-called formation factor. The parameter n, or the saturation exponent, ranges around 2 (Frohlich and Parke, 1989; Schön, 1996; Reynolds, 1997; Amente et al., 2000; Amidu and Dunbar, 2007) and is mostly larger than the cementation exponent m. This suggests that for the same volumetric water content, σ_b should increase with decreasing porosity.

In Eq. [1], σ_b is linearly related with σ_w and approaches 0 when σ_w reaches 0, suggesting that all electrical conduction takes place in the pore fluid. In soils or sediments with a considerable clay content and a high cation exchange capacity (CEC), cations in the electrical double layer that develops at the surface of charged solid particles make a significant contribution to the electrical conductivity of the bulk soil volume. Therefore, an extra surface conductivity term, $\sigma_{surface}$, is added to right hand side of Eq. [1] (Waxman and Smits, 1968):

$$\sigma_b = a'\phi^m \left(\frac{\theta}{\phi}\right)^n \sigma_w + \sigma_{surface}\left(\phi, \theta, \sigma_w\right) \qquad [2]$$

The surface conductivity depends on the characteristics of the mineral surface, the chemical composition of the pore water solution, and the soil water content (Waxman and Smits, 1968; Revil and Glover, 1997; Revil and Glover, 1998). For relatively large σ_w, $\sigma_{surface}$ becomes independent of σ_w and represents the intercept of the linear σ_b–σ_w relation.

An often used model for the σ_b–σ_w–θ relation in the soil science community is the Rhoades model (Rhoades et al., 1976), which assumes a constant surface conductivity:

$$\sigma_b = \left[a\theta^2 + b\theta \right] \sigma_w + \sigma_{surface} \tag{3}$$

where a and b are empirical parameters that are a function of the soil type. For clay soils, Rhoades et al. (1976) found $a = 2.1$ and $b = -0.25$ and for loam soils: $1.3 \leq a \leq 1.4$ and $-0.11 \leq b \leq -0.06$. Both Eq. [2] and [3] are based on a parallel conductor concept that decouples the electrical current through the bulk pore fluid from that along the fluid grain surfaces. This may lead to an overestimation of the conductivity in the pore fluid and its contribution to the bulk electrical conductivity (Friedman, 1998). A more sophisticated model that considers electrical current through a series of conductors that represent bulk fluid and solid–fluid interfaces was proposed by Rhoades et al., (1989). This model adds flexibility to the σ_b–σ_w–θ relation as it leads to a nonlinear relation between σ_b and σ_w for small σ_w and to a surface conductivity that depends on θ (Mallants et al., 1996; Mallants and Vanderborght, 1998).

The semi-empirical σ_b–σ_w–θ relations Eq. [1], [2], and [3] contain parameters that depend on soil properties such as soil porosity, bulk density, soil structure, and cation exchange capacity. The parameters m and n are a measure of the tortuosity of the electrical flow paths in the porous medium and are related to the structure of the porous medium and the spatial arrangement of the conducting and non-conducting regions. Larger m and n values were found for aggregated soils in which the tortuosity increases considerably when intra-aggregate pores drain and disconnect the more conducting aggregates whereas smaller m and n were found in compacted soils (Seladji et al., 2010). Changes in soil structure due to the formation of cracks in drying clay soils were found to decrease the bulk conductivity or increase the resistivity dramatically (Samouelian et al., 2003; Amidu and Dunbar, 2007) and led to an anisotropy of the bulk conductivity/resistivity (Samouelian et al., 2004).

The soil properties affecting the parameters of the petrophysical models may vary considerably with depth (e.g., between different soil layers) and within a certain soil layer. Therefore, site- and depth-specific calibration relations are required to relate ERT derived σ_b to θ. The petrophysical models may be parameterized using lab experiments on soil cores taken from the field site. Application of such core-scale derived petrophysical models to interpret ERT derived σ_b images may, however, be problematic for the following reasons: (i) the ERT and core-scale derived σ_b have a different support volume and soil heterogeneities lead to a scale dependent σ_b–θ–σ_w relation, (ii) the ill-posed and smoothness constrained inversion problem leads to lower and spatially variable sensitivities in the ERT image so that "apparent" petrophysical relations are needed to relate spatiotemporal fluctuations of σ_b to θ and σ_w (Singha and Gorelick, 2006) and to avoid water balance errors (Binley et al., 2002a), and (iii) soil cores are conditioned in the lab using a solution with a known σ_w, whereas σ_w varies with time and depth in field soils (Heimovaara et al., 1995). Furthermore, differences in σ_w between different land-use types due to differences in transpiration and concurrent salt accumulation were observed (Jayawickreme et al., 2010). However, these authors found that the effect of spatial and temporal variations of σ_w on σ_b was small compared with variations in σ_b caused by temperature and soil moisture

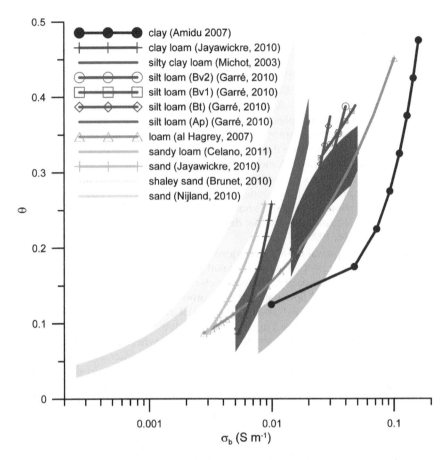

Fig. 12–1. Relation between volumetric water content θ and bulk electrical conductivity that was derived from electrical resistivity tomography for different soil types. The soil texture range between the different studies is represented by the color range from gray and/or yellow (sand) over orange and red (loam and silt) to dark brown (clay). The colored bands for a few relations represent the scatter of the soil moisture and correspond with the range: $(\theta-\sigma_\theta,\theta+\sigma_\theta)$ where σ_θ is the standard deviation of the water content, i.e., the root mean square error of the regression.

fluctuations. Experiments and sensitivity analyses by Brunet et al. (2010) also suggest that the effect of spatiotemporal variations in σ_w on σ_b are small.

Alternatively, point-scale measurements of soil moisture obtained with, for example, time domain reflectometry or neutron probes or from soil sampling can be used to derive petrophysical models in situ. The scatter in in situ calibration relations often leads to very simple linear or power law calibration relations between θ and σ_b. In Fig. 12–1, a few in situ derived relations between σ_b and θ are exemplarily shown. For some relations, we also added the range of the soil moisture (mean ± standard deviation) which gives an indication of the variation that can be expected. Although there seems to be a general trend of increasing σ_b with finer texture, for a certain texture class there is a considerable variation of

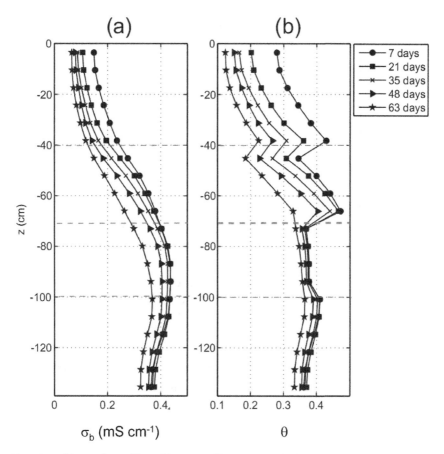

Fig. 12-2. (a) Depth profiles of horizontally averaged bulk electrical conductivity, σ_b, derived from electrical resistivity tomography measurements in a lysimeter on which Barley was grown, and (b) corresponding depth profiles of the soil water content that were estimated from σ_b using soil layer specific σ_b–θ relations (see also Fig. 12-1) (from Garré et al., 2011).

the σ_b–θ relation between different sites. Besides the scatter in the calibration or petrophysical relationship, the spatial variability of this relationship adds to the uncertainty in the derived spatial distribution of θ (Yeh et al., 2002). In Fig. 12-2, inverted profiles of σ_b at different times during a growing season when the soil dried out progressively are shown together with derived profiles of θ (Garré et al., 2011). The effect of the soil-layer-dependent σ_b–θ relation becomes very clear in the θ profiles, and it is evident that information about the soil layer depths is crucial to infer θ profiles from σ_b profiles. At a larger scale (approximately 100-m-wide and 10-m-deep transects), Schwartz et al. (2008) found no correlation between local time domain reflectometry (TDR), measured soil moisture contents, and ERT-derived σ_b. To infer θ from σ_b, they had to use a petrophysical relationship of which the parameters depended on the clay content and the local pore water conductivity, which was estimated from the extractable Ca and Mg. Maps

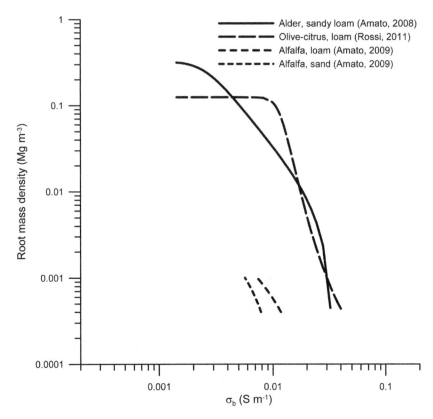

Fig. 12–3. Relations between root mass density and bulk electrical conductivity, σ_b, for different plant species and different soils.

of clay content and extractable Ca and Mg were obtained by spatial interpolation using these point scale measurements. In such an approach, it is evident that the accuracy of the derived θ maps depends on the accuracy of the spatial distribution of the clay content and Ca–Mg contents.

Besides depending on water content, pore water electrical conductivity, and soil properties, the bulk soil electrical conductivity was also found to depend on the presence of plant roots. In Fig. 12–3, the change in σ_b due to changes in root mass density (RMD) and changes in soil moisture are shown. Since woody material has a lower conductivity than bulk soil, the presence of roots was found to decrease the soil conductivity. Amato et al. (2008) and Rossi et al. (2011) found a negative correlation between the RMD of trees and σ_b. Since no correlation was found with root length density, which is mainly determined by fine roots, and since RMD is mainly determined by thick roots, σ_b depends on the thick root distribution in which the suberized cells of woody roots act as insulators. However, Rossi et al. (2011) note that the increase in bulk soil resistivity is more than proportional with volume fraction of the more resistive woody root material. This indicates that the root structure also influences the structure and tortuosity of the electric flow field. The impact of RMD on the bulk soil conductivity depends on the conductivity of soil without roots and is more outspoken in soils with a higher conductivity (e.g., wet loam soil) than in soils with a lower conductivity (e.g., dry sandy soil). Moreover, for herbaceous plants, a decrease of bulk soil electrical conductivity with increasing RMD was observed for alfalfa by Amato et al. (2009), whereas Werban et al. (2008) found an increase in σ_b in a soil with lupine roots. Thus, fine roots may also act as electric conductors or their exudates may influence the electrical conductivity of the soil around the roots. Similarly, al Hagrey (2007) also found higher bulk electrical conductivities in the root zone of a 5-yr-old poplar tree and attributed this to the presence of conductive "soft" roots. For older poplar trees, Zenone et al. (2008) reported lower σ_b at locations with higher RMD. Therefore, the effect of RMD on the bulk soil electrical conductivity seems to depend on the plant type, age, and background electrical conductivity. It is important to note that the presence of roots also influences other soil properties and variables, such as θ, which are also related to σ_b. As a consequence, observed relations between σ_b, θ, and RMD may be indirect relations.

Application of ERT for Investigating Root Zone Processes

ERT Monitoring of Water Movement

ERT can be used to monitor the movement of water that was applied at the soil surface or injected into the soil. The two-dimensional or three-dimensional distributions of the inverted resistivity provide information about the water content distribution that cannot be obtained with point-scale measurements. Daily et al. (1992) monitored point source and line source infiltration of water in vertical transects of approximately 6-m width and 20-m depth in a deep vadose zone. The effect of capillary barriers and the geology at the site on the water flow could be inferred from the ERT images. al Hagrey and Michaelson (1999) imaged infiltration in 6.2- and 9.4-m wide transects and traced preferential flow paths in the dense subsoil that were generated by a compacted plow pan on top of which water ponded and was laterally redistributed. Nimmo et al. (2009) used ERT to image the water infiltration and redistribution under 1.0-m-diameter infiltration ponds. These experiments were performed in desert soils of different age to investigate the hydrology of these soils and its influence on the vegetation that develops in these soils. In young soils with little or no pedological development, the infiltration capacity was much higher and more homogeneous than in older soils with distinct features of pedogenic development such as soil aggregation, horizonation, and macropore development. In the younger soils, the infiltrated water was not retained for a long time in the surface layers but drained rapidly to deeper layers, which supports the development of deep-rooting vegetation. The larger water retention in the soil surface layer and the lateral variation of water retention in the older soils is more supportive of shallow-rooted vegetation that distributes itself around the more retentive parts.

ERT can also be used to monitor water infiltration under natural rain conditions at the field-plot scale [3.5 by 3.5. by 1.5 m (Zhou et al., 2001)] and to quantify preferential flow phenomena in heterogeneous loamy soils (Zhou et al., 2002) or cracking clay soils in 17-m-long and 1.25-m-deep transects (Amidu and Dunbar, 2007). Brunet et al. (2010) used ERT to monitor the spatiotemporal evolution of the water deficit in soil profiles to a depth of 1.2 m along a transect of 17.5 m over a period of 22 mo. The ERT-derived soil moisture contents and profile water deficits were comparable to TDR measurements, illustrating the potential of ERT to monitor soil water balances on a spatial and temporal scale at which temporal dynamics and spatial variation of vegetation as well as spatial variability in soil pedology become important. A similar monitoring study was carried by Celano et al. (2011), who investigated the soil water balance in 11.75-m-wide and 3-m-deep vertical transects in olive groves with different soil management (cover-cropped versus tilled plots) during the autumn–spring period. In the cover-cropped plots, more rainwater could infiltrate and be stored in the soil profile than in the tilled plots.

ERT can also be used for landscape-scale tomography to investigate ecohydrological processes and their relation to spatial structures in the subsurface, land use, and vegetation. Koch et al. (2009) imaged σ_b along a 260-m-long transect comprising both a steep valley slope and the valley bottom down to a depth of 18 m. In the flood plane, an ERT grid array was used to obtain three-dimensional information about the subsurface. ERT surveys provided insight in the hydrology of the headwater catchment by identifying the spatial structure of flow paths such as connectivity or lack of connectivity between groundwater and the stream channel and the connectivity of deep groundwater with the surface in the flood plane generating exfiltration and surface runoff. This type of spatial information is valuable to interpret traditional hydrological measurements such as stream hydrographs, groundwater and stream water chemistry, and piezometric head measurements. Berthold et al. (2004) used ERT to investigate depression-focused groundwater recharge in the Canadian prairies and how this might be altered by recent land-use changes. Therefore, 496-m- and 192-m-long transects and areas of 112 by 112 m and 84 by 84 m were investigated and imaged to approximately 30-m depth. Spatial patterns in groundwater electrical conductivity, σ_w, and its effect on ERT imaged σ_b, were used to identify zones of leaching and groundwater recharge under depressions and zones of net upward water flow and salt accumulation under adjacent uplands. A similar hydrological situation can be found in sinkholes that develop in karst systems. Sinkholes are believed to be points of local and focused rapid groundwater recharge. Using ERT, Schwartz and Schreiber (2009) monitored and imaged soil water content along 72-m-long and 9-m-deep profiles across mantled sinkholes. By comparing the temporal evolution of the water storage in the profile, precipitation, and potential evapotranspiration, groundwater recharge could be estimated. The ERT data revealed that a considerable amount of water can be stored in mantled sinkholes and slowly released to the underlying aquifer, which is in contrast with the generally assumed functioning of sinkholes. However, the water storage was found to depend strongly on the soil properties and soil layer thickness.

ERT Monitoring of Root Water Uptake

ERT has also been used to monitor root water uptake and to identify the depth of the root zone. These studies illustrate that ERT can be used to assess depths of zones from which water can be extracted by plant roots, how these zones vary laterally, and how plant roots influence water infiltration patterns. Due to the lateral and vertical extent of the imaged soil region, the spatial variability of the soil and soil moisture content, and the inaccessibility of deeper soil or bedrock layers, this type of information is mostly inaccessible with other methods. Using ERT, information on these ecohydrological processes could be obtained over larger lateral and vertical scales than what could be obtained using classical point scale soil moisture measurements.

Barker and Moore (1998) imaged the change in σ_b along a 36-m-long and 5-m-deep transect with time-lapse ERT after a water infiltration experiment and attributed the larger decreases in σ_b in a region below a tree to the water uptake by the tree. al Hagrey (2007) monitored changes of σ_b in a cork oak forest along two 20-m-long and 3-m-deep transects during 15 d and observed a decrease in σ_b within the upper 2 m of the profile in regions with cork trees, whereas no decrease in σ_b was observed in regions without trees. ERT along 27-m-long transects was used by Nijland et al. (2010) to investigate the effect of lithology on root water uptake by Mediterranean natural vegetation in regions characterized by shallow soils of 0.6 to 1.5 m thickness. Using time-lapse ERT measurements at the beginning and end of the summer season, they could demonstrate root water uptake by the natural vegetation from fractured bedrock below the shallow soil layer down to 6-m depth. This type of information is hard if not impossible to obtain from local- or point-scale measurements. Jayawickreme et al. (2008) monitored soil moisture for 1 yr using ERT along a 128-m-long transect that crossed a forest and grassland ecotone. This study illustrates the effect of vegetation type on the temporal dynamics of the soil moisture and its impact on groundwater recharge and evapotranspiration. Under forest, soil moisture was depleted to a greater extent and to a greater depth than under grassland, which reflects differences in root depth, rain interception, and transpiration between the two ecotones.

ERT was also used to derive information about the lateral spatial variability of root water uptake and preferential infiltration facilitated by plants. Zenone et al. (2008) derived from time-lapse ERT images that water infiltrated along vertically oriented pine tree roots whereas horizontal main roots acted as a kind of filter and impeded vertical water movement. Michot et al. (2003) monitored the spatiotemporal soil moisture distribution in a corn field along a 6.2-m-long and 0.8-m-deep transect perpendicular to the corn rows during a 10 d period starting just before an irrigation event. The ERT data clearly demonstrated a different hydrological regime between the row and inter-row locations. Water interception by the corn leaves and stem flow led to a stronger wetting of the soil below the corn rows whereas root water uptake and concurrent water depletion was also larger at the row than inter-row locations. Srayeddin and Doussan (2009) monitored σ_b along six 9.3-m-long transects during a three month period in corn and sorghum fields that obtained different amounts of irrigation together with water content profiles using neutron probes. Larger water depletion was observed in the poorly irrigated than in the fully irrigated plots and in the sorghum than in the corn fields. For the plots receiving less irrigation, water depletion was observed to a depth of 1.6 m below the surface. The conductivity distribution obtained with ERT was in general agreement with local-scale water content measurements. However, due to a loss of sensitivity and resolution of ERT with depth, the depth of the drying front could not be derived from the ERT measurements. A larger spatial heterogeneity or patchiness of resistivity was observed in those transects where the soil dried out more. This could indicate a more irregular root water uptake under drier soil conditions, but could also be the result of a stronger sensitivity of resistivity to water content changes under dry soil conditions due to the nonlinear relation between θ and ρ or $1/\sigma_b$, or due to the higher measurement error because of a bad soil-electrode contact under dry soil conditions.

Differences in resistivity between 5 d in the fully irrigated plot were persistent for different time intervals; this reflect a persistent root water uptake pattern in a well-watered soil. Looking at 67-d differences in electrical resistivity, larger differences that extended to a greater depth were observed in the poorly irrigated plots. The spatial variation in the resistivity was the largest in transects with an intermediate irrigation amount where dried out regions coexisted with wetter patches. This reflects the spatial variation in rooting pattern and root water uptake efficiency. In the plots receiving less irrigation, the crops were forced to dry out the soil stronger and extract all the available water, which led to a more uniform soil moisture pattern.

The water storage in a soil profile and its temporal evolution is an important term in the soil water balance. Images of the two- or three-dimensional water distribution in a soil that are derived from ERT could be used to calculate the water storage in a soil profile. Garré et al. (2011) demonstrated that three-dimensional ERT-derived soil moisture distributions in a weighted 1.5-m-long undisturbed soil monolith with a diameter of 1.16 m reproduced the water storage in the lysimeter well. This study demonstrated that ERT data can be used to obtain quantitative information about water storage in a soil profile from which evaporation rates and local water depletion rates associated with root water uptake can be estimated.

ERT can also be used to image σ_b distributions at the centimeter scale, for instance to monitor water infiltration and root water uptake of plants grown in pots. Werban et al. (2008) monitored σ_b changes in a vertical transect of 220 cm^2 in the upper part of the root zone of a lupine plant that was grown in a pot of 30 cm diameter and 50 cm height. Therefore, they used a 30-cm-long electrode profile at the top of the pot and a 14-cm-long profile that was buried at 10-cm depth in the pot, both with 2-cm electrode spacing. Water infiltration was found to be heterogeneous with preferential infiltration around the main roots. Differences in σ_b between 2 h were monitored during two consecutive days of plant transpiration. Surprisingly, an increase of σ_b or water content was observed in the upper part of the pot from dawn until noon, when plant transpiration was largest. This increase in water content was also observed with a TDR probe but contradicts previous observations and expectations. The spatial variability of σ_b was the largest during the day when root water uptake occurred. During night, water redistribution averaged out these variations. It must, however, be noted that the data-acquisition time was also 2 h and that the data inversion did not account for changes occurring during data acquisition. As a consequence, part of the spatial variability in the σ_b differences may be artifacts that come from temporal variations.

Challenges

The soil state variable that is derived from ERT depends simultaneously on a number of variables and soil properties that may vary in space and/or time. For the interpretation of ERT data, it is therefore common to assume that only one of the variables changes over time whereas the other variables and properties remain constant but may vary with location. In most studies related to soil–plant interactions, it is assumed that soil water content varies with time. If this assumption is valid, the large spatial variation of the relation between water content and bulk electrical conductivity poses a challenge to quantify water contents and soil profile water balances from ERT-derived bulk electrical conductivities. Therefore,

ERT surveys must be accompanied by local soil water content measurements to derive site specific calibration relationships.

Soil water content changes over time go along with root water uptake and root growth. Since RMD also influences the bulk soil conductivity, this may add uncertainty to the relation between water content and bulk electrical conductivity.

Finally, root water uptake may lead to an accumulation of salts in the root zone, especially when irrigation water with a considerable amount of dissolved salts is used. The increased salinity affects the relation between soil moisture content and bulk electrical conductivity and therefore poses a challenge to the interpretation of the measured bulk electrical conductivity in terms of water content.

Electrical Impedance Spectroscopy and Tomography
General Principles

Electrical impedance spectroscopy (EIS) is a natural extension of electrical resistivity methods in which the complex electrical conductivity of a sample is determined for a broad range of frequencies (mHz to kHz). The complex electrical conductivity, $\sigma^*(\omega)$, is a measure of both electrolytic conductivity (real part, σ') and polarization effects (imaginary part, σ''). To measure $\sigma^*(\omega)$, an electrical excitation voltage with known amplitude and frequency (f) is applied at two current electrodes and the associated voltages are measured at two potential electrodes. The recorded voltages, $V(\omega)$, and current, $I(\omega)$, are converted to $\sigma^*(\omega)$ using the geometric factor, k, for the electrode arrangement. $\sigma^*(\omega)$ can also be expressed as conductivity magnitude ($|\sigma|$) and phase angle (φ) between the excited sinusoidal current and measured voltage signal:

$$\sigma^*(\omega) = k\frac{I(\omega)}{V(\omega)} = |\sigma(\omega)| \cdot e^{i\varphi(\omega)} = \sigma'(\omega) + i\sigma''(\omega) \qquad [4]$$

with $i = \sqrt{-1}$ and $\omega = 2\pi f$. It should be noted that this method is more widely known as Spectral Induced Polarization in the field of geophysics. Alternatively, measurements of $\sigma^*(\omega)$ can also be made in the time domain, where the voltage decay after switching off the current is measured and typically transformed into a measure of polarizability. Although time domain and frequency domain measurements are directly related in principle, frequency domain measurements of $\sigma^*(\omega)$ are of superior accuracy and cover a much broader frequency range than time domain measurements.

Following the same procedures as in the case of ERT, tomographic images of the subsurface can be obtained when different combinations of current and potential electrodes are used to measure $\sigma^*(\omega)$. In addition to imaged distributions of σ', which are similar to those obtained with ERT, this type of measurement also provides imaged distributions of σ'' (or the phase angle), both for a range of frequencies (Kemna et al., 2000; Hordt et al., 2007). To distinguish this tomographic method from ERT, it is commonly referred to as electrical impedance tomography (EIT), where the use of impedance implies the consideration of the complex nature of the electrical conductivity. In geophysics, it is also referred to as induced polarization (IP) imaging. In contrast to ERT, it is important to mention that EIT is far from being an established imaging method. Most important, a consistent inversion strategy that not only considers spatial regularization but also regularization along the frequency axis still needs development (Brandstätter et al., 2003; Son et al., 2007).

Relation between σ″ and Soil and Root Properties

In soils, polarization is caused by the presence of charged interfaces and constrictions in the pore space, which lead to zones of unequal ionic transport (Lesmes and Morgan, 2001; Titov et al., 2002). Generally, $\sigma^*(\omega)$ varies considerably with frequency. Frequencies associated with characteristic maxima in the measured phase angle (so-called relaxation frequencies) are thought to relate to characteristic length scales of the porous medium, such as pore size (Titov et al., 2002) or grain size (Leroy and Revil, 2009). Because of this link, several studies have investigated the link between relaxation frequency and soil hydraulic conductivity (Binley et al., 2005; Breede et al., 2011). Nevertheless, polarization is generally low (<15 mrad) in metal-free soil, which has led to the development of dedicated EIS and EIT equipment to measure the complex electrical conductivity of unconsolidated porous media (Zimmermann et al., 2008a, 2008b).

Plant and root materials are expected to polarize much stronger in the presence of an alternating electrical field because of the presence of capacitances associated with thin resistive cell membranes separating the highly conductive inner cell from the less conductive material outside of the cell membranes. EIS measurements have been used to study physiological processes in plants, such as responses to freezing (Zhang and Willison, 1993; Repo and Pulli, 1996; Repo et al., 2000). These studies have shown that the merit of EIS measurements depends on the ability to relate the electrical measurements correctly to physical or physiological properties of the plant through an appropriate equivalent circuit model (Zhang et al., 1990). A simple equivalent circuit model of plant tissue considers the extracellular conductivity in parallel with the intracellular conductivity and the capacitance of the membranes (Hayden et al., 1969; Dvorak et al., 1981). EIS measurements have shown that this model is too simplified to adequately describe the electrical measurements (Zhang et al., 1990). More complicated circuit model have been proposed and evaluated to take in the account the presence of additional capacitances in the cell (Zhang et al., 1990; Zhang et al., 1995), air spaces in pine needles (Zhang and Willison, 1993), and the presence of different plant tissue type, for example, bark and wood (Repo and Zhang, 1993).

Dalton (1995) proposed an equivalent electrical circuit model for the entire root system. In this model, each individual root is represented by a resistance and a capacitance in parallel and all root segments are connected in parallel. This electrical circuit model suggests that a measurement of the effective capacitance of a root system, which is directly related to σ″, is simply the sum of the capacitance of each individual root segment. In addition, it can be shown that the capacitance of a root segment is directly proportional to the root surface area, which in turn depends on the diameter and length of the roots. Therefore, it depends on the morphology of the root system (e.g., distribution of root diameters) whether the capacitance can be meaningfully related to root mass. Several more recent studies have also proposed equivalent electrical circuit models for the entire root system (Ozier-Lafontaine and Bajazet, 2005; Repo et al., 2005; Cao et al., 2011).

Applications of EIS to Characterize Root Properties

Many applications of EIS have focused on relating measured electrical capacitance to properties of the root system. These studies have all relied on a very similar experimental set-up. Typically, the electrical capacitance was measured by an impedance bridge with two electrodes. One electrode was inserted into the stem of the plant as close as possible to the surface of the growing medium (soil or nutrient solution), and the second electrode was inserted in the growing medium directly. Chloupek (1972) and Chloupek (1977) were among the first to apply this method and reported strong linear relationships between fresh root weight and electrical capacitance at 1 kHz for a range of plants (see example in Fig. 12–4). The intercepts of these linear relations were attributed to parasitic capacitances associated with the wiring and the soil, and the slopes of the linear relations depended on plant type, soil type, measurement frequency, and applied voltage. This dependence on applied voltage is most likely related to the use of a two-point electrode set-up that introduces nonlinear effects because of the high current densities associated with the electrode inserted in the stem of the plant. Such effects can be greatly reduced when a four-point electrode set-up is used.

Several studies have attempted to apply the methodology outlined above, some with positive results (Preston et al., 2004; Pitre et al., 2010) and others with more mixed results. Kendall et al. (1982) attempted to relate capacitance measurements on the root system to the dry root weight of alfalfa and red clover. They found significant relations in three out of seven tests and attributed the lack of correlation at four out of seven occasions to variable soil moisture content that affected the contacts between electrodes, plant, and soil. A similar finding in which root mass and capacitance were only correlated at certain stages of the growth cycle was reported by van Beem et al. (1998) for maize. However, they did

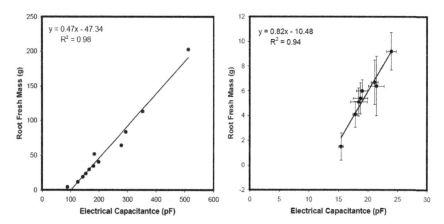

Fig. 12–4. Relationship between electrical capacitance and fresh root mass for two case studies. (Left) Relationship between electrical capacitance and fresh root mass for spherical carrot roots grown in a loam soil. Data points represent the mean of capacitance and root mass measurements on twenty carrots (data from Chloupek, 1977). (Right) Relationship between electrical capacitance and fresh root mass for eight maize genotypes grown under greenhouse conditions. Data points represent the mean of capacitance and root mass measurements on four individual plants, and the error bars indicate the standard error of the mean (data from van Beem et al., 1998).

find a good relationship between root mass and electrical capacitance across different maize genotypes, despite potential differences in physiological and electrical properties of the root system between genotypes (Fig. 12–4). In a more controlled experiment with tomato cultivars growing in nutrient solutions, Dalton (1995) found that 77% of the variance in root mass at the time of harvest could be explained with capacitance measurements. In a second experiment, Dalton (1995) monitored the temporal development of roots of tomato seedlings growing in nutrient solutions and the associated temporal development of the capacitance. Between Day 51 and 57, the total root capacitance dropped considerably from 12 to 8 pF despite an increase in dry root weight from 6 to 11 g. This sudden decrease was attributed to an effectively reduced active root mass (e.g., reduced activity, suberization, or death) and changes in the root electrical properties during plant maturation (Dalton, 1995). Finally, Dalton (1995) noted the strong influence of the height of the plant electrode above the surface of the growing medium, which highlights the importance of a consistent electrode insertion procedure. In a field trial with maize seedlings growing in soil, van Beem et al. (1998) found a significant relationship between fresh root mass and capacitance. Interestingly, differences in the regression equations for greenhouse and field experiments indicated that root capacitance was not an absolute measurement of root mass, but instead should be considered as a relative measurement that depends on a range of other factors affecting the electrical measurements (van Beem et al., 1998).

In the context of the measurement set-up discussed above, it is also worth mentioning the earth impedance measurement method for studying tree roots as proposed by Aubrecht et al. (2006) and Cermak et al. (2006). The use of "impedance" is not entirely appropriate for this method because it implies the consideration of the complex electrical conductivity, which was not the case in these studies. In fact, a procedure similar to ERT was used where one current electrode is inserted in the stem of the plant and the second current electrode in the soil. Two other electrodes are used to measure the potential difference. One of the potential electrodes is inserted at a fixed position in the root collar, whereas a second potential electrode is repeatedly inserted in the soil along the line connecting the two current electrodes. For each potential difference measurement, the grounding resistance is calculated and plotted as a function of the position of the second moving potential electrode. The resulting curve typically consists of a nonlinear and a linear part, and the transition between these two regions is assumed to correspond to the effective root length, which can then be used to calculate the "active" root surface area from the grounding resistance and the conductivity of the root (typically assumed to equal the conductivity of the stem). In a first evaluation, Cermak et al. (2006) showed that this method gave plausible results. Recently, Cao et al. (2010) and Urban et al. (2011) revisited key assumptions in this approach, such as (i) the resistance of stem xylem and soil are low and can be neglected, and (ii) the radial conductivity of stem and "nonactive" roots is negligible, among others. In a series of dedicated experiments, Urban et al. (2011) came to the conclusion that none of these key assumptions was fulfilled, thus strongly invalidating previous results where the earth impedance measurement method was used to estimate root surface area.

Many studies derived empirical relations between capacitance and root properties at a single measurement frequency. Interestingly, most of these studies have also pointed out several interfering effects that could deteriorate the quality of the empirical relationship. When using a single measurement frequency, the prospects of separating interfering effects are limited. Therefore, a number of recent studies have focused on the use of EIS for a broad range of frequencies and attempted to interpret the electrical measurement using equivalent circuit models for the entire root system (Ozier-Lafontaine and Bajazet, 2005; Repo et al., 2005; Cao et al., 2011). For example, Ozier-Lafontaine and Bajazet (2005) included the soil conductivity in their equivalent electrical circuit model, thus potentially removing the interfering effect of variable soil water content on the electrical measurements, as reported by Kendall et al. (1982) and van Beem et al. (1998). They also showed that one of the fitted capacitances of the equivalent electrical circuit model was a better predictor of dry root weight than the measured capacitance at 1 kHz used in many previous studies. Nevertheless, it was still difficult to separate changes in the soil–root interface and in the root itself (Ozier-Lafontaine and Bajazet, 2005). Using a more elaborate equivalent electrical circuit and willow cuttings grown in nutrient solutions, Cao et al. (2011) were able to separate bulk root and interface effects. In addition, they found that root capacitance was a better proxy of root size than root resistance.

EIS measurements on the entire root system have mostly relied on a set-up where one of the current electrodes was inserted in the plant stem. This has the advantage that individual plants can be investigated and that the current passes at least through part of the root system. However, it also adds the complicating factor of capacitances and resistances associated with the stem and the stem-electrode interface (Dalton, 1995; Ozier-Lafontaine and Bajazet, 2005). Recently, Zanetti et al. (2011) used an alternative EIS set-up where all four electrodes were installed in the soil. Laboratory measurements on boxes with three soil materials (i.e., gravel, sand, silt) with and without freshly cut root samples showed that the conductivity ratio between soil and soil with roots was much more favorable for the imaginary part than for the real part of the conductivity. The ratio for the real part of the conductivity varied from 0.78 to 1.79 for a range of soil materials and roots of different trees, thus indicating that successful detection of roots with ERT as reported by several researchers (Amato et al., 2008; Zenone et al., 2008) depends on the type of root and soil. In contrast, the ratio for the imaginary part of the conductivity ranged from 1.11 to 9.81 for the same soil and tree roots. Therefore, field EIT measurements are expected to be more successful than ERT measurements for in situ investigations of roots.

Challenges and Future Research

In summary, empirical relationships between root properties (mass, length, area) and electrical capacitance and resistance measured at a single measurement frequency allow noninvasive investigations of root processes. However, these empirical calibrations were mostly successful when standardized growing media and genotypes were used that minimized variations in soil and root conductivity. The prospects of using capacitance and resistance measurements at a single measurement frequency in a more natural setting seem troubled by a wide range of additional processes that affect the measured electrical properties, such as soil and plant water content, soil textural heterogeneity, soil temperature and solute content, and plant cell anatomy and tissue integrity (Urban et al., 2011).

A more promising research avenue seems to be the use of broadband electrical measurement in combination with the use of equivalent electrical circuit models that attempt to directly represent biophysical processes in the root–soil system.

The development of such models would greatly benefit from measurements and modeling of the individual components of the root–soil system, where in particular more knowledge on the electrical properties of the roots and their variability in time and space and across plants and genotypes seems to be required. This will require new experimental approaches, perhaps in the spirit of earlier work of Zhang and Willison (1993). Finally, EIS work seems biased to measurement setups with current electrodes installed in the plants. More conventional EIT measurements with current and potential electrodes in the soil also seem worthwhile to explore. Studies in controlled conditions have indicated that contrasts in the imaginary part of the electrical conductivity are more favorable than contrast in the real part of the electrical conductivity, thus indicating that the potential of electrical imaging methods in root studies can be further increased by considering the complex electrical conductivity.

Ground Penetrating Radar
General Principles

Ground penetrating radar (GPR) uses high-frequency (50–5000 MHz) electromagnetic waves to determine dielectric and conductivity properties of the shallow subsurface or to detect contrasts in these properties. Electromagnetic wave propagation is described by Maxwell's equations. The complex propagation constant $\gamma = \alpha + i\omega/v$ describes the propagation in an electrically conducting dielectric medium, where the velocity v and the attenuation coefficient α are given by Balanis (1989):

$$v = \frac{c}{\left\{ \frac{\varepsilon_r}{2}\left[\sqrt{1 + \left(\frac{\sigma}{2\pi f \varepsilon_r \varepsilon_0} \right)^2} + 1 \right] \right\}^{\frac{1}{2}}} \approx \frac{c}{\sqrt{\varepsilon_r}} \text{ for } \sigma = 0 \tag{5}$$

$$\alpha = 2\pi f \sqrt{\mu_0 \varepsilon_r \varepsilon_0} \left\{ \frac{1}{2}\left[\sqrt{1 + \left(\frac{\sigma}{2\pi f \varepsilon_r \varepsilon_0} \right)^2} - 1 \right] \right\}^{\frac{1}{2}} \approx \frac{\sigma}{2} \sqrt{\frac{\mu_0}{\varepsilon_r \varepsilon_0}} \text{ for large } f, \tag{6}$$

where c is the speed of light (0.3m ns^{-1}), f is the frequency (Hz), ε_r is the relative permittivity, ε_0 is the permittivity in free space (8.854 × 10^{-12} F m^{-1}), μ_0 is the magnetic permeability in free space (4π × 10^{-7} H m^{-1}), and σ is the electrical conductivity (S m^{-1}).

When the emitted electromagnetic waves reach a discontinuity with different electromagnetic properties, part of the energy is reflected back to the receiving antenna. For a wave with normal incidence on an interface between two media, the reflected amplitude is expressed by

$$r = \frac{\eta_2 - \eta_1}{\eta_2 + \eta_1} \tag{7}$$

where

$$\eta = \sqrt{\frac{2i\pi f \mu_0}{\sigma + 2i\pi f \varepsilon_r \varepsilon_0}}. \qquad\qquad [8]$$

The propagation velocity, the wave attenuation, and the wave reflection coefficient are used to interrogate the subsurface and to detect and locate anomalous targets in the subsurface, such as soil layer boundaries, water tables, pipes, and roots, among others.

A GPR measurement is typically made using one emitting antenna and one (or more) receiving antenna. Different measurement setups with varying antenna configurations exist: (i) off-ground GPR, where antennas are suspended in the air and the main reflection that comes from the soil surface is analyzed; (ii) surface GPR, where both antennas are placed on the soil surface and the main energy is emitted into the subsurface, thus allowing detection of deeper reflectors; and (iii) crosshole GPR, where antennas are lowered into the ground in boreholes and the area between two (or more) boreholes can be investigated.

Common processing steps for off-ground GPR are based on the analysis of the air–ground reflection coefficient (see Eq. [7]) using a reference measurement over a metal plate. From these measurements, the dielectric permittivity and conductivity of the soil surface layer can be derived (Huisman et al., 2003; Serbin and Or, 2005). More sophisticated analyses include the full-waveform inversion of these measurements (Lambot et al., 2006b) so that profiles of dielectric permittivity and electrical conductivity can also be inferred.

For surface and crosshole GPR, the propagation velocity of an electromagenetic (EM) pulse is analyzed. For surface GPR, two categories of methods to determine the propagation velocity v and the mean dielectric permittivity, ε_r, (Eq. [5]) can be recognized. First, the velocity of the ground wave can be determined using GPR measurements with different antenna separations. The ground wave is a wave that propagates directly from the emitter to the receiver antenna through the top of the soil (Huisman et al., 2001). When strong vertical heterogeneities in permittivity are present, the ground wave might get trapped and result in a dispersive waveguide that requires dedicated interpretation methods to obtain subsurface information (van der Kruk et al., 2006). A second category relies on reflected waves to determine the propagation velocity. For flat reflectors, such as soil layers and the ground water table, GPR measurements with a single fixed antenna separation can be used to estimate propagation velocity if the depth to the reflector is known. When reflector depth is also not known, GPR measurements with multiple antenna offsets can be used to estimate both simultaneously (Huisman et al., 2003). Because the sending GPR antenna emits energy in a wide beam, the receiving antenna also registers reflections of objects that are not directly below the antennas. Small point reflectors such as stones, pipes, and plant roots will, therefore, show up as reflection hyperbola when a transect of GPR measurements with a fixed antenna separation is made. The apex of this hyperbola indicates the position of the reflecting object, and the slopes of the hyperbola can be used to determine the GPR wave propagation velocity. Such measurements can also be used to obtain information about the geometry of the reflecting object using a processing step called migration (also imaging), which properly decomposes and compacts the geometry of the hyperbolas into a form that is close to the actual feature (Fisher et al., 1992).

In the case of crosshole GPR, the vertical offset between emitter antenna in one borehole and receiver antenna in a second borehole are typically systematically varied. The distance between boreholes depends on the attenuation of the EM signal in the soil under investigation. When the area between the boreholes is discretized in rectangular cells of constant velocity or permittivity ε_r, the velocity of each cell can be estimated by minimizing the difference between the measured arrival times and arrival times calculated for raypaths passing through these cells. The simplest inversion algorithms assume straight raypaths and neglect the bending or refraction of the EM wave by spatial variation of ε_r and σ. Contemporary common practice includes inversion algorithms that use curved rays. Recently, so-called full-waveform inversion strategies have been developed. Full waveform inversion relies on the simulation of EM wave propagation by solving Maxwell's equations. Using full-waveform inversions, the spatial ε_r distribution can be obtained with higher spatial resolution. It also allows obtaining more accurate information in parts of the image domain that are prone to interferences with surface reflections of the EM waves, that is, close to the soil–atmosphere boundary. Finally, full-waveform inversion also considers GPR signal attenuation and can, therefore, be used to image the electrical conductivity distribution in addition to the dielectric permittivity distribution.

The ability of GPR to detect layers or anomalous target objects embedded in a medium depends on the (i) electromagnetic contrast between the medium and the target, (ii) attenuation of the medium, (iii) antenna frequency, and (iv) target depth. An indication for the vertical resolution or the thinnest layer that can be detected is given by the Rayleigh criterion, which states that the top and bottom of a layer can only be distinguished when it has a thickness greater than one-fourth wavelength. The wavelength is determined by the ratio of the velocity and the frequency of the electromagnetic wave: $\lambda = v/f$. A limit on the horizontal resolution of a GPR survey is described by the Fresnel zone $w = \sqrt{2z\lambda}$ that depends on the dominant wavelength λ and the reflector depth z. Using full-waveform inversion, it is, however, possible to obtain sub-wavelength horizontal resolution. The penetration depth of GPR is largely determined by antenna frequency and the attenuation caused by the soil electrical conductivity (see Eq. [6]). Soils having high electrical conductivity rapidly attenuate radar energy, and limit the effectiveness of GPR (Doolittle et al., 2010). Waves with lower frequency penetrate deeper, but have a lower vertical and horizontal resolution. It is clear that one must make a compromise between penetration depth and resolution and select an optimum system frequency.

Relation between ε and σ and Soil and Root Properties

The relative permittivity of moist to water-saturated substances is dominated by the relative permittivity and volume fraction of water. To convert relative permittivity into volumetric soil water content, an appropriate petrophysical relationship is needed that can be based on empirical relationships, volumetric mixing formulae, or effective medium approximations (Topp et al., 1980; Roth et al., 1990; Robinson et al., 2003). A commonly used mixing formula is the refractive index model for an unsaturated porous material (three-phase mixture):

$$\sqrt{\varepsilon_{eff}} = \theta\sqrt{\varepsilon_w} + (n-\theta)\sqrt{\varepsilon_a} + (1-n)\sqrt{\varepsilon_s} \qquad [9]$$

where n is the porosity, θ is the water content, the relative permittivities of water and air are $\varepsilon_w = 80$ and $\varepsilon_a = 1$, and ε_s is the relative permittivity of the mineral grains. Alternatively, several empirical equations between ε and θ have been developed. The most widely used $\varepsilon_{eff}(\theta)$ relationship for soils is the empirical formula of Topp et al. (1980):

$$\varepsilon_{eff} = 3.03 + 9.3\theta + 146\theta^2 - 76.7\theta^3 \qquad [10]$$

which yields good predictions for coarse- and medium-textured soils and gives similar predictions as more theoretically based models.

In contrast to the relation between the bulk electrical conductivity and the water content, the relation between water content and the dielectric permittivity is more universal and less dependent on other soil properties and soil temperature. However, for soil with high clay content, there might be issues associated with the accuracy of these relationships due to interactions of water molecules with electrically charged clay particles that reduce the dielectric permittivity of water (Heimovaara et al., 1994). These interactions typically introduce a dependence of the permittivity on frequency in the lower MHz range. For example, West et al. (2003) found that sand–clay mixtures showed a frequency-dependent dielectric permittivity up to 350 MHz. In the lower MHz frequency range, the electrical conductivity of the pore fluid, the microscopic spatial arrangement or configuration of the air, water and solid phases, and the soil temperature influence the dielectric permittivity so that the relation between dielectric permittivity and soil moisture content becomes less unequivocal in this frequency range (Chen and Or, 2006a, 2006b). Most relationships between water content and dielectric permittivity were derived with TDR, which operates in the frequency range between 300 and 1000 MHz, which should be remembered when applying them to estimate water content with low-frequency GPR antennas.

Applications of GPR for Investigating Soil and Root Zone Properties

GPR Monitoring of Water Movement

Because GPR waves are significantly influenced by the presence of water, GPR surveys can be used to address numerous hydrogeological questions, ranging from investigating the geological structure to estimating material properties and water content. Repeated GPR measurements can be used to monitor changes in the relative permittivity that correspond to moisture content. A review of studies that determined soil water content using GPR was provided by Huisman et al. (2003). Using surface GPR antennas that are mounted on a sledge and moved over a field, several hectares can be monitored within a day. The use of multi-channel GPR acquisition systems has emerged in recent years and is particularly promising in this context, because it improves the interpretation in case of the ground wave method and allows the simultaneous determination of reflector depth and the associated average dielectric permittivity (Gerhards et al., 2008). In the context of investigations of root–soil interactions, there are two distinct disadvantages of surface GPR measurements. First, there is no control on the resolution of the soil water content measurements because this is determined by the presence of reflecting layers and objects. Second, the instruments need to be driven or pulled over the investigation area. Therefore, these methods are less suited to map water contents in cropped fields and were used mainly in bare fields (Weihermüller et al., 2007), orchards (Grote et al., 2003), or grassland (Huisman et al., 2001).

The acquisition and interpretation of off-ground GPR measurements has been considerably improved in recent years. Full-waveform inversion of off-ground GPR can now accurately measure the water content of the upper centimeters of the soils (Lambot et al., 2006a). Fields of tens of hectares can be surveyed within a day using an off-ground GPR system mounted on a small vehicle (Weihermüller et al., 2007; Minet et al., 2011). A comparison between off-ground and surface GPR with local soil moisture measurements (Weihermüller et al., 2007) revealed considerable differences between the different methods which were attributed to different soil layer depths that were observed by the different methods and the effect of the soil surface roughness. In the context of investigations of root–soil interactions, the investigation depth of off-ground GPR might be too shallow.

Using borehole antennas, the water contents and changes in water content were imaged in vertical transects with a scale up to a few tens of meters in both vertical and horizontal direction. Changes in volumetric water content were obtained using time-lapse zero offset profile (ZOP) radar data sets (Binley et al., 2002b). Permeable pathways and moisture migration were delineated using time-lapse tomographic multi offset profile (MOG) radar data (Hubbard et al., 1997; Eppstein and Dougherty, 1998; Deiana et al., 2007; Kuroda et al., 2009). Despite the more invasive nature of borehole GPR measurements, because of the need to install access tubes, this type of GPR measurements seems most promising to investigate soil water content changes associated with root–soil interactions because of the control on the spatial resolution of the obtained soil water content data.

Root Detection and Biomass Estimation Using GPR

In Table 12–1, an overview of detected root diameters and depths of root detection as a function of the antenna frequency is given. This table illustrates that both detected root diameter and the depth of root detection generally decrease with increasing antenna frequency. Hyperbolic reflectors in GPR profiles indicate roots or other point anomalies. al Hagrey (2007) showed that the relative dielectric permittivity of wood ranges between 4.5 and 22. It depends on the contrast in dielectric permittivity between roots and soil whether a characteristic reflection pattern will be observed in GPR. The larger the difference in water content between roots and soil, the larger the reflection coefficient (e.g., Eq. [7]), and the clearer the roots can potentially be observed.

Roots can have a tremendous diversity of reflector shapes and orientations such that sophisticated three-dimensional imaging of GPR data is necessary to obtain a realistic image of the tree root distribution. Until now, most studies only investigated roots present in the direction perpendicular to the measurement transect. In such a measurement set-up, two-dimensional migration of the GPR data collapses the hyperbola, which potentially provides valuable information about the roots. However, successful GPR application for root detection has been site-specific, and

Table 12–1. Antenna frequencies and diameter and depth estimation of root detection (from Hirano et al., 2009).

Frequency	Detected root diameter		Detected root depth		Reference
	Min.	Max.	Min.	Max.	
MHz	cm				
400	3.7	10	–	130	(Butnor et al., 2001)
450	3–4	–	–	200	(Hruska et al., 1999)
500,800	1	10	15	155	(Barton and Montagu, 2004)
900	2.5	8.2	11	114	(Cox et al., 2005)
900	1.1	5.2	–	–	(Dannoura et al., 2008)
900	1.9	7.8	30	80	(Hirano et al., 2009)
1500	0.6	–	–	45	(Butnor et al., 2001)
1500	0.5	–	–	50	(Butnor et al., 2001)

numerous factors can affect the success of root detection with GPR (Hirano et al., 2009). The most important factors were root diameter, root water content, and the horizontal and vertical separation between the roots. Moreover, the classic hyperbolic anomaly shape of GPR targets is lost when the direction of the root is not perfectly perpendicular to the profile. This limits the ability to identify the true root position and associated root properties.

One of the first studies that investigated the mapping of tree root systems in forest stands using three-dimensional GPR measurements is described by Hruska et al. (1999). Here, a grid of GPR measurements was made using 450 MHz antennas. The grid consisted of lines of GPR measurements with a separation of 0.25 m. GPR measurements within each line were separated by 0.05 m. The data were processed using two-dimensional migration and manual redrawing of the final images. Roots with a diameter greater than 2 cm were identified down to a depth of 2.5 m. Objects running parallel to the transmitted electromagnetic waves could not be reliably detected. To obtain information on root diameter under field conditions, Butnor et al. (2001) investigated the signal amplitude of reflected waves. Roots of known size were buried at 15- and 30-cm depths and a correlation between return signal strength and root diameter was observed. However, the quality of the correlation deteriorated rapidly with depth (correlation coefficient of $r = 0.81$ and 0.55 at 15- and 30-cm depths, respectively). This was probably due to the amplitude attenuation of the GPR signals by geometrical spreading and the electrical conductivity (Eq. [6]). GPR has also been used to determine total root biomass in a forest stand (Butnor et al., 2003). In this study, Hilbert transformed and migrated 1.5 GHz two-dimensional GPR data were related to tree root biomass up to a depth of 30 cm, which was estimated by root samples using soil cores. The resulting correlation was quite strong with a correlation coefficient of $r = 0.86$ based on 60 samples (Butnor et al., 2003).

An alternative method to estimate root diameter from GPR data was presented by Barton and Montagu (2004). Instead of focusing on the signal amplitude, they proposed to relate the length of the GPR pulse of the reflected wave to root diameter, which has the advantage of being independent of the waveform and the signal attenuation. To test this idea, tree roots were buried under optimal conditions and migrated 500 MHz two-dimensional GPR data were analyzed and correlated with the actual root diameter. Root diameters were predicted with a root mean squared error of 0.6 cm, allowing detection and quantification of roots as small as 1 cm in diameter. To take advantage of the different polarizations, Wielopolski et al. (2002) measured two datasets that were orthogonal to each other, that is, the antenna was rotated 90° while measuring along the same GPR lines. This improved the detection of tree roots. Another improvement for the detection of roots is the use of circular GPR scans, which allows a quasi-perpendicular intersection with a larger part of the roots, thus generating more hyperbolic features that can be interpreted (Zenone et al., 2008).

Most of the former studies were performed without appreciable changes in soil and root moisture content. Hirano et al. (2009) investigated the influence of volumetric root water content and observed that roots with high water content were easily detected but roots with less than 20% water content were not detected in a relatively dry soil with a volumetric water content of 13%. Zenone et al. (2008) reported that occasional rain events resulted in higher water content of the shallow soil layer which resulted in a smaller reflection of the GPR signal between shallow and deeper soil layer making the roots in the deeper layer clearer visible. A larger water content of the roots could as well have caused an increasing contrast in electrical properties between the root and the surrounding soil so that the hyperbolas relative to root targets resulted more clearly.

Challenges and Future Research

Many researchers have shown that GPR can be used to detect tree roots in a three-dimensional space. However, many of the data processing have been performed in a two-dimensional sense. Future work should focus on three-dimensional GPR measurements and processing to fully explore the information content present in the GPR data. Recent developments in GPR surveying techniques include the use of accurate positioning systems, like laser theodolite (Lehmann and Green, 1999; Heincke et al., 2005), laser positioning sensors (Grasmueck and Viggiano, 2007), differential GPS (Streich et al., 2006), and self-tracking total station methods (Boniger and Tronicke, 2010). Nowadays, every GPR system is equipped with the possibility to store GPS positioning data directly for every recorded trace such that high-quality and high-precision GPR data can be kinematically acquired using off-the-shelf instrumentation without further hardware modifications. These techniques should be tested and applied for the acquisition of full three-dimensional blocks of GPR data on a smaller scale with a very small line spacing (a few centimeters when using frequencies above 500 MHz), although this might be challenging for rough ground surfaces. Three-dimensional topographic processing and migration of three-dimensional GPR data has already been performed on larger scales (i.e., 30 by 50 m, Lehmann and Green, 2000; Heincke et al., 2005). These approaches need to be extended to enable the conversion of a dense series of circular scans (Zenone et al., 2008) into a full three-dimensional model.

Most GPR measurements are performed by moving along a field or in boreholes manually. Therefore, monitoring can generally not be done automatically so is accompanied with a high work load, which limits the temporal resolution and the number of measurement times. With the development of multi-array emitter and receiver antennas, automatic monitoring may become possible.

Until now, mainly empirical approaches have been used to estimate information about tree root systems. Due to the wide diversity of reflector shapes and orientations and the many factors that influence the GPR response, these empirical methods could only be used after calibration. More quantitative images can be obtained by full-waveform inversion methods (Klotzsche et al., 2010; Meles et al., 2010) that have the potential to image roots at the subwavelength scale. These full-waveform methods also consider orientational effects that have been found to strongly influence reflection properties (Radzevicius and Daniels, 2000).

With current GPR systems and the highest antenna frequency of 1500 MHz, the smallest tree roots that could be detected had a diameter of 0.5 cm (Butnor et al., 2001). The use of GPR to detect crop roots will be difficult, since crop roots have diameters that are too small to directly detect with the current high frequency antennas. For example, fine roots smaller than 0.2 mm in diameter represent up to more than 50% of root length for canola (Pierret et al., 2005) and maize roots smaller than 0.175 mm in diameter can also account for more than 56% of the root length (Pallant et al., 1993). The use of higher frequency antennas to detect smaller roots does not seem to be promising because of increased conductivity and scattering losses for high frequencies. A possible solution to indirectly measure the influence of fine roots is to perform time-lapse three-dimensional measurement of GPR with natural or artificial changes in soil water content that can be related to root water uptake. Several time-lapse three-dimensional surveys have been used to determine subsurface fluid migration (Birken and Versteeg, 2000; Truss et al., 2007) and can perhaps be applied in studies of soil–root interactions.

Conclusions and Outlook

In Table 12–2, a summary of the characteristics of the three methods: ERT, EIS, and GPR relevant for their use to investigate soil–plant interactions is given. This table demonstrates the differences between the methods indicating that their combination could lead to important synergisms. Since different soil electrical properties are measured by the different methods, a combination of these measurements will enable a better separation of the soil and root parameters that influence the electrical properties simultaneously. The different spatial and temporal resolution and extent that could be achieved by the methods also offers opportunities for synergisms when the methods are combined. In the following section, we provide a short overview of procedures that could be used to combine methods and to integrate them with process models.

The quantitative use of multiple streams of geophysical data is still a major challenge. An overview of different methods to combine various geophysical data in the context of hydrological modeling was provided by Ferre et al. (2009) and Hinnell et al. (2010). They distinguished two broad classes of inversion strategies: uncoupled and coupled hydrogeophysical inversion. In the uncoupled approach, the geophysical data are first inverted to obtain images of geoelectrical properties (e.g., dielectric permittivity, electrical conductivity). In the next step, the geoelectrical properties are converted to model states or parameters, which are then used in the model parameterization or calibration process (e.g., Chen et al., 2001; Hubbard et al., 2001; Binley et al., 2002a; Kemna et al., 2002; Binley and Beven, 2003; Cassiani and Binley, 2005; Muller et al., 2010). Independent geo-

Table 12–2. Overview of characteristics of different methods: Electrical Resistivity Tomography (ERT), Electrical Impedance Spectroscopy (EIS) and Ground Penetrating Radar (GPR).

	ERT	EIS	GPR
Derived electrical soil property	Bulk electrical conductivity (resistivity)	Bulk electrical conductivity and polarization or complex electrical conductivity (possible at different frequencies)	Dielectric permittivity (also complex dielectrical permittivity, i.e. including bulk electrical conductivity is possible)
Relation to soil parameters/properties	Soil water content, salt concentration in pore water, clay content, porosity, soil temperature. Relation between soil water content and bulk electrical conductivity varies considerable between different soil types	see ERT for real part of the conductivity. Imaginary part of the conductivity depends on grain/pore size, chemical composition of pore water, and cation exchange capacity.	Soil water content. For higher frequencies (> 200 MHz), relatively universal calibration relation between dielectric permittivity and water content. At lower frequencies, this relation is less certain.
Relation to plant parameters	Root mass density [(RMD) may decrease or increase the bulk electrical conductivity depending on soil electrical conductivity]	Root mass, length and area (electrical model for the soil–root system and its spectral electrical properties requires additional research)	Individual roots with diameter larger than 0.5 cm can be detected.
Experimental setup and consequences for surveying and monitoring	Line or grid arrays of surface electrodes, or buried electrodes. Permanent electrodes and automated data acquisition allow automated continuous monitoring with high temporal resolution. Electrodes can be installed and left without disturbance of the plant/crop growth. Spatial extent of the area or transect that can be monitored or surveyed is limited by the number of electrodes that can be connected to the measurement device during a survey.	Electrodes in plants and soil. Standard measurement equipment is not available yet. Potentially, the method would be similar to ERT.	Off-ground, surface, or borehole antennas. Automated data acquisition and positioning of antennas allow efficient surveys of large areas. Using surface antennas, orchards, vineyards, grassland can be surveyed but surveying of cropped fields or densely vegetated areas is difficult. Installation of fixed and permanent antennas is difficult, hampering continuous automated monitoring.
Spatial extent and resolution	Spatial extent ranges from three-dimensional imaging of lab-scale studies with cm resolution to vertical transects of tens of meters depth and kilometers width.	So far, mainly used at the single plant scale to derive spatially integrated characteristic of the root system (e.g. total root mass). Imaging is possible but requires further development.	Using off-ground and surface antennas, areas of several hectares can be surveyed. Using borehole antennas, vertical transects of tens of meters depth and up to 10 m width can be measured.

physical inversion is generally ill-posed and underdetermined, and regularization constraints need to be applied to condition and stabilize the inverse problem. These regularization constraints, in combination with inherent spatial variability in resolution, can lead to uncertain and sometimes erroneous inversion results (Day-Lewis et al., 2005), which are directly propagated to the model parameterization or calibration process in the uncoupled inversion approach. Many studies have attempted to alleviate the shortcomings of this sequential and uncoupled inversion approach. For example, time-lapse geophysical inversion approaches (Day-Lewis et al., 2002) now allow a consistent treatment of geophysical data obtained during subsurface monitoring, which reduces some of the imaging artifacts. Alternatively, spatially variable apparent petrophysical relations can be applied to obtain images of geoelectrical properties corrected for variable spatial resolution and smoothing (Moysey et al., 2005; Singha and Moysey, 2006). Finally, joint inversion approaches (Gallardo and Meju, 2003; Linde et al., 2006) have been proposed to reduce the inversion artifacts by using multiple sets of preferably complimentary geophysical data to obtain improved subsurface images. In view of imaging soil–plant interactions, a combination of different geophysical data is promising, especially to disentangle the effect of soil and plant variables or properties (soil moisture, soil salinity, clay content, root mass, or root length density) on the imaged geo-electrical properties (dielectric permittivity and complex electrical conductivity).

A second inversion strategy, the coupled inversion approach, has received considerable attention in recent years for the interpretation of geophysical monitoring data (Kowalsky et al., 2004; Lambot et al., 2006a; Hinnell et al., 2010). In this approach, geophysical data are directly interpreted using a forward geophysical model coupled to a model describing the process that is being monitored. The model parameters are perturbed to minimize the difference between simulated and measured geophysical and otherwise available ancillary data. If the model is an adequate representation of the process under investigation, the coupled inversion is expected to provide a much stronger constraint than the regularization in the uncoupled inversion approach. Despite the potential advantages of coupled hydrogeophysical inversion, the number of studies that have used this approach to interpret actual geophysical data has started to increase only recently (Kowalsky et al., 2005; Deiana et al., 2007; Looms et al., 2008; Lambot et al., 2009; Huisman et al., 2010; Rings et al., 2010; Mboh et al., 2011). For soil–plant investigations, coupled inversion approaches could be used to determine key parameters of soil–vegetation atmosphere transfer models such as rooting depths, root length density profiles or two- or three-dimensional distributions, and functions that describe root water uptake in terms of soil moisture or soil water potential. So far, local soil moisture measurements have been used to derive these parameters (Coelho and Or, 1999; Musters and Bouten, 1999; Vrugt et al., 2001; Hupet et al., 2003). However, root and soil hydraulic parameters could hardly be determined simultaneously or independently (Musters and Bouten, 2000; Hupet et al., 2002), rooting depths and root water uptake vary considerably with location in a field (Hupet and Vanclooster, 2002) and for some applications, a three-dimensional mapping of the root water uptake is required. Combining geophysical methods that provide information about the spatial distribution of both root and soil moisture distributions directly with models that describe water flow and root water uptake processes is therefore expected to be promising.

References

al Hagrey, S.A. 2007. Geophysical imaging of root-zone, trunk, and moisture heterogeneity. J. Exp. Bot. 58:839–854. doi:10.1093/jxb/erl237

al Hagrey, S.A., and J. Michaelson. 1999. Resistivity and percolation study of preferential flow in vadose zone at Bokhorst, Germany. Geophysics 64:746–753. doi:10.1190/1.1444584

Amato, M., B. Basso, G. Celano, G. Bitella, G. Morelli, and R. Rossi. 2008. In situ detection of tree root distribution and biomass by multi-electrode resistivity imaging. Tree Physiol. 28:1441–1448.

Amato, M., G. Bitella, R. Rossi, J.A. Gomez, S. Lovelli, and J.J.F. Gomes. 2009. Multi-electrode 3d resistivity imaging of alfalfa root zone. Eur. J. Agron. 31:213–222 10.1016/j.eja.2009.08.005. doi:10.1016/j.eja.2009.08.005

Amente, G., J.M. Baker, and C.F. Reece. 2000. Estimation of soil solution electrical conductivity from bulk soil electrical conductivity in sandy soils. Soil Sci. Soc. Am. J. 64:1931–1939. doi:10.2136/sssaj2000.6461931x

Amidu, S.A., and J.A. Dunbar. 2007. Geoelectric studies of seasonal wetting and drying of a texas vertisol. Vadose Zone J. 6:511–523 10.2136/vzj2007.0005. doi:10.2136/vzj2007.0005

Andreu, L., J.W. Hopmans, and L.J. Schwankl. 1997. Spatial and temporal distribution of soil water balance for a drip-irrigated almond tree. Agric. Water Manage. 35:123–146. doi:10.1016/S0378-3774(97)00018-8

Archie, G.E. 1942. The electrical resistivity log as an aid in determining some reservoir characteristics. Trans. Am. Inst. Min. Metall. Eng. 146:54–61.

Aubrecht, L., Z. Stanek, and J. Koller. 2006. Electrical measurement of the absorption surfaces of tree roots by the earth impedance method: 1. Theory. Tree Physiol. 26:1105–1112. doi:10.1093/treephys/26.9.1105

Balanis, C.A. 1989. Advanced engineering electromagnetics. John Wiley & Sons, New York.

Barker, R., and J. Moore. 1998. The application of time-lapse electrical tomography in groundwater studies. Lead. Edge (Tulsa Okla.) 17:1454–1458. doi:10.1190/1.1437878

Barton, C.V.M., and K.D. Montagu. 2004. Detection of tree roots and determination of root diameters by ground penetrating radar under optimal conditions. Tree Physiol. 24:1323–1331. doi:10.1093/treephys/24.12.1323

Berthold, S., L.R. Bentley, and M. Hayashi. 2004. Integrated hydrogeological and geophysical study of depression-focused groundwater recharge in the Canadian prairies. Water Resour. Res. 40:W06505. doi:10.1029/2003WR002982

Binley, A., and K. Beven. 2003. Vadose zone flow model uncertainty as conditioned on geophysical data. Ground Water 41:119–127. doi:10.1111/j.1745-6584.2003.tb02576.x

Binley, A., G. Cassiani, R. Middleton, and P. Winship. 2002a. Vadose zone flow model parameterisation using cross-borehole radar and resistivity imaging. J. Hydrol. 267:147–159. doi:10.1016/S0022-1694(02)00146-4

Binley, A., L.D. Slater, M. Fukes, and G. Cassiani. 2005. Relationship between spectral induced polarization and hydraulic properties of saturated and unsaturated sandstone. Water Resour. Res. 41:W12417. doi:10.1029/2005WR004202

Binley, A., P. Winship, L.J. West, M. Pokar, and R. Middleton. 2002b. Seasonal variation of moisture content in unsaturated sandstone inferred from borehole radar and resistivity profiles. J. Hydrol. 267:160–172. doi:10.1016/S0022-1694(02)00147-6

Birken, R., and R. Versteeg. 2000. Use of four-dimensional ground penetrating radar and advanced visualization methods to determine subsurface fluid migration. J. Appl. Geophys. 43:215–226. doi:10.1016/S0926-9851(99)00060-9

Boniger, U., and J. Tronicke. 2010. On the potential of kinematic gpr surveying using a self-tracking total station: Evaluating system crosstalk and latency. IEEE Trans. Geosci. Rem. Sens. 48:3792–3798. doi:10.1109/TGRS.2010.2048332

Brandstätter, B., K. Hollaus, H. Hutten, M. Mayer, R. Merwa, and H. Scharfetter. 2003. Direct estimation of cole parameters in multifrequency eit using a regularized gauss-newton method. Physiol. Meas. 24:437–448. doi:10.1088/0967-3334/24/2/355

Breede, K., A. Kemna, O. Esser, E. Zimmermann, H. Vereecken, and J.A. Huisman. 2011. Joint measurement setup for determining spectral induced polarization and soil hydraulic properties. Vadose Zone J. 10:716–726. doi:10.2136/vzj2010.0110

Brunet, P., R. Clement, and C. Bouvier. 2010. Monitoring soil water content and deficit using electrical resistivity tomography (ert)- a case study in the cevennes area, France. J. Hydrol. 380:146–153. doi:10.1016/j.jhydrol.2009.10.032

Butler, D.K. 2005. Near surface geophysics. Investigations in geophysics series. Society of Exploration Geophysics, Tulsa, OK.

Butnor, J.R., J.A. Doolittle, K.H. Johnsen, L. Samuelson, T. Stokes, and L. Kress. 2003. Utility of ground-penetrating radar as a root biomass survey tool in forest systems. Soil Sci. Soc. Am. J. 67:1607–1615. doi:10.2136/sssaj2003.1607

Butnor, J.R., J.A. Doolittle, L. Kress, S. Cohen, and K.H. Johnsen. 2001. Use of ground-penetrating radar to study tree roots in the southeastern United States. Tree Physiol. 21:1269–1278. doi:10.1093/treephys/21.17.1269

Cao, Y., T. Repo, R. Silvennoinen, T. Lehto, and P. Pelkonen. 2010. An appraisal of the electrical resistance method for assessing root surface area. J. Exp. Bot. 61:2491–2497. doi:10.1093/jxb/erq078

Cao, Y., T. Repo, R. Silvennoinen, T. Lehto, and P. Pelkonen. 2011. Analysis of the willow root system by electrical impedance spectroscopy. J. Exp. Bot. 62:351–358. doi:10.1093/jxb/erq276

Cassiani, G., and A. Binley. 2005. Modeling unsaturated flow in a layered formation under quasi-steady state conditions using geophysical data constraints. Adv. Water Resour. 28:467–477. doi:10.1016/j.advwatres.2004.12.007

Celano, G., A.M. Palese, A. Ciucci, E. Martorella, N. Vignozzi, and C. Xiloyannis. 2011. Evaluation of soil water content in tilled and cover-cropped olive orchards by the geoelectrical technique. Geoderma 163:163–170. doi:10.1016/j.geoderma.2011.03.012

Cermak, J., R. Ulrich, Z. Stanek, J. Koller, and L. Aubrecht. 2006. Electrical measurement of tree root absorbing surfaces by the earth impedance method: 2. Verification based on allometric relationships and root severing experiments. Tree Physiol. 26:1113–1121. doi:10.1093/treephys/26.9.1113

Chen, J.S., S. Hubbard, and Y. Rubin. 2001. Estimating the hydraulic conductivity at the south oyster site from geophysical tomographic data using bayesian techniques based on the normal linear regression model. Water Resour. Res. 37:1603–1613. doi:10.1029/2000WR900392

Chen, Y.P., and D. Or. 2006a. Effects of maxwell-wagner polarization on soil complex dielectric permittivity under variable temperature and electrical conductivity. Water Resour. Res. 42:W06424. doi:10.1029/2005WR004590

Chen, Y.P., and D. Or. 2006b. Geometrical factors and interfacial processes affecting complex dielectric permittivity of partially saturated porous media. Water Resour. Res. 42:W06423. doi:10.1029/2005WR004744

Chloupek, O. 1972. Relationship between electric capacitance and some other parameters of plant roots. Biol. Plant. 14:227–230. doi:10.1007/BF02921255

Chloupek, O. 1977. Evaluation of size of a plants root-system using its electrical capacitance. Plant Soil 48:525–532. doi:10.1007/BF02187258

Coelho, E.F., and D. Or. 1999. Root distribution and water uptake patterns of corn under surface and subsurface drip irrigation. Plant Soil 206:123–136. doi:10.1023/A:1004325219804

Constable, S.C., R.L. Parker, and C.G. Constable. 1987. Occams inversion- a practical algorithm for generating smooth models from electromagnetic sounding data. Geophysics 52:289–300. doi:10.1190/1.1442303

Cox, K.D., H. Scherm, and N. Serman. 2005. Ground-penetrating radar to detect and quantify residual root fragments following peach orchard clearing. Horttechnology 15:600–607.

Daily, W., A. Ramirez, D. Labrecque, and J. Nitao. 1992. Electrical-resistivity tomography of vadose water-movement. Water Resour. Res. 28:1429–1442. doi:10.1029/91WR03087

Dalton, F.N. 1995. In-situ root measurements by electrical capacitance methods. Plant Soil 173:157–165. doi:10.1007/BF00155527

Dannoura, M., Y. Hirano, T. Igarashi, M. Ishii, K. Aono, K. Yamase, and Y. Kanazawa. 2008. Detection of cryptomeria japonica roots with ground penetrating radar. Plant Biosyst. 142:375–380. doi:10.1080/11263500802150951

Dawson, T.E. 1996. Determining water use by trees and forests from isotopic, energy balance and transpiration analyses: The roles of tree size and hydraulic lift. Tree Physiol. 16:263–272. doi:10.1093/treephys/16.1-2.263

Day-Lewis, F.D., J.M. Harris, and S.M. Gorelick. 2002. Time-lapse inversion of crosswell radar data. Geophysics 67:1740–1752. doi:10.1190/1.1527075

Day-Lewis, F.D., K. Singha, and A.M. Binley. 2005. Applying petrophysical models to radar travel time and electrical resistivity tomograms: Resolution-dependent limitations. J. Geophys. Res. Solid Earth 110:B08206. doi:10.1029/2004JB003569

Deiana, R., G. Cassiani, A. Kemna, A. Villa, V. Bruno, and A. Bagliani. 2007. An experiment of non-invasive characterization of the vadose zone via water injection and cross-hole time-lapse geophysical monitoring. Near Surf. Geophys. 5:183–194.

Doolittle, J., R. Dobos, S. Peaslee, S. Waltman, E. Benham, and W. Tuttle. 2010. Revised ground-penetrating radar soil suitability maps. J. Environ. Eng. Geophys. 15:111–118. doi:10.2113/JEEG15.3.111

Dvorak, M., J. Cernohorska, and K. Janacek. 1981. Characteristics of current passage through plant-tissue. Biol. Plant. 23:306–310. doi:10.1007/BF02895374

Eppstein, M.J., and D.E. Dougherty. 1998. Efficient three-dimensional data inversion: Soil characterization and moisture monitoring from cross-well ground-penetrating radar at a Vermont test site. Water Resour. Res. 34:1889–1900.

Feddes, R.A., H. Hoff, M. Bruen, T. Dawson, P. de Rosnay, O. Dirmeyer, R.B. Jackson, P. Kabat, A. Kleidon, A. Lilly, and A.J. Pitman. 2001. Modeling root water uptake in hydrological and climate models. Bull. Am. Meteorol. Soc. 82:2797–2809. doi:10.1175/1520-0477(2001)082<2797:MRWUIH>2.3.CO;2

Ferre, T., L. Bentley, A. Binley, N. Linde, A. Kemna, K. Singha, K. Holliger, J.A. Huisman, and B. Minsley. 2009. Critical steps for the continuing advancement of hydrogeophysics. Eos 90:200. doi:10.1029/2009EO230004

Fisher, E., G.A. McMechan, A.P. Annan, and S.W. Cosway. 1992. Examples of reverse-time migration of single-channel, ground-penetrating radar profiles. Geophysics 57:577–586. doi:10.1190/1.1443271

Friedman, S.P. 1998. Simulation of a potential error in determining soil salinity from measured apparent electrical conductivity. Soil Sci. Soc. Am. J. 62:593–599. doi:10.2136/sssaj1998.03615995006200030006x

Frohlich, R.K., and C.D. Parke. 1989. The electrical-resistivity of the vadose zone- field survey. Ground Water 27:524–530. doi:10.1111/j.1745-6584.1989.tb01973.x

Gallardo, L.A., and M.A. Meju. 2003. Characterization of heterogeneous near-surface materials by joint 2d inversion of dc resistivity and seismic data. Geophys. Res. Lett. 30:1658. doi:10.1029/2003GL017370

Garré, S., M. Javaux, J. Vanderborght, L. Pagès, and H. Vereecken. 2011. Three-dimensional electrical resistivity tomography to monitor root zone water dynamics. Vadose Zone J. 10:412–424. doi:10.2136/vzj2010.0079

Gerhards, H., U. Wollschlager, Q.H. Yu, P. Schiwek, X.C. Pan, and K. Roth. 2008. Continuous and simultaneous measurement of reflector depth and average soil-water content with multichannel ground-penetrating radar. Geophysics 73:J15–J23. doi:10.1190/1.2943669

Grasmueck, M., and D.A. Viggiano. 2007. Integration of ground-penetrating radar and laser position sensors for real-time 3-d data fusion. IEEE Trans. Geosci. Rem. Sens. 45:130–137. doi:10.1109/TGRS.2006.882253

Green, S., and B. Clothier. 1998. The root zone dynamics of water uptake by a mature apple tree. Plant Soil 206:61–77. doi:10.1023/A:1004368906698

Green, S.R., and B.E. Clothier. 1995. Root water-uptake by kiwifruit vines following partial wetting of the root-zone. Plant Soil 173:317–328. doi:10.1007/BF00011470

Grote, K., S. Hubbard, and Y. Rubin. 2003. Field-scale estimation of volumetric water content using ground-penetrating radar ground wave techniques. Water Resour. Res. 39:1321. doi:10.1029/2003WR002045

Hayden, R.I., C.A. Moyse, F.W. Calder, D.P. Crawford, and D.S. Fensom. 1969. Electrical impedance studies on potato and alfalfa tissue. J. Exp. Bot. 20:177–200.doi: 10.1093/jxb/20.2.177.

Heimovaara, T.J., W. Bouten, and J.M. Verstraten. 1994. Frequency-domain analysis of time-domain reflectometry wave-forms.2. A 4-component complex dielectric mixing model for soils. Water Resour. Res. 30:201–209. doi:10.1029/93WR02949

Heimovaara, T.J., A.G. Focke, W. Bouten, and J.M. Verstraten. 1995. Assessing temporal variations in soil-water composition with time-domain reflectometry. Soil Sci. Soc. Am. J. 59:689–698. doi:10.2136/sssaj1995.03615995005900030009x

Heincke, B., A.G. Green, J. van der Kruk, and H. Horstmeyer. 2005. Acquisition and processing strategies for 3d georadar surveying a region characterized by rugged topography. Geophysics 70:K53–K61. doi:10.1190/1.2122414

Hinnell, A.C., T.P.A. Ferre, J.A. Vrugt, J.A. Huisman, S. Moysey, J. Rings, and M.B. Kowalsky. 2010. Improved extraction of hydrologic information from geophysical data through coupled hydrogeophysical inversion. Water Resour. Res. 46:W00d40. doi: 10.1029/2008wr007060.

Hirano, Y., M. Dannoura, K. Aono, T. Igarashi, M. Ishii, K. Yamase, N. Makita, and Y. Kanazawa. 2009. Limiting factors in the detection of tree roots using ground-penetrating radar. Plant Soil 319:15–24. doi:10.1007/s11104-008-9845-4

Hordt, A., R. Blaschek, A. Kemna, and N. Zisser. 2007. Hydraulic conductivity estimation from induced polarisation data at the field scale- the Krauthausen case history. J. Appl. Geophys. 62:33–46. doi:10.1016/j.jappgeo.2006.08.001

Hruska, J., J. Cermak, and S. Sustek. 1999. Mapping tree root systems with ground-penetrating radar. Tree Physiol. 19:125–130. doi:10.1093/treephys/19.2.125

Hubbard, S.S., J.S. Chen, J. Peterson, E.L. Majer, K.H. Williams, D.J. Swift, B. Mailloux, and Y. Rubin. 2001. Hydrogeological characterization of the south oyster bacterial transport site using geophysical data. Water Resour. Res. 37:2431–2456. doi:10.1029/2001WR000279

Hubbard, S.S., J.E. Peterson Jr., E.L. Majer, P.T. Zawislanski, K.H. Williams, J. Roberts, and F. Wobber. 1997. Estimation of permeable pathways and water content using tomographic radar data. Leading Edge16:1623–1630.

Hubbard, S., and Y. Rubin. 2004. Hydrogeophysics overview. In Y. Rubin and S. Hubbard, editors, Hydrogeophysics. Springer, New York. p. 3–21.

Huisman, J.A., S.S. Hubbard, J.D. Redman, and A.P. Annan. 2003. Measuring soil water content with ground penetrating radar: A review. Vadose Zone J. 2:476–491. doi:10.2113/2.4.476.

Huisman, J.A., J. Rings, J.A. Vrugt, J. Sorg, and H. Vereecken. 2010. Hydraulic properties of a model dike from coupled bayesian and multi-criteria hydrogeophysical inversion. J. Hydrol. 380:62–73. doi:10.1016/j.jhydrol.2009.10.023

Huisman, J.A., C. Sperl, W. Bouten, and J.M. Verstraten. 2001. Soil water content measurements at different scales: Accuracy of time domain reflectometry and ground-penetrating radar. J. Hydrol. 245:48–58. doi:10.1016/S0022-1694(01)00336-5

Hupet, F., S. Lambot, R.A. Feddes, J.C. van Dam, and M. Vanclooster. 2003. Estimation of root water uptake parameters by inverse modeling with soil water content data. Water Resour. Res. 39:1312–1328. doi:10.1029/2003WR002046

Hupet, F., S. Lambot, M. Javaux, and M. Vanclooster. 2002. On the identification of macroscopic root water uptake parameters from soil water content observations. Water Resour. Res. 38:1300–1314. doi:10.1029/2002WR001556

Hupet, F., and A. Vanclooster. 2005. Micro-variability of hydrological processes at the maize row scale: Implications for soil water content measurements and evapotranspiration estimates. J. Hydrol. 303:247–270. doi:10.1016/j.jhydrol.2004.07.017

Hupet, F., and M. Vanclooster. 2002. Intraseasonal dynamics of soil moisture variability within a small agricultural maize cropped field. J. Hydrol. 261:86–101. doi:10.1016/S0022-1694(02)00016-1

Jarvis, N.J. 1989. A simple empirical-model of root water-uptake. J. Hydrol. 107:57–72. doi:10.1016/0022-1694(89)90050-4

Jayawickreme, D.H., R.L. Van Dam, and D.W. Hyndman. 2008. Subsurface imaging of vegetation, climate, and root-zone moisture interactions. Geophys. Res. Lett. 35:L18404. doi:10.1029/2008GL034690

Jayawickreme, D.H., R.L. Van Dam, and D.W. Hyndman. 2010. Hydrological consequences of land-cover change: Quantifying the influence of plants on soil moisture with time-lapse electrical resistivity. Geophysics 75:WA43–WA50. doi:10.1190/1.3464760

Keller, G.V., and F.C. Frischknecht. 1966. Electrical methods in geophysical prospecting. Pergamon Press, Oxford.

Kemna, A., A. Binley, A. Ramirez, and W. Daily. 2000. Complex resistivity tomography for environmental applications. Chem. Eng. J. 77:11–18. doi:10.1016/S1385-8947(99)00135-7

Kemna, A., J. Vanderborght, B. Kulessa, and H. Vereecken. 2002. Imaging and characterisation of subsurface solute transport using electrical resistivity tomography (ert) and equivalent transport models. J. Hydrol. 267:125–146. doi:10.1016/S0022-1694(02)00145-2

Kendall, W.A., G.A. Pederson, and R.R. Hill. 1982. Root size estimates of red-clover and alfalfa based on electrical capacitance and root diameter measurements. Grass Forage Sci. 37:253–256. doi:10.1111/j.1365-2494.1982.tb01604.x

Klotzsche, A., J. van der Kruk, G.A. Meles, J. Doetsch, H. Maurer, and N. Linde. 2010. Full-waveform inversion of cross-hole ground-penetrating radar data to characterize a gravel aquifer close to the Thur river, Switzerland. Near Surf. Geophys. 8:635–649.doi:10.3997/1873–0604.2010054.

Koch, K., J. Wenninger, S. Uhlenbrook, and M. Bonell. 2009. Joint interpretation of hydrological and geophysical data: Electrical resistivity tomography results from a process hydrological research site in the Black Forest Mountains, Germany. Hydrol. Processes 23:1501–1513. doi:10.1002/hyp.7275

Kowalsky, M.B., S. Finsterle, J. Peterson, S. Hubbard, Y. Rubin, E. Majer, A. Ward, and G. Gee. 2005. Estimation of field-scale soil hydraulic and dielectric parameters through joint inversion of GPR and hydrological data. Water Resour. Res. 41:W11425. doi:10.1029/2005WR004237.

Kowalsky, M.B., S. Finsterle, and Y. Rubin. 2004. Estimating flow parameter distributions using ground-penetrating radar and hydrological measurements during transient flow in the vadose zone. Adv. Water Resour. 27:583–599. doi:10.1016/j.advwatres.2004.03.003

Kuroda, S., H. Jang, and H.J. Kim. 2009. Time-lapse borehole radar monitoring of an infiltration experiment in the vadose zone. J. Appl. Geophys. 67:361–366.doi: 10.1016/j.jappgeo.2008.07.005.

Lai, C.T., and G. Katul. 2000. The dynamic role of root-water uptake in coupling potential to actual transpiration. Adv. Water Resour. 23:427–439. doi:10.1016/S0309-1708(99)00023-8

Lambot, S., E. Slob, J. Rhebergen, O. Lopera, K.Z. Jadoon, and H. Vereecken. 2009. Remote estimation of the hydraulic properties of a sand using full-waveform integrated hydrogeophysical inversion of time-lapse, off-ground GPR data. Vadose Zone J. 8:743–754. doi:10.2136/vzj2008.0058

Lambot, S., E.C. Slob, M. Vanclooster, and H. Vereecken. 2006a. Closed loop GPR data inversion for soil hydraulic and electric property determination. Geophys. Res. Lett. 33:L21405. doi:10.1029/2006GL027906

Lambot, S., L. Weihermuller, J.A. Huisman, H. Vereecken, M. Vanclooster, and E.C. Slob. 2006b. Analysis of air-launched ground-penetrating radar techniques to measure the soil surface water content. Water Resour. Res. 42:W11403. doi:10.1029/2006WR005097

Lehmann, F., and A.G. Green. 1999. Semiautomated georadar data acquisition in three dimensions. Geophysics 64:719–731. doi:10.1190/1.1444581

Lehmann, F., and A.G. Green. 2000. Topographic migration of georadar data: Implications for acquisition and processing. Geophysics 65:836–848. doi:10.1190/1.1444781

Leroy, P., and A. Revil. 2009. A mechanistic model for the spectral induced polarization of clay materials. J. Geophys. Res. 114:B10202. doi:10.1029/2008JB006114

Lesmes, D.P., and F.D. Morgan. 2001. Dielectric spectroscopy of sedimentary rocks. J. Geophys. Res. 106:13329–13346. doi:10.1029/2000JB900402

Li, Y., R. Wallach, and Y. Cohen. 2002. The role of soil hydraulic conductivity on the spatial and temporal variation of root water uptake in drip-irrigated corn. Plant Soil 243:131–142. doi:10.1023/A:1019911908635

Linde, N., A. Binley, A. Tryggvason, L.B. Pedersen, and A. Revil. 2006. Improved hydrogeophysical characterization using joint inversion of cross-hole electrical resistance and ground-penetrating radar traveltime data. Water Resour. Res. 42:W12404. doi:10.1029/2006WR005131.

Looms, M.C., A. Binley, K.H. Jensen, L. Nielsen, and T.M. Hansen. 2008. Identifying unsaturated hydraulic parameters using an integrated data fusion approach on cross-borehole geophysical data. Vadose Zone J. 7:238–248. doi:10.2136/vzj2007.0087

Mallants, D., M. Vanclooster, N. Toride, J. Vanderborght, M.T. vanGenuchten, and J. Feyen. 1996. Comparison of three methods to calibrate tdr for monitoring solute movement in undisturbed soil. Soil Sci. Soc. Am. J. 60:747–754. doi:10.2136/sssaj1996.03615995006000030010x

Mallants, D., and J. Vanderborght. 1998. Comment on 'comparison of three methods to calibrate tdr for monitoring solute movement in undisturbed soil- reply. Soil Sci. Soc. Am. J. 62:490–492. doi:10.2136/sssaj1998.03615995006200020030x

Mboh, C.M., J.A. Huisman, and H. Vereecken. 2011. Feasibility of sequential and coupled inversion of time domain reflectometry data to infer soil hydraulic parameters under falling head infiltration. Soil Sci. Soc. Am. J. 75:775–786. doi:10.2136/sssaj2010.0285

Meles, G.A., J. Van der Kruk, S.A. Greenhalgh, J.R. Ernst, H. Maurer, and A.G. Green. 2010. A new vector waveform inversion algorithm for simultaneous updating of conductivity and permittivity parameters from combination crosshole/borehole-to-surface gpr data. IEEE Trans. Geosci. Rem. Sens. 48:3391–3407. doi:10.1109/TGRS.2010.2046670

Michot, D., Y. Benderitter, A. Dorigny, B. Nicoullaud, D. King, and A. Tabbagh. 2003. Spatial and temporal monitoring of soil water content with an irrigated corn crop cover using surface electrical resistivity tomography. Water Resour. Res. 39:1138–1158. doi:10.1029/2002WR001581

Minet, J., A. Wahyudi, P. Bogaert, M. Vanclooster, and S. Larnbot. 2011. Mapping shallow soil moisture profiles at the field scale using full-waveform inversion of ground penetrating radar data. Geoderma 161:225–237. doi:10.1016/j.geoderma.2010.12.023

Moysey, S., K. Singha, and R. Knight. 2005. A framework for inferring field-scale rock physics relationships through numerical simulation. Geophys. Res. Lett. 32:L08304. doi:10.1029/2004GL022152

Muller, K., J. Vanderborght, A. Englert, A. Kemna, J.A. Huisman, J. Rings, and H. Vereecken. 2010. Imaging and characterization of solute transport during two tracer tests in a shallow aquifer using electrical resistivity tomography and multilevel groundwater samplers. Water Resour. Res. 46:W03502. doi:10.1029/2008WR007595

Musters, P.A.D., and W. Bouten. 1999. Assessing rooting depths of an Austrian pine stand by inverse modeling soil water content maps. Water Resour. Res. 35:3041–3048. doi:10.1029/1999WR900173

Musters, P.A.D., and W. Bouten. 2000. A method for identifying optimum strategies of measuring soil water contents for calibrating a root water uptake model. J. Hydrol. 227:273–286. doi:10.1016/S0022-1694(99)00187-0

Nijland, W., M. van der Meijde, E.A. Addink, and S.M. de Jong. 2010. Detection of soil moisture and vegetation water abstraction in a mediterranean natural area using electrical resistivity tomography. Catena 81:209–216. doi:10.1016/j.catena.2010.03.005

Nimmo, J.R., K.S. Perkins, K.M. Schmidt, D.M. Miller, J.D. Stock, and K. Singha. 2009. Hydrologic characterization of desert soils with varying degrees of pedogenesis: 1. Field experiments evaluating plant-relevant soil water behavior. Vadose Zone J. 8:480–495. doi:10.2136/vzj2008.0052

Ozier-Lafontaine, H., and T. Bajazet. 2005. Analysis of root growth by impedance spectroscopy (eis). Plant Soil 277:299–313. doi:10.1007/s11104-005-7531-3

Pallant, E., R.A. Holmgren, G.E. Schuler, K.L. McCracken, and B. Drbal. 1993. Using a fine-root extraction device to quantify small-diameter corn roots (less-than-or-equal-to-0.025 mm) in-field soils. Plant Soil 153:273–279. doi:10.1007/BF00013000

Pierret, A., C.J. Moran, and C. Doussan. 2005. Conventional detection methodology is limiting our ability to understand the roles and functions of fine roots. New Phytol. 166:967–980. doi:10.1111/j.1469-8137.2005.01389.x

Pitre, F.E., N.J.B. Brereton, S. Audoire, G.M. Richter, I. Shield, and A. Karp. 2010. Estimating root biomass in salix viminalis x salix schwerinii cultivar oolofo using the electrical capacitance method. Plant Biosyst. 144:479–483. doi:10.1080/11263501003732092

Preston, G.M., R.A. McBride, J. Bryan, and M. Candido. 2004. Estimating root mass in young hybrid poplar trees using the electrical capacitance method. Agrofor. Syst. 60:305–309. doi:10.1023/B:AGFO.0000024439.41932.e2

Radzevicius, S.J., and J.J. Daniels. 2000. Ground penetrating radar polarization and scattering from cylinders. J. Appl. Geophys. 45:111–125. doi:10.1016/S0926-9851(00)00023-9

Repo, T., J. Laukkanen, and R. Silvennoinen. 2005. Measurement of the tree root growth using electrical impedance spectroscopy. Silva Fennica 39:159–166.

Repo, T., and S. Pulli. 1996. Application of impedance spectroscopy for selecting frost hardy varieties of English ryegrass. Ann. Bot. (Lond.) 78:605–609. doi:10.1006/anbo.1996.0167

Repo, T., G. Zhang, A. Ryyppo, and R. Rikala. 2000. The electrical impedance spectroscopy of scots pine (pinus sylvestris l.) shoots in relation to cold acclimation. J. Exp. Bot. 51:2095–2107. doi:10.1093/jexbot/51.353.2095

Repo, T., and M.I.N. Zhang. 1993. Modeling woody plant-tissues using a distributed electrical circuit. J. Exp. Bot. 44:977–982. doi:10.1093/jxb/44.5.977

Revil, A., and P.W.J. Glover. 1997. Theory of ionic-surface electrical conduction in porous media. Phys. Rev. B 55:1757–1773. doi:10.1103/PhysRevB.55.1757

Revil, A., and P.W.J. Glover. 1998. Nature of surface electrical conductivity in natural sands, sandstones, and clays. Geophys. Res. Lett. 25:691–694. doi:10.1029/98GL00296

Reynolds, J.M. 1997. An introduction to applied environmental geophysics. John Wiley & Sons, Chichester, UK.

Rhoades, J.D., N.A. Manteghi, P.J. Shouse, and W.J. Alves. 1989. Soil electrical-conductivity and soil-salinity- new formulations and calibrations. Soil Sci. Soc. Am. J. 53:433–439. doi:10.2136/sssaj1989.03615995005300020020x

Rhoades, J.D., P.A.C. Raats, and R.J. Prather. 1976. Effects of liquid-phase electrical-conductivity, water-content, and surface conductivity on bulk soil electrical-conductivity. Soil Sci. Soc. Am. J. 40:651–655. doi:10.2136/sssaj1976.03615995004000050017x

Rings, J., J.A. Huisman, and H. Vereecken. 2010. Coupled hydrogeophysical parameter estimation using a sequential bayesian approach. Hydrol. Earth Syst. Sci. 14:545–556. doi:10.5194/hess-14-545-2010

Robinson, D.A., S.B. Jones, J.M. Wraith, D. Or, and S.P. Friedman. 2003. A review of advances in dielectric and electrical conductivity measurement in soils using time domain reflectometry. Vadose Zone J. 2:444–475.

Rossi, R., M. Amato, G. Bitella, R. Bochicchio, J.J.F. Gomes, S. Lovelli, E. Martorella, and P. Favale. 2011. Electrical resistivity tomography as a non-destructive method for mapping root biomass in an orchard. Eur. J. Soil Sci. 62:206–215. doi:10.1111/j.1365-2389.2010.01329.x

Roth, K., R. Schulin, H. Fluhler, and W. Attinger. 1990. Calibration of time domain reflectometry for water-content measurement using a composite dielectric approach. Water Resour. Res. 26:2267–2273. doi: 10.1029/90wr01238

Samouelian, A., I. Cousin, G. Richard, A. Tabbagh, and A. Bruand. 2003. Electrical resistivity imaging for detecting soil cracking at the centimetric scale. Soil Sci. Soc. Am. J. 67:1319–1326. doi:10.2136/sssaj2003.1319

Samouelian, A., G. Richard, I. Cousin, R. Guerin, A. Bruand, and A. Tabbagh. 2004. Three-dimensional crack monitoring by electrical resistivity measurement. Eur. J. Soil Sci. 55:751–762. doi:10.1111/j.1365-2389.2004.00632.x

Schön, J.H. 1996. Physical properties of rocks: Fundamentals and principles of petrophysics. Handbook of geophysical exploration–section I seismic exploration. Elsevier Science Ltd., Amsterdam, The Netherlands.

Schwartz, B.F., and M.E. Schreiber. 2009. Quantifying potential recharge in mantled sinkholes using ert. Ground Water 47:370–381. doi:10.1111/j.1745-6584.2008.00505.x

Schwartz, B.F., M.E. Schreiber, and T. Yan. 2008. Quantifying field-scale soil moisture using electrical resistivity imaging. J. Hydrol. 362:234–246. doi:10.1016/j.jhydrol.2008.08.027

Seladji, S., P. Cosenza, A. Tabbagh, J. Ranger, and G. Richard. 2010. The effect of compaction on soil electrical resistivity: A laboratory investigation. Eur. J. Soil Sci. 61:1043–1055. doi:10.1111/j.1365-2389.2010.01309.x

Serbin, G., and D. Or. 2005. Ground-penetrating radar measurement of crop and surface water content dynamics. Remote Sens. Environ. 96:119–134. doi:10.1016/j.rse.2005.01.018

Simunek, J., and J.W. Hopmans. 2009. Modeling compensated root water and nutrient uptake. Ecol. Modell. 220:505–521. doi:10.1016/j.ecolmodel.2008.11.004

Singha, K., and S.M. Gorelick. 2006. Hydrogeophysical tracking of three-dimensional tracer migration: The concept and application of apparent petrophysical relations. Water Resour. Res. 42:W06422. doi:10.1029/2005WR004568

Singha, K., and S. Moysey. 2006. Accounting for spatially variable resolution in electrical resistivity tomography through field-scale rock-physics relations. Geophysics 71:A25–A28. doi:10.1190/1.2209753

Son, J.S., J.H. Kim, and M.J. Yi. 2007. A new algorithm for sw parameter estimation from multi-frequency ip data: Preliminary results. Explor. Geophys. 38:60–68. doi:10.1071/EG07009

Srayeddin, I., and C. Doussan. 2009. Estimation of the spatial variability of root water uptake of maize and sorghum at the field scale by electrical resistivity tomography. Plant Soil 319:185–207. doi:10.1007/s11104-008-9860-5

Streich, R., J. van der Kruk, and A.G. Green. 2006. Three-dimensional multicomponent georadar imaging of sedimentary structures. Near Surf. Geophys. 4:39–48.

Tikhonov, A.N., and V.Y. Arsenin. 1977. Solutions of ill-posed problems. W.H. Winston and Sons, Washington.

Timlin, D., Y. Pachepsky, and V.R. Reddy. 2001. Soil water dynamics in row and interrow positions in soybean (Glycine max L.). Plant Soil 237:25–35. doi:10.1023/A:1013385026500

Titov, K., V. Komarov, V. Tarasov, and A. Levitski. 2002. Theoretical and experimental study of time domain-induced polarization in water-saturated sands. J. Appl. Geol. 50:417–433. doi:10.1016/S0926-9851(02)00168-4

Topp, G.C., J.L. Davis, and A.P. Annan. 1980. Electromagnetic determination of soil-water content-measurements in coaxial transmission-lines. Water Resour. Res. 16:574–582. doi:10.1029/WR016i003p00574

Truss, S., M. Grasmueck, S. Vega, and D.A. Viggiano. 2007. Imaging rainfall drainage within the miami oolitic limestone using high-resolution time-lapse ground-penetrating radar. Water Resour. Res. 43:W03405. doi:10.1029/2005WR004395

Urban, J., R. Bequet, and R. Mainiero. 2011. Assessing the applicability of the earth impedance method for in situ studies of tree root systems. J. Exp. Bot. 62:1857–1869. doi:10.1093/jxb/erq370

van Beem, J., M.E. Smith, and R.W. Zobel. 1998. Estimating root mass in maize using a portable capacitance meter. Agron. J. 90:566–570. doi:10.2134/agronj1998.00021962009000040021x

van der Kruk, J., R. Streich, and A.G. Green. 2006. Properties of surface waveguides derived from separate and joint inversion of dispersive TE and TM GPR data. Geophysics 71:K19–K29. doi:10.1190/1.2168011

Vereecken, H., A. Binley, G. Cassiani, A. Revil, and K. Titov. 2006. Applied hydrogeophysics. Springer, Dordrecht, The Netherlands.

Vrugt, J.A., J.W. Hopmans, and J. Simunek. 2001. Calibration of a two-dimensional root water uptake model. Soil Sci. Soc. Am. J. 65:1027–1037. doi:10.2136/sssaj2001.6541027x

Waxman, M.H., and L.J.M. Smits. 1968. Electrical conductivities in oil-bearing shaly sands. Soc. Pet. Eng. J. 8:107.

Weihermüller, L., J.A. Huisman, S. Lambot, M. Herbst, and H. Vereecken. 2007. Mapping the spatial variation of soil water content at the field scale with different ground penetrating radar techniques. J. Hydrol. 340:205–216. doi:10.1016/j.jhydrol.2007.04.013

Werban, U., S.A. al Hagrey, and W. Rabbel. 2008. Monitoring of root-zone water content in the laboratory by 2d geoelectrical tomography. J. Plant Nutr. Soil Sci.-. Z. Pflanzenernaehr. Bodenkd. 171:927–935. doi:10.1002/jpln.200700145

West, L.J., K. Handley, Y. Huang, and M. Pokar. 2003. Radar frequency dielectric dispersion in sandstone: Implications for determination of moisture and clay content. Water Resour. Res. 39:1026–1038. doi:10.1029/2001WR000923

Wielopolski, L., G. Hendrey, M. McGuigan, and J. Daniels. 2002. Imaging tree root systems in situ. Proccedings of the GPR 2002: Ninth International Conference on Ground Penetrating Radar. p. 58–62.

Yeh, T.C.J., S. Liu, R.J. Glass, K. Baker, J.R. Brainard, D. Alumbaugh, and D. LaBrecque. 2002. A geostatistically based inverse model for electrical resistivity surveys and its applications to vadose zone hydrology. Water Resour. Res. 38:1278–1291. doi:10.1029/2001WR001204

Zanetti, C., A. Weller, M. Vennetier, and P. Meriaux. 2011. Detection of buried tree root samples by using geoelectrical measurements: A laboratory experiment. Plant Soil 339:273–283. doi:10.1007/s11104-010-0574-0

Zenone, T., G. Morelli, M. Teobaldelli, F. Fischanger, M. Matteucci, M. Sordini, A. Armani, C. Ferre, T. Chiti, and G. Seufert. 2008. Preliminary use of ground-penetrating radar and electrical resistivity tomography to study tree roots in pine forests and poplar plantations. Funct. Plant Biol. 35:1047–1058. doi:10.1071/FP08062

Zhang, M.I.N., T. Repo, J.H.M. Willison, and S. Sutinen. 1995. Electrical-impedance analysis in plant-tissues- on the biological meaning of cole-cole-alpha in scots pine needles. Eur. Biophys. J. 24:99–106. doi:10.1007/BF00211405

Zhang, M.I.N., D.G. Stout, and J.H.M. Willison. 1990. Electrical-impedance analysis in plant-tissues- symplasmic resistance and membrane capacitance in the hayden model. J. Exp. Bot. 41:371–380. doi:10.1093/jxb/41.3.371

Zhang, M.I.N., and J.H.M. Willison. 1993. Electrical-impedance analysis in plant-tissues- impedance measurement in leaves. J. Exp. Bot. 44:1369–1375. doi:10.1093/jxb/44.8.1369

Zhou, Q.Y., J. Shimada, and A. Sato. 2001. Three-dimensional spatial and temporal monitoring of soil water content using electrical resistivity tomography. Water Resour. Res. 37:273–285. doi:10.1029/2000WR900284

Zhou, Q.Y., J. Shimada, and A. Sato. 2002. Temporal variations of the three-dimensional rainfall infiltration process in heterogeneous soil. Water Resour. Res. 38:1030–1045. doi:10.1029/2001WR000349

Zimmermann, E., A. Kemna, J. Berwix, W. Glaas, H.M. Munch, and J.A. Huisman. 2008a. A high-accuracy impedance spectrometer for measuring sediments with low polarizability. Meas. Sci. Technol. 19:105603. doi:10.1088/0957-0233/19/10/105603

Zimmermann, E., A. Kemna, J. Berwix, W. Glaas, and H. Vereecken. 2008b. EIT measurement system with high phase accuracy for the imaging of spectral induced polarization properties of soils and sediments. Meas. Sci. Technol. 19:094010. doi:10.1088/0957-0233/19/9/094010

Index

Printed and bound by CPI Group (UK) Ltd, Croydon, CR0 4YY

27/10/2024

14580267-0002